# Geospatial Applications for Natural Resources Management

# Geospatial Applications for Natural Resources Management

Edited by
Chander Kumar Singh

CRC Press
Taylor & Francis Group
Boca Raton London New York

CRC Press is an imprint of the
Taylor & Francis Group, an **informa** business

CRC Press
Taylor & Francis Group
6000 Broken Sound Parkway NW, Suite 300
Boca Raton, FL 33487-2742

© 2018 by Taylor & Francis Group, LLC
CRC Press is an imprint of Taylor & Francis Group, an Informa business

No claim to original U.S. Government works

Printed on acid-free paper

International Standard Book Number-13: 978-1-1386-2628-7 (Hardback)

This book contains information obtained from authentic and highly regarded sources. Reasonable efforts have been made to publish reliable data and information, but the author and publisher cannot assume responsibility for the validity of all materials or the consequences of their use. The authors and publishers have attempted to trace the copyright holders of all material reproduced in this publication and apologize to copyright holders if permission to publish in this form has not been obtained. If any copyright material has not been acknowledged please write and let us know so we may rectify in any future reprint.

Except as permitted under U.S. Copyright Law, no part of this book may be reprinted, reproduced, transmitted, or utilized in any form by any electronic, mechanical, or other means, now known or hereafter invented, including photocopying, microfilming, and recording, or in any information storage or retrieval system, without written permission from the publishers.

For permission to photocopy or use material electronically from this work, please access www.copyright.com (http://www.copyright.com/) or contact the Copyright Clearance Center, Inc. (CCC), 222 Rosewood Drive, Danvers, MA 01923, 978-750-8400. CCC is a not-for-profit organization that provides licenses and registration for a variety of users. For organizations that have been granted a photocopy license by the CCC, a separate system of payment has been arranged.

**Trademark Notice:** Product or corporate names may be trademarks or registered trademarks, and are used only for identification and explanation without intent to infringe.

**Visit the Taylor & Francis Web site at**
http://www.taylorandfrancis.com

**and the CRC Press Web site at**
http://www.crcpress.com

Printed and bound in the United States of America by Sheridan

# Contents

List of Figures ................................................................................................................vii
List of Tables ................................................................................................................ xiii
Preface............................................................................................................................xvii
Editor ............................................................................................................................. xix
Contributors ................................................................................................................. xxi

**Chapter 1** Spatiotemporal Analysis of Urban Expansion and Its Impact on Surface
Temperature and Water Bodies ................................................................. 1

*Chander Kumar Singh, M. Kumari, N. Kikon, and R.K. Tomar*

**Chapter 2** Landscape Pattern and Dynamics in a Fast-Growing City, Khamis-Mushyet,
Saudi Arabia, Using Geoinformation Technology ..................................... 11

*Javed Mallick and Hoang Thi Hang*

**Chapter 3** Understanding the Spatio-Temporal Monitoring of Glaciers: Application
of Geospatial Technology ........................................................................... 27

*Shruti Dutta and AL. Ramanathan*

**Chapter 4** Urban Imprints on City's Environment—Unfolding Four Metro Cities of India ...... 51

*Richa Sharma and P.K. Joshi*

**Chapter 5** Predictive Modeling of a Metropolitan City in India Using a Land Change
Modeling Approach..................................................................................... 73

*Akanksha Balha and Chander Kumar Singh*

**Chapter 6** Performance Analysis of Different Predictive Algorithms for the Land
Features Modeling....................................................................................... 87

*Pradeep Kumar, Rajendra Prasad, Arti Choudhary, and Sudhir Kumar Singh*

**Chapter 7** Urban Growth and Management in Lucknow City, the Capital of Uttar Pradesh ...... 109

*Akanksha Balha and Chander Kumar Singh*

**Chapter 8** Change in Volume of Glaciers and Glacierets in Two Catchments of
Western Himalayas, India since 1993–2015 ............................................. 123

*Mohd Soheb, AL. Ramanathan, Manish Pandey, and Sarvagya Vatsal*

**Chapter 9** Analysis of Drainage Morphometry and Tectonic Activity in the Dehgolan Basin Kurdistan, Iran, Using Remote Sensing and Geographic Information System ............... 131

*Payam Sajadi, Amit Singh, Saumitra Mukherjee, Harshita Asthana, P Pingping. Luo, and Kamran Chapi*

**Chapter 10** Fog—A Ground Observation-Based Climatology and Forecast over North India ...... 151

*Sanjay Kumar Srivastava, Rohit Sharma, Kamna Sachdeva, and Anu Rani Sharma*

**Chapter 11** Estimation of Evapotranspiration through Open Access Earth Observation Data Sets and Its Validation with Ground Observation ........................................... 173

*Kishan Singh Rawat, Sudhir Kumar Singh, and Anju Bala*

**Chapter 12** Use of Hydrological Modeling Coupled with Geographical Information System for Plotting Sustainable Management Framework ....................................... 191

*Pankaj Kumar and Chander Kumar Singh*

**Chapter 13** CERES-Rice Model to Define Management Strategies for Rice Production; Soil Moisture and Evapotranspiration Estimation during Drought Years—A Study over Parts of Madhya Pradesh, India .......................................... 207

*Sourabh Shrivastava, S.C. Kar, and Anu Rani Sharma*

**Chapter 14** Simulation of Hydrologic Processes through Calibration of SWAT Model with MODIS Evapotranspiration Data for an Ungauged Basin in Western Himalaya, India ........................................................................................................ 223

*Pratik Dash*

**Chapter 15** Impact of Land Use and Land Cover Changes on Nutrient Concentration in and around Kabar Tal Wetland, Begusarai (Bihar), India .................................... 243

*Rajesh Kumar Ranjan and Priyanka Kumari*

**Chapter 16** Evaluation of Spectral Mapping Methods of Mineral Aggregates and Rocks along the Thrust Zones of Uttarakhand Using Hyperion Data ............................... 251

*Soumendu Shekhar Roy and Chander Kumar Singh*

**Chapter 17** Assessment of Flood-Emanated Impediments to Kaziranga National Park Grassland Ecosystem—A Binocular Vision with Remote Sensing and Geographic Information System ............................................................................. 275

*Surajit Ghosh, Subrata Nandy, Debarati Chakraborty, and Raj Kumar*

**Supplementary Information** ........................................................................................................... 291

**Index** ............................................................................................................................................... 301

# List of Figures

**Figure 1.1** Flowchart showing the detailed methodology adopted in the study ........................ 3
**Figure 1.2** Map of the study area ........................ 4
**Figure 1.3** Map of LULC of Kuttanad for the years 2005 and 2014 ........................ 6
**Figure 1.4** Map of NDWI of Kuttanad for the years 2005 and 2014 ........................ 7
**Figure 1.5** Map of LST of Kuttanad for the years 2005 and 2014 ........................ 7
**Figure 1.6** Map of NDVI of Kuttanad for the years 2005 and 2014 ........................ 8
**Figure 2.1** Study area ........................ 14
**Figure 2.2** Digital elevation model map of Khamis-Mushyet ........................ 14
**Figure 2.3** Flow diagram of methodology adopted ........................ 16
**Figure 2.4** LU/LC Map 1990 ........................ 17
**Figure 2.5** LU/LC Map 2002 ........................ 18
**Figure 2.6** LU/LC Map 2014 ........................ 19
**Figure 2.7** Urban Expansion Map during 1990–2002–2014 ........................ 25
**Figure 3.1** Location map of Chhota Shigri Glacier, Himachal Pradesh. (a) Outline of India, (b) route map, and (c) outline map of Chhota Shigri ........................ 29
**Figure 3.2** Lateral moraines in the inactive zone of the glacier ........................ 30
**Figure 3.3** Identification of morainic loop ........................ 30
**Figure 3.4** Comparative lithological characteristics and variation in reflectance of Chhota Shigri and Bada Shigri glaciers ........................ 31
**Figure 3.5** Debris cover in the ablation zone of the glacier ........................ 31
**Figure 3.6** Identification of break in the topography, indicating sites for crevasse development ........................ 32
**Figure 3.7** Position of the snout ........................ 32
**Figure 3.8** Identification of glacier boundary ........................ 33
**Figure 3.9** Glacier boundary delineation through manual digitization in the Parbati sub-basin, Beas ........................ 34
**Figure 3.10** Graphical representation of the deglaciation pattern in the Beas basin, 1972–2006 ........................ 35
**Figure 3.11** Glacier boundary delineation through manual digitization in the Ravi Basin ........................ 35
**Figure 3.12** Mapping of debris-covered glaciers through satellite images for the glaciers encompassing the Great Himalayan Range. (a) TM4/TM5 ratio image, (b) thresholding applied to TM4/TM5 ratio image showing snow- and ice-covered parts of the glaciers (black), (c) hue component from Intensity Hue Saturation (IHS) image derived after transformation of RGB composite using TM bands 3, 4, and 5, (d) thresholding of hue component of IHS image for demarcation of vegetation-free areas (black), (e) slope image derived from the Cartosat-1 DEM, and (f) slope map showing all slope values of <15 deg (black) ........................ 39

| | | |
|---|---|---|
| **Figure 4.1** | The location of the four cities on a map of India | 53 |
| **Figure 4.2** | LULC maps for Chennai | 58 |
| **Figure 4.3** | LST zones and their areas under different LULC in Chennai for 1991, 2000, and 2014 | 59 |
| **Figure 4.4** | LST statistics for various LULC classes over the years for the city of Chennai | 60 |
| **Figure 4.5** | LULC maps for Delhi | 60 |
| **Figure 4.6** | LST zones and their areas under different LULC in Delhi for 1998, 2003, and 2011 | 62 |
| **Figure 4.7** | LST statistics for various LULC classes over the years for the city of Delhi | 62 |
| **Figure 4.8** | LULC maps for Kolkata | 63 |
| **Figure 4.9** | LST zones and their areas under different LULC in Kolkata for 1989, 2006, and 2010 | 64 |
| **Figure 4.10** | LST statistics for various LULC classes over the years for the city of Kolkata | 65 |
| **Figure 4.11** | LULC maps of Mumbai for the years 1998 and 2009 | 66 |
| **Figure 4.12** | LST maps for 1998 and 2009 | 67 |
| **Figure 4.13** | LST statistics for various LULC classes over the years for the city of Mumbai | 68 |
| **Figure 5.1** | Study area | 81 |
| **Figure 5.2** | Land use and land cover (LULC) maps for Lucknow district for time period 2000–2031, depicting the changing land use on a temporal scale | 82 |
| **Figure 6.1** | Location map of the study area shown by LISS-IV imagery | 90 |
| **Figure 6.2** | Methodology adopted | 91 |
| **Figure 6.3** | The structure of a three-layer ANN | 94 |
| **Figure 6.4** | Representation of training pixels in 2D space between bands 3 and 4 | 97 |
| **Figure 6.5** | (a) SVMs with linear kernel, (b) SVMs with polynomial of degree 2 kernel, (c) SVMs with radial basis kernel, and (d) SVMs with sigmoid function kernel-based classification | 98 |
| **Figure 6.6** | ANN-based classification | 101 |
| **Figure 6.7** | RF-based classification | 102 |
| **Figure 7.1** | Study area | 115 |
| **Figure 7.2** | Land use land cover (LULC) maps for Lucknow district for years 2000, 2007, and 2017 | 116 |
| **Figure 7.3** | Increasing built-up (urban) area from 2000 to 2017 | 117 |
| **Figure 7.4** | Variation of spatial metrics derived using FRAGSTATS for built-up class over the period of time | 118 |
| **Figure 8.1** | Study area for the present study. Stok and Matoo catchments are presented in pink and blue boundary, and yellow boundary presents glacier/glacieret outlines | 124 |

# List of Figures

| | | |
|---|---|---|
| **Figure 8.2** | Catchmentwide change in surface area of glaciers/glacierets of Stok and Matoo village catchments | 127 |
| **Figure 8.3** | Total ice reserve of Glacier and Glacieret of Stok and Matoo village catchments in 1993 and 2015 | 128 |
| **Figure 9.1** | Location map of the study area | 134 |
| **Figure 9.2** | Stream order of drainage network in the Qorveh–Dehgolan basin | 138 |
| **Figure 9.3** | Log stream number and stream order relationship for individual watersheds | 139 |
| **Figure 9.4** | Log stream length and stream order relationship | 140 |
| **Figure 9.5** | Drainage density map for all watersheds | 140 |
| **Figure 9.6** | Hypsometric curve (*HC*) and hypsometric integral (*HI*) for eight watersheds | 141 |
| **Figure 9.7** | *Sl* values overlaid on 3D-DEM | 143 |
| **Figure 9.8** | Elongation map for eight watersheds | 145 |
| **Figure 9.9** | Basin slope map for eight watersheds | 146 |
| **Figure 9.10** | Lithology map of study area overlaid on watershed boundary | 147 |
| **Figure 10.1** | Study area—IGP and Ghaziabad | 153 |
| **Figure 10.2** | Ground meteorological observatories in IGP | 154 |
| **Figure 10.3** | Monthly mean frequency of fog over Ghaziabad (October–February) | 157 |
| **Figure 10.4** | Decadal frequency of fog | 157 |
| **Figure 10.5** | Average monthly visibility (less than 1 km in days) | 159 |
| **Figure 10.6** | (a) Percentage frequency of duration, (b) diurnal variability of fog (January and December), (c) diurnal variability of fog (clusterwise), and (d) percentage frequency of intensity | 159 |
| **Figure 10.7** | Average daily fog persistence (h) | 160 |
| **Figure 10.8** | Correlation between the time series and number of fog days. (a) Mann–Kendall trend (October) and (b) linear trend (October) | 162 |
| **Figure 10.9** | Time series analysis of fog frequency over IGP considered independently for the months of December and January over the last 40 years. (a) Time series analysis (January and December) and (b) time series analysis (Peak winter) | 162 |
| **Figure 10.10** | Trend analysis of winter fog | 163 |
| **Figure 10.11** | Decadewise trend analysis | 164 |
| **Figure 10.12** | Spatial variability of fog over the entire IGP during the months of December and January. (a) Spatial cluster analysis (Z-score), (b) spatial cluster analysis (*p*-value), and (c) spatial cluster analysis (% average fog frequency) | 166 |
| **Figure 10.13** | ARIMA model forecast | 168 |
| **Figure 10.14** | AOD for the duration 2000–2010 | 169 |
| **Figure 11.1** | Location of study area. (a) Lysimeter location in WTC experimental field and (b) WTC experimental field location over New Delhi | 176 |

| | | |
|---|---|---|
| **Figure 11.2** | Net radiation over New Delhi | 178 |
| **Figure 11.3** | Soil heat flux over New Delhi | 179 |
| **Figure 11.4** | Sensible heat flux over New Delhi | 180 |
| **Figure 11.5** | Evaporative fraction over New Delhi | 180 |
| **Figure 11.6** | $ET_{24h}$ during December 23, 2011, over New Delhi | 181 |
| **Figure 11.7** | $ET_{24h}$ during February 9, 2011, over New Delhi | 182 |
| **Figure 11.8** | $ET_{24h}$ during March 29, 2011, over New Delhi | 182 |
| **Figure 11.9** | $ET_{24h}$ during April 14, 2011, over New Delhi | 183 |
| **Figure 11.10** | Variation of LAI and NDVI with cropping period | 184 |
| **Figure 11.11** | LAI–NDVI relationship | 184 |
| **Figure 11.12** | Correlation between measured and predicted $ET_{24h}$ | 186 |
| **Figure 11.13** | Seasonal variation of $ET_{24h}$ over the study area | 187 |
| **Figure 12.1** | Location map of the study area | 193 |
| **Figure 12.2** | Types of aquifer in Sri Lanka | 194 |
| **Figure 12.3** | Distribution of the confined aquifers in the various basins | 196 |
| **Figure 12.4** | Rainfall time series used in the model (in m/day) | 197 |
| **Figure 12.5** | Horizontal hydraulic conductivity distribution map of the area—From the groundwater flow model (m/day) | 197 |
| **Figure 12.6** | Rainfall and EVT for the simulated years (1984–1986) | 198 |
| **Figure 12.7** | Major land use classes of the Killinochi district | 199 |
| **Figure 12.8** | Major soil types of the Killinochi district | 201 |
| **Figure 12.9** | Head contours in meters. (a) At the beginning of the simulation (SP-1, TS-1) and (b) at the end of the simulation (SP-12, TS-365). SP = Stress period, TS = Time step | 202 |
| **Figure 12.10** | Head contours in meters. (a) At the beginning of the simulation (SP-1, TS-1) and (b) at the end of the simulation (SP-36, TS-1095). SP = Stress period, TS = Time step | 202 |
| **Figure 12.11** | Rainfall time-series (in mm/month) | 202 |
| **Figure 12.12** | Recharge rate for the simulated consecutive years in cubic meter (1984–1986) | 203 |
| **Figure 12.13** | (a) Zonal budget recharge rates for the different basins, simulated, and observed (MCM) and (b) major basin land use distribution. N.N1 and N.N2 are the two basins with no name | 203 |
| **Figure 13.1** | Study area Balaghat, Jabalpur, Natshinghpur, and Seoni districts | 209 |
| **Figure 13.2** | Rainfall (mm/day), soil moisture ($m^3/m^3$), and evapotranspiration over Madhya Pradesh. (a) Rain June climatology, (b) rain July climatology, (c) rain August climatology, (d) SM June climatology, (e) SM July climatology, (f) SM August climatology, (g) ET June climatology, (h) ET July climatology, and (i) ET August climatology | 211 |

List of Figures ................................................................................................................ xi

**Figure 13.3** Observed rainfall and mean temperature for the four stations in Madhya Pradesh, India. (a) Climatology of rainfall and mean temperature during January to December, (b) seasonal mean (JJAS) rainfall anomaly for the Balaghat, Jabalpur, Narsinghpur, and Seoni districts ............................................. 212

**Figure 13.4** Composite of break phases during different RMM phases. (a) RMM phase 1, (b) RMM phase 2, (c) RMM phase 6, (d) RMM phase 7, and (e) composite of all phases (1, 2, 6, and 7) .................................................................................. 213

**Figure 13.5** Year-to-year variations in the observed rice yield. (a) Observed yield and (b) simulated yield from 1990 to 2011 for the Balaghat, Jabalpur, Narsinghpur, and Seoni districts ...................................................................... 214

**Figure 13.6** Difference between the observed and simulated yield for (a) control runs (b) by changing the genetic coefficients for IR36 rice crop .................................. 215

**Figure 13.7** Observed anomaly and detrend anomaly of rice yield the (a) Balaghat, (b) Jabalpur, (c) Narsinghpur, and (d) Seoni districts ............................................. 217

**Figure 13.8** Model-simulated and MODIS evapotranspiration from 2000 to 2011 (per data availability) for (a) Balaghat, (b) Jabalpur, (c) Narsinghpur, and (d) Seoni districts .................................................................................................. 218

**Figure 13.9** RMSE of SM from remote sensing and model simulations during drought years ................................................................................................................. 219

**Figure 14.1** Location of Sirsa River basin including towns and subbasins (numbers indicate subbasin ID) .................................................................................... 226

**Figure 14.2** Sensitivity of water balance components to groundwater parameters. (a) GWQMN, (b) REVAPMN, (c) GW_REVAP, and (d) RCHR_DP .................... 231

**Figure 14.3** Sensitivity of water balance components to soil parameters. (a) SOL_Z, (b) SOL_AWC, and (c) SOL_K ......................................................................... 232

**Figure 14.4** Sensitivity of water balance components to HRU parameters. (a) EPCO, (b) ESCO, and (c) CANMX ............................................................................ 234

**Figure 14.5** Comparison of actual evapotranspiration ($ET_a$) between SWAT simulation and MODIS data at (a) daily (8-day composite) and (b) monthly time step during calibration (2004–2006) and validation (2007–2008) periods ................... 235

**Figure 14.6** Box-and-whisker plot of monthly deviation in evapotranspiration, that is, $\Delta ET$ (SWAT ET—MODIS ET) for (a) daily (8-day cumulative), and (b) monthly comparisons during the period of 2004–2008. The box-and-whisker plots show the median, first, and third quartiles. The caps at the end of the boxes show the extreme values .................................................................................. 236

**Figure 14.7** Average monthly $ET_a$ simulated from SWAT and MODIS data for the period 2004–2008. (a) Comparison of average monthly ET simulated from SWAT model with MODIS ET and (b) scatter plot of SWAT simulated monthly ET against monthly MODIS ET values ................................................................. 237

**Figure 14.8** Mean monthly value of selected hydrologic components of Sirsa basin for the period of 2003–2008 ........................................................................................ 238

**Figure 15.1** Map of the study area (Kabar Tal wetland) .......................................................... 245

**Figure 15.2** Land use cover map of Kabar Tal wetland 1990, 2010, and 2015 .......................... 246

| | | |
|---|---|---|
| **Figure 15.3** | Graph showing nutrients concentration (log scale) in and around Kabar Tal in 1990, 2012, and 2015 | 247 |
| **Figure 16.1** | Lithological map and thrust zones of Lesser Himalaya-Kumaun | 254 |
| **Figure 16.2** | The minimum noise fraction image obtained from the Hyperion scene | 257 |
| **Figure 16.3** | n-dimensional space for endmember extraction (pure pixel) at 10,000 iterations, 2.5 threshold for PPI over the Hyperion image with the axes of MNF bands, which are the dimensions of this subspace (ENVI n-dimension Visualizer) | 257 |
| **Figure 16.4** | Schematic diagram of the methodology of the SAM method | 261 |
| **Figure 16.5** | Spectral curve matching of endmembers and the most suitable rock spectra from the spectral library according to the equal weightage defined to the spectral mapping methods. (a) and (b) Image end members matching with serpentine marble, (c) image end members matching with chloritic gneiss and (d) image end members matching with green quartzite | 263 |
| **Figure 16.6** | Classified image of Hyperion data according to the equal weightage provided to SAM and BE classifier spectral mapping methods | 264 |
| **Figure 16.7** | Classified image of Hyperion data according to the 100% weightage provided to BE classifier spectral mapping method | 267 |
| **Figure 16.8** | Classified image of Hyperion data according to a weightage factor of 0.6 (60%) of binary encoding and 0.4 (40%) of SAM classifier | 269 |
| **Figure 16.9** | Classified image of Hyperion data according to field spectral data of phyllite and quartzite data collected using ASD-Spectroradiometer | 271 |
| **Figure 17.1** | Study area of Kaziranga National Park | 278 |
| **Figure 17.2** | Topography of Kaziranga National Park | 278 |
| **Figure 17.3** | Rainfall trend of the Kaziranga National Park region | 279 |
| **Figure 17.4** | Rainfall distribution of the Nagaon, Sonitpur, Golaghat, and Kaziranga districts | 279 |
| **Figure 17.5** | Land use land cover map | 284 |
| **Figure 17.6** | Flood map of Brahmaputra River | 285 |
| **Figure 17.7** | Flood area statistics of grassland | 285 |
| **Figure 17.8** | Water level anomalies derived from SARAL/AltiKa | 286 |
| **Figure 17.9** | Interface of *River App* android application | 287 |
| **Figure A.1** | Surface albedo ($\alpha$) | 295 |
| **Figure A.2** | Normalized difference vegetation index (NDVI) | 296 |
| **Figure A.3** | Leaf area index (LAI) | 296 |
| **Figure A.4** | Soil-adjusted vegetation index (SAVI) | 297 |
| **Figure A.5** | Surface emissivity | 298 |
| **Figure A.6** | Surface temperature (Kelvin) | 299 |

# List of Tables

| | | |
|---|---|---|
| **Table 1.1** | Specifications of Landsat TM and OLI Data | 3 |
| **Table 1.2** | Table Showing the Change in the Area of the Various Land Use Categories between 2005 and 2014 for Kuttanad (in km$^2$) | 6 |
| **Table 2.1** | Classification Description | 15 |
| **Table 2.2** | Land Use and Land Covers Distribution (1990) | 17 |
| **Table 2.3** | Land Use and Land Cover Distribution (2002) | 18 |
| **Table 2.4** | Land Use and Land Cover Distribution (2014) | 19 |
| **Table 2.5** | Gains and Losses and Net Change | 21 |
| **Table 2.6** | Land Transformation | 22 |
| **Table 3.1** | A Tabular Representation of the Literature for Automated Delineation Methods of Debris Covered Glaciers | 36 |
| **Table 3.2** | Parameters of the Applied V–A Relations. Cogely (2011) Used the Same Relationships for the Glacier Volume Estimations of the Same Region. Volumes Are Calculated in m$^3$ for Input Glacier Areas Measured in m$^3$. $c$ is in m$^{3-2\gamma}$; $\gamma$ is Dimensionless | 40 |
| **Table 4.1** | Basic Information on Demography and Climate of the Cities | 53 |
| **Table 4.2** | Details and Characteristics of the Satellite Data Used | 54 |
| **Table 4.3** | List of Landsat Datasets Used for LULC Classification | 55 |
| **Table 5.1** | A Brief Description of the Review Studies on Various Land Change Models by Different Authors | 74 |
| **Table 5.2** | Description of Broad Characteristics (Defined by Verburg et al., 2006) to Classify Different Land Change Models | 76 |
| **Table 5.3** | Description of the Land Use and Land Cover Classes Identified in the Study Area, Mentioning the Different Features | 82 |
| **Table 5.4** | Area (in km$^2$) Covered by Different Land Use and Land Cover Classes in Lucknow District over the Period of 2000–2031 | 83 |
| **Table 6.1** | Training and Testing Pixels Used in the Classification | 91 |
| **Table 6.2** | Number of Correctly and Incorrectly Classified Pixels for the Two Algorithms | 96 |
| **Table 6.3** | Separability Analysis between Diverse Crop and Noncrop Classes Using the TD Method | 97 |
| **Table 6.4** | Separability Analysis between Diverse Crop and Noncrop Classes Using the JM Distance Method | 98 |
| **Table 6.5** | Selected Measures (SVMs with Linear Kernel) | 99 |
| **Table 6.6** | Selected Measures (SVMs with Polynomial Kernel) | 99 |
| **Table 6.7** | Selected Measures (SVMs with Radial Basis Kernel) | 100 |

| | | |
|---|---|---|
| Table 6.8 | Selected Measures (SVMs with Sigmoid Kernel) | 100 |
| Table 6.9 | Selected Measures (ANN Classification) | 102 |
| Table 6.10 | Selected Measures (RF Classification) | 103 |
| Table 6.11 | Classification Algorithms OA, κ Results before and after Postprocessing Steps | 104 |
| Table 6.12 | Statistical Significance in the Accuracy between Two Different Algorithms by Z-Test | 104 |
| Table 7.1 | Brief Description of the Spatial Metrics Used in the Study | 113 |
| Table 7.2 | Description of the Land Use/Land Cover Classes Identified in the Study Area, Mentioning the Different Features | 116 |
| Table 7.3 | Area (in $km^2$) Covered by Different Land Use/Land Cover Classes in Lucknow District over the Period of 2000–2017 | 117 |
| Table 7.4 | Change Detection in Area ($km^2$) between Land Use/Land Cover Classes from 2000 to 2017 | 118 |
| Table 8.1 | Brief Details of the Study Area | 124 |
| Table 8.2 | Details of the Satellite Data Used for the Present Study | 125 |
| Table 8.3 | Scaling Parameters Used in Equations 8.5 and 8.6 | 126 |
| Table 8.4 | Area Change and Surface Area of Glaciers/Glacierets in 1993 and 2015 | 127 |
| Table 8.5 | Volume Change of Glaciers and Glacierets between 1993 and 2015 in Stok and Matoo Catchment | 128 |
| Table 9.1 | Morphometric and Morphotectonic Parameters and Their Equation | 135 |
| Table 9.2 | Stream Number in Given Order | 138 |
| Table 9.3 | Stream Length in Given Order | 139 |
| Table 9.4 | $HI$ for Eight Watersheds and Their Classes | 141 |
| Table 9.5 | $Sl$ Value for Eight Watersheds and Their Classes | 142 |
| Table 9.6 | Basin Asymmetry ($Af$) Value for Eight Watersheds and Their Classes | 144 |
| Table 9.7 | Schumm (1956) Classification of Elongation Ratio | 144 |
| Table 9.8 | Elongation Ratio ($R_e$) of Eight Watersheds and Their Classes | 144 |
| Table 9.9 | River Sinuosity ($K$) Value in Given Order for All Watersheds | 145 |
| Table 10.1 | Climatological Data over Ghaziabad | 158 |
| Table 10.2 | Fog Statistics over IGP | 158 |
| Table 10.3 | Mann–Kendall Trend Statistics | 161 |
| Table 10.4 | Kendall's Tau Correlation and Sen's Slope Magnitude | 163 |
| Table 10.5 | Change in Fog Frequency by Decade | 165 |
| Table 10.6 | ARIMA Model Fit Statistics | 168 |
| Table 10.7 | Relationship with AOD 550 nm | 170 |

# List of Tables

| | | |
|---|---|---|
| Table 11.1 | Detail Remote Sensing (RS) Data Description | 176 |
| Table 11.2 | Quantified Surfaces Parameter Value over Lysimeter Point | 183 |
| Table 11.3 | Land Surface Heat, Heat Fluxes, and Daily ET over Lysimeter Point | 185 |
| Table 11.4 | Comparison of ET (from Wheat Crop) from SEBAL with Lysimeter | 185 |
| Table 12.1 | Major Land Use Classes in the Basins of Killinochi | 196 |
| Table 12.2 | Extinction Depth Estimates Based on Land Use Classes | 198 |
| Table 12.3 | Evapotranspiration Rate Estimated Based on the Data from WRB (mm/Month; Killinochi Basin) | 200 |
| Table 12.4 | Saturated Hydraulic Conductivity Values of the Various Soil Groups | 201 |
| Table 13.1 | Soil Physical Characteristics of 0–60 cm Soil Depth at the Experimental Area | 209 |
| Table 13.2 | Description of Data Sets | 210 |
| Table 13.3 | Calibrated Genetic Coefficients for Different Varieties of Rice | 210 |
| Table 13.4 | Excess and Deficit Rainfall Years for the Four Districts of Madhya Pradesh | 212 |
| Table 13.5 | RMSE of Simulated Yield for Both Genetic Coefficients (GC) | 215 |
| Table 13.6 | Genetic Coefficients for Different Sensitivity Experiments | 216 |
| Table 13.7 | RMSE of Simulated Yield for All Genetic Coefficients (IR36) Experiments | 216 |
| Table 13.8 | RMSE and SD of MODIS and Simulated Evapotranspiration: Difference of ET between MODIS and Simulated Values during Excess and Deficit Years | 219 |
| Table 14.1 | Description and Statistics of Sensitive Parameters of SWAT Including Sensitivity Rank and Optimal Parameterization Values | 229 |
| Table 14.2 | Description and Results of Criteria Used for Testing Statistical Performance of SWAT Model during Calibration | 229 |
| Table 15.1 | Data Source | 245 |
| Table 15.2 | Areas of Land Use/Land Cover Classes | 246 |
| Table 15.3 | Land Use/Land Cover Change: Trend, Rate, and Magnitude | 247 |
| Table 15.4 | Change Detection in the Nutrient Concentration with Reference to Reports Published in 2002, 2012, and 2015 for the Kabar Tal | 247 |
| Table 15.5 | Correlation between Land Use and Nutrient Concentration | 248 |
| Table 16.1 | Data Used in This Study and Their Sources | 256 |
| Table 16.2 | End Members Extracted by n-Dimensional Visualizer on PPI Band | 258 |
| Table 16.3 | Result of the Spectral Mapping Methods for Corresponding Spectral Endmember Classes for Equal Weight to the Three Spectral Mapping Methods (SAM, BE, and SFF) | 262 |
| Table 16.4 | Percentage of Endmember Classes in the Study Area (Equal Proportion in Weight for SAM, BE, SFF, and Single Range) | 265 |
| Table 16.5 | Result of the Spectral Mapping Methods with 100% Weight on BE | 266 |

| | | |
|---|---|---|
| **Table 16.6** | Percentage of Spectral Endmember Classes in BE Classification | 268 |
| **Table 16.7** | Percentage of Endmember Considering SAM and BE Methods | 270 |
| **Table 16.8** | The Class Distribution Result of the Field Spectral Mapping for the Study Area between North and South Almora Thrusts | 272 |
| **Table 17.1** | Key Studies in Kaziranga National Park Using Remote Sensing and GIS | 283 |
| **Table 17.2** | Landsat 8 Band Description | 283 |
| **Table 17.3** | Land Use Land Cover Area in Percentage | 284 |
| **Table A.1** | Landsat7-ETM+ Solar Spectral Irradiances ($ESUN_\lambda$) | 294 |

# Preface

Space is everywhere and so are spatial problems. Geoinformatics deals with acquiring, processing, examining, and visualizing spatial data. With the advancement of computer technology and availability of earth observation satellite data, the application domain of geospatial technologies has increased many times. It has begun contributing significantly to monitoring and managing natural resources. The increased effectiveness of geospatial technology in natural resources management can be understood from the fact that geospatial technologies have begun providing decision-making tools in scientific research as well as policy and planning.

Today geospatial technology is being used in applications in several domains of society, be it natural resources, ecosystem services, urban planning, or climate change. However, although many books on the subject deal with very specific topics of natural resources, we did not find a book dealing with different applications for remotely sensed data. This book has been written to provide people who deal with geoinformatics a comprehensive view of the applications of remote sensing in multiple domains of natural resources ranging from urban, glacier, and forest land change modeling to the use of remotely sensed spectroscopic techniques in tectonic applications.

This book comprises work by eminent scientists in the world in the area of management of natural resources and is bound to generate interest in students, researchers, and academicians. The book's uniqueness lies in the case studies in each chapter and should help the reader develop a comprehensive overview of what and how geospatial applications can be used in the management of natural resources.

The editor wishes to thank the authors who have contributed to the book and have emended the chapters to suit its needs.

*Geospatial Applications for Natural Resources Management* has been written to introduce its readers to the vast areas where geospatial technology has been used in the best possible way in managing the natural resources.

The use of the following features covered in the book make its subject matter easy to understand:

- Each chapter contains an abstract providing a brief summary of its content for quick reading and understanding.
- Each chapter includes a particular case study to demonstrate the application of geospatial technology in analyzing the real-world problems, for example, urban expansion, glaciers, fog, and hydrological processes.
- Each chapter includes figures and tables for easy analysis of concepts discussed.
- Each chapter has a considerable reference list.

The book is written for researchers working in several domains of geoinformatics who are interested in knowing how geospatial technology can be used to analyze real-world processes. For better understanding, the reader needs practical knowledge of geospatial technology and real-world phenomena (e.g., urban expansion, land use change, drainage processes, fog) to which the technology can be applied. The overall purpose of the book is to demonstrate, by the means of particular case study in each chapter, the vast applications of geospatial technologies in analyzing real world phenomenon and managing the resources. Hence, the book caters the needs of researchers working in several domains of geoinformatics.

# Editor

**Dr. Chander Kumar Singh** is assistant professor at the Department of Energy and Environment at TERI School of Advanced Studies, New Delhi, India. He earned his PhD from Jawaharlal Nehru University (2011) and was awarded the Sat Pal Mittal Fellowship by Indian Association of Parliamentarians on Population and Development for his doctoral research. Dr. Singh's research ranges across several domains including the field of Geogenic Contamination, Climate Change and Impact on Hydrological System, Remote sensing and GIS applications in Ground Water Exploration, Geochemistry and Hydrogeology, and Public Health. Dr. Singh received the Young Scientist Award by the International Union of Geological Sciences at the Euro Conference 2009, Switzerland. His collaborations on research front ranges across several leading institutions of the world such as the Columbia University of New York, New York; Massachusetts Institute of Technology, Cambridge, Massachusetts; United Nations University, Tokyo, Japan; and many more. Dr. Singh has authored more than 80 research papers in journals and conferences and has supervised and trained doctoral scholars. He leads one of the most prestigious grants from USAID and National Science Foundation, on Targeting Low fluoride and Low Arsenic aquifers in rural areas of Punjab, India. Pertaining to his research, Dr. Singh was invited speaker at MIT, Boston University, and Columbia University under HESN and Superfund research program. He has received grants from several international and national funding agencies such as USAID, National Science Foundation, International Growth Centre, UKAID, Ministry of Environment and Forest, and Ministry of Science and Technology, Govt. of India. He is also on the editorial board of several renowned journals and is reviewer of several high-quality peer-reviewed journals.

# Contributors

**Harshita Asthana**
School of Environmental Sciences
Jawaharlal Nehru University (JNU)
New Delhi, India

**Anju Bala**
Department of Civil Engineering
World College of Technology and Management (WCTM)
Gurgaon, India

**Akanksha Balha**
Department of Energy and Environment
TERI School of Advanced Studies
New Delhi, India

**Debarati Chakraborty**
Department of Molecular Biology and Biotechnology
University of Kalyani
and
Centre for Interdisciplinary Studies
Kolkata, India

**Kamran Chapi**
Department of Rangeland and Watershed Management
College of Natural Resources
University of Kurdistan
Sanandaj, Iran

**Arti Choudhary**
Department of Civil Engineering
Indian Institute of Technology
Guwahati, India

**Pratik Dash**
Department of Geography
School of Science
Adamas University
Kolkata, India

**Shruti Dutta**
School of Earth & Environmental Sciences
Amity University Haryana
Gurgaon, India

and

Glacier Research Group
School of Environmental Sciences
Jawaharlal Nehru University (JNU)
New Delhi, India

**Surajit Ghosh**
Department of Civil Engineering
National Institute of Technology
Durgapur, India

and

IORA Ecological Solutions Pvt. Ltd.
New Delhi, India

**Hoang Thi Hang**
Department of Geography
Jamia Millia Islamia
New Delhi, India

**P.K. Joshi**
Department of Natural Resources
TERI School of Advanced Studies
and
School of Environmental Sciences
Jawaharlal Nehru University (JNU)
Delhi, India

**S.C. Kar**
National Centre for Medium Range Weather Forecasting
Noida, India

**N. Kikon**
Nagaland State Disaster Management Authority
Nagaland Civil Secretariat
Nagaland, India

**Pankaj Kumar**
Institute for the Advanced Study of
   Sustainability (UNU-IAS)
United Nations University
Tokyo, Japan

**Pradeep Kumar**
Department of Physics
Indian Institute of Technology (BHU)
Varanasi, India

**Raj Kumar**
IORA Ecological Solutions Pvt. Ltd.
New Delhi, India

**M. Kumari**
Department of Energy and Environment
TERI School of Advanced Studies
New Delhi, India

and

Department of Civil Engineering
Amity School of Engineering & Technology
Amity University Uttar Pradesh
Noida, India

**Priyanka Kumari**
Centre for Environmental Sciences
Central University of South Bihar
Patna, India

**P Pingping. Luo**
Disaster Prevention Research Institute
Kyoto University
Kyoto, Japan

**Javed Mallick**
Department of Civil Engineering
College of Engineering
King Khalid University
Abha, Saudi Arabia

**Saumitra Mukherjee**
School of Environmental Sciences
Jawaharlal Nehru University
New Delhi, India

**Subrata Nandy**
Indian Institute of Remote Sensing
ISRO
Dehradun, India

**Manish Pandey**
School of Environmental Sciences
Jawaharlal Nehru University
New Delhi, India

**Rajendra Prasad**
Department of Physics
Indian Institute of Technology (BHU)
Varanasi, India

**AL. Ramanathan**
Glacier Research Group
School of Environmental Sciences
Jawaharlal Nehru University (JNU)
New Delhi, India

**Rajesh Kumar Ranjan**
Centre for Environmental Sciences
Central University of South Bihar
Patna, India

**Kishan Singh Rawat**
Centre for Remote Sensing and
   Geoinformatics
Sathyabama University (A Joint Initiative of
   ISRO & Sathyabama University)
Chennai, India

**Soumendu Shekhar Roy**
Department of Energy and Environment
TERI School of Advanced Studies
New Delhi, India

**Kamna Sachdeva**
Department of Natural Resources
TERI School of Advanced Studies
New Delhi, India

**Payam Sajadi**
School of Environmental Sciences
Jawaharlal Nehru University
New Delhi, India

## Contributors

**Anu Rani Sharma**
Department of Natural Resources
TERI School of Advanced Studies
New Delhi, India

**Richa Sharma**
Flemish Institute of Technological Research (VITO)
Mol, Belgium

and

Department of Natural Resources
TERI School of Advanced Studies
New Delhi, India

**Rohit Sharma**
Department of Natural Resources
TERI School of Advanced Studies
New Delhi, India

**Sourabh Shrivastava**
National Centre for Medium Range Weather Forecasting
Noida, India

and

Department of Natural Resources
TERI School of Advanced Studies
New Delhi, India

**Amit Singh**
School of Environmental Sciences
Jawaharlal Nehru University
New Delhi, India

**Chander Kumar Singh**
Department of Energy and Environment
TERI School of Advanced Studies
New Delhi, India

**Sudhir Kumar Singh**
K. Banerjee Centre of Atmospheric & Ocean Studies
IIDS, Nehru Science Centre
University of Allahabad
Allahabad, India

**Mohd Soheb**
School of Environmental Sciences
Jawaharlal Nehru University
New Delhi, India

**Sanjay Kumar Srivastava**
Department of Natural Resources
TERI School of Advanced Studies
and
Indian Air Force
New Delhi, India

**R.K. Tomar**
Department of Civil Engineering
Amity School of Engineering & Technology
Amity University Uttar Pradesh
Noida, India

**Sarvagya Vatsal**
School of Environmental Sciences
Jawaharlal Nehru University
New Delhi, India

# 1 Spatiotemporal Analysis of Urban Expansion and Its Impact on Surface Temperature and Water Bodies

*Chander Kumar Singh, M. Kumari, N. Kikon, and R.K. Tomar*

## CONTENTS

1.1 Introduction ........................................................................................................2
1.2 Materials and Methods .......................................................................................2
    1.2.1 Study Area ..............................................................................................3
    1.2.2 Data Set...................................................................................................4
    1.2.3 Preprocessing of Image ..........................................................................4
    1.2.4 Image Classification................................................................................4
    1.2.5 Derivation of Normalized Difference Water Index ................................4
    1.2.6 Derivation of Normalized Difference Vegetation Index ........................5
    1.2.7 Mono-Window Algorithm for the Estimation of Land Surface Temperature ............5
1.3 Result and Discussions .......................................................................................6
    1.3.1 Spatiotemporal Analysis of Normalized Difference Water Index and Land Use/Land Cover .................6
    1.3.2 Spatiotemporal Analysis of Normalized Difference Vegetation Index and Land Surface Temperature ............7
1.4 Conclusion ..........................................................................................................8
References...................................................................................................................8

**ABSTRACT** During last few decades, the Kuttanad district of Kerala, India, experienced exponential urban expansion. This has several environmental impacts, the major ones being the decrease in water bodies and the rise in land surface temperature (LST). The main objective of this study was to understand the impact of urbanization on surface temperature and water bodies. Landsat 5 data of February 10, 2005, and Landsat 8 data of February 11, 2014, were used in this study. To validate the classification result for water bodies, the Normalized Difference Water Index (NDWI) was performed. The Normalized Difference Vegetation Index and the LST were also computed to quantify the impact on vegetation spread and temperature change. Results indicated that the area covered by water bodies decreased from 2005 to 2014. Likewise, a prominent increase in the area occupied by built up was noted from 2005 to 2014. The overall vegetation spread was found to be increasing as a result of the different agricultural practices. The LST was found to be low in the year 2014 due to the increase in vegetation cover. However, local analysis indicated that the LST surrounding the dense built-up region was high as compared to that of the less dense built-up area.

## 1.1 INTRODUCTION

Over the past few decades, urbanisation has rapidly increased as the result of the population increase. The urbanization has had a detrimental impact on the environment and natural resources and has become an important area of research. Remote sensing (RS) technology is extensively used now in diverse applications for monitoring changes such as urban expansion,[1,2] change in green cover,[3,4] hydrology,[4,5] and change in the pattern of land use/land cover.[6,7,8] Water bodies are the most essential parts on earth; they uphold all manifestations of life on Earth. One of the most irreplaceable natural resource essential for human survival is the surface water.[9] It is necessary for all living beings, vegetation and the ecosystem.[6] It is important to have proper information regarding the distribution of surface water for the assessment of current and future water resources.[10] Hence, there is a need to understand the cause of depletion in water bodies so that appropriate steps can be taken to preserve them. The accessibility of remotely sensed data have enabled the wide use of the data for identifying and extracting information regarding surface water bodies, thus making it possible to view substantial regions of Earth's surface at temporal scale and to perform spatial comparisons of change in land use.[11–14] In recent years, a number of image-processing techniques have been introduced for the extraction of information regarding water bodies from satellite data. For the extraction of water bodies from Landsat imagery, the NDWI was developed.[15,16]

This study highlights the importance of water bodies in the Kuttanad region of Kerala which is under continuous threat from urban expansion and as a consequence, water bodies are disappearing, and some have become totally contaminated by waste from processing plants, pesticides that are flushed out from the rice fields, and transfer of local sewage from homes. The exploitation of water bodies and an increased water demand for cultivation and domestic use were the main cause of the decrease of water bodies. It is therefore important to keep a continuous track on the declining water bodies to develop a proper understanding of the human impact on the water bodies and to manage them in a more efficient manner.

This study aims to model the spatiotemporal analysis of the Kuttanad district of Kerala, India, for the years 2005 and 2014 using Landsat 5 and Landsat 8 data. The Kuttanad district has experienced exponential urban expansion during recent years. This has several environmental impacts, the major one being the decrease in water bodies and the rise in LST. Thus, the main objective of this study was to understand the impact of urbanization on surface temperature and water bodies using RS and the geographic information system as analysis tools.

## 1.2 MATERIALS AND METHODS

The following steps were performed to achieve the objectives of the study: gathering data, defining the study area, preprocessing images, change detection, assessment of two different satellite images for the detection of water bodies, and extraction of water bodies from both images. Figure 1.1 shows the detailed methodology adopted in this study for monitoring the changes taking place in the area of water bodies with respect to urbanization. The satellite images were acquired and preprocessed, which served as input for the extraction of required information. The vector layers of water bodies and built-up area were extracted after processing the image. The reflectance and radiance calculation was done, and Normalized Difference Vegetation Index (NDVI), NDWI, and LST were performed. The analysis was then performed to identify the major causes responsible for the diminishing water bodies as well as their impact on water bodies and LST. Table 1.1 lists the data used. For the preprocessing of image and additional analysis, Leica Geosystems ERDAS IMAGINE 9.1 and ESRI ArcGIS 10.2.1 software were used. The layer stacking, georectification, and subsetting were the preprocessing steps performed before image analysis.

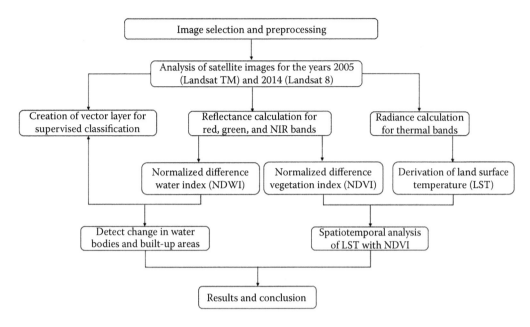

**FIGURE 1.1** Flowchart showing the detailed methodology adopted in the study.

**TABLE 1.1**
**Specifications of Landsat TM and OLI Data**

| Satellite | Sensor | Date | Resolution (m) | Wavelength (μm) |
|---|---|---|---|---|
| Landsat 5 | TM | February 10, 2005 | 30 | Band 1: 0.45–0.52 |
| | | | | Band 2: 0.52–0.60 |
| | | | | Band 3: 0.63–0.69 |
| | | | | Band 4: 0.76–0.90 |
| | | | | Band 5: 1.55–1.75 |
| | | | | Band 7: 2.08–2.35 |
| Landsat 8 | OLI | February 11, 2014 | 30 | Band 1: 0.435–0.451 |
| | | | | Band 2: 0.452–0.512 |
| | | | | Band 3: 0.533–0.590 |
| | | | | Band 4: 0.636–0.673 |
| | | | | Band 5: 0.851–0.879 |
| | | | | Band 6: 1.566–1.651 |
| | | | | Band 7: 2.107–2.294 |
| | | | | Band 9: 1.363–1.384 |

TM: Thematic Mapper; OLI: Operational Land Imager.

### 1.2.1 Study Area

Kuttanad is located in the district of Alappuzha in the state of Kerala. It lies between 9°25′30″N latitude and 76°27′50′E longitude, covering an area of 301.33 km² (Figure 1.2). The agriculture in this area occurs below sea level and at the lowest altitude in India. Kuttanad is generally characterized by a warm and humid climate for most of the year. The weather remains hot during the summer season with the temperature ranging in between 32°C to 38°C. Winter temperatures are relatively cooler with a minimum temperature of 21°C and can rise to 32°C. Kuttanad receives an average rainfall of 3000 mm.[17] During the past three decades, the town has had a rapid urban expansion due to various socioeconomic factors.

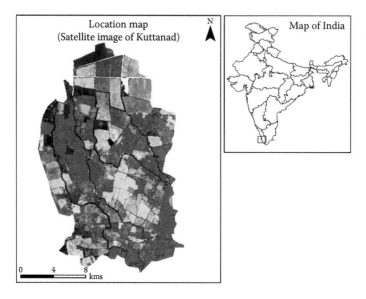

**FIGURE 1.2** Map of the study area.

### 1.2.2 Data Set

Two time period satellite images of Landsat 5 data acquired in February 10, 2005, and Landsat 8 data acquired in February 11, 2014, were obtained from U.S. Geological Survey (USGS) Earth explorer. Both images acquired are of the same season (early spring). Specifications of Landsat TM and Landsat 8 images are presented in Table 1.1.

### 1.2.3 Preprocessing of Image

For further processing of the satellite images, the following preprocessing techniques were implemented: atmospheric correction, radiometric calibration, resampling, and mosaicking. In doing so, the digital number (DN) of band 6 (TIR) of Landsat 5 and band 10 of Landsat 8 were converted into radiance values that were then used for the retrieval of LST. The bands within the solar reflectance spectral range were used for extracting the vegetation and water indexes.[18] The satellite images of both the years were mosaicked to produce new images to cover the entire study area, and finally, the study region was clipped by the method of subsetting.

### 1.2.4 Image Classification

Supervised classification was performed for the Landsat images to extract the water bodies and built-up areas. In supervised classification, the pixels with similar signatures were first vectorized and merged into unique classes, and the classification was completed by utilizing the method of maximum likelihood classifier, a method in which the pixels were distributed based on the class of the highest probability. Four classes, namely, water bodies, built up, vegetation, and cultivated and open land were classified, and the total area covered by each land use class was then calculated.

### 1.2.5 Derivation of Normalized Difference Water Index

NDWI is used to extract water bodies from a satellite image. The green wavelength is used to maximize the reflectance of water features, and the NIR is used to minimize the reflectance of the water features due to which the vegetation has been suppressed because of negative or zero values, and the water bodies are highlighted from the positive values.[15,16] NDWI is calculated using the reflectance measurements of the green and near infrared (NIR) portion of the spectrum as given in Equation 1.1.

$$\text{NDWI} = \frac{\text{Green} - \text{NIR}}{\text{Green} + \text{NIR}} \tag{1.1}$$

where:
  NIR is the Band 4 (for Landsat 5) and Band 5 (for Landsat 8)
  Green is the Band 2 (for Landsat 5) and Band 3 (for Landsat 8)

### 1.2.6 Derivation of Normalized Difference Vegetation Index

The Normalized Difference Vegetation Index (NDVI) is used to monitor the vegetation and biomass generation.[19,20] The value of NDVI is between −1 and 1 in which the value of thick vegetation ranges between 0.3 to 0.8, the value of soil lies in the range of 0.1 to 0.2, and clear water has a low positive or even marginally negative range because of the low reflectance in both the visible and infrared radiation. NDVI is calculated using the reflectance measurements of the red and NIR portion of the spectrum as given in Equation 1.2.

$$\text{NDVI} = \frac{\text{NIR} - \text{Red}}{\text{NIR} + \text{Red}} \tag{1.2}$$

where:
  NIR is the Band 4 (for Landsat 5) and Band 5 (for Landsat 8)
  Red is the Band 3 (for Landsat 5) and Band 4 (for Landsat 8)

### 1.2.7 Mono-Window Algorithm for the Estimation of Land Surface Temperature

Once the radiance values were calculated using the DNs of the thermal bands, the reverse of the Plank function was applied to derive the values of brightness temperature.[21]

$$T = \frac{K_2}{\ln\left(\dfrac{K_1 \times \varepsilon}{CV_{R1}} + 1\right)} \tag{1.3}$$

where:
  $T$ is the degree (in Kelvin)
  $CV_{R1}$ is the cell value as radiance
  $K_1 = 774.89$ (for Landsat 8) and $666.09$ (for Landsat 5 TM)
  $K_2 = 1321.08$ (for Landsat 8) and $1282.71$ (for Landsat 5 TM)

LST is then estimated by using mono-window algorithm proposed by Qin et al.[22] The mono-window algorithm is based on three principle parameters: emissivity, transmittance, and the mean atmospheric temperature. The thermal band, that is, band 6 of Landsat TM and band 10 of Landsat 8, records the radiation with the spectral range from 10.40 to 12.50 for Landsat TM and 10.60 to 11.19 for Landsat 8. The following formula is then used for deriving LST.[23]

$$T_s = \{a(1-C-D)+[b(1-C-D)+C+D]T_i - D*T_a\}/C$$

where:
 $a = -67.355351$
 $b = 0.4558606$
 $C = \varepsilon_i * \tau_i$
 $D = (1 - \tau_i)[1 + (1 - \varepsilon_i) * \tau_i]$
 $\varepsilon_i$ is the emissivity
 $\tau_i$ is the transmissivity
 $T_i$ is the brightness temperature
 $T_a$ is the mean atmospheric temperature

## 1.3 RESULT AND DISCUSSIONS

### 1.3.1 Spatiotemporal Analysis of Normalized Difference Water Index and Land Use/Land Cover

The changes in the land use and land cover (LULC) of Kuttanad are shown in the map in Figure 1.3, and the change in an area of each LULC category form 2005 to 2014 is shown in Table 1.2. During 2005, the water bodies covered a total area of 34.60 km², which decreased to 16.56 km² in 2014. The area covered by built up increased from 17.48 km² in 2005 to 38.43 km² in 2014. Vegetative area

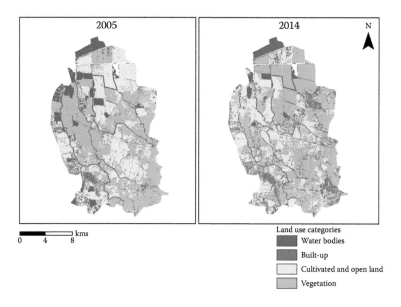

**FIGURE 1.3** Map of LULC of Kuttanad for the years 2005 and 2014.

**TABLE 1.2**
Table Showing the Change in the Area of the Various Land Use Categories between 2005 and 2014 for Kuttanad (in km²)

| Date | Water Bodies | Built Up | Vegetation | Cultivated and Open Land |
|---|---|---|---|---|
| February 10, 2005 | 34.60 | 17.48 | 125.38 | 123.85 |
| February 11, 2014 | 16.56 | 38.43 | 130.28 | 116.05 |

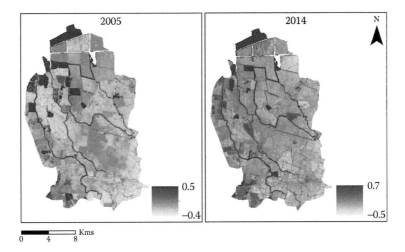

**FIGURE 1.4** Map of NDWI of Kuttanad for the years 2005 and 2014.

was 125.38 km² in 2005, which increased to 130.28 km² in 2014. A decline in the area of cultivated and open land was also observed in 2005 to be 123.85 km² and in 2014 was 116.05 km². Also, for confirming the result of water bodies derived from supervised classification, NDWI–based analysis was carried out. The NDWI values ranged from −0.4 to 0.5 in 2005 and −0.5 to 0.7 in 2014. It is evident from the map shown in Figure 1.4 that few locations of Kuttanad, which were covered by water bodies during 2005, have been converted to built-up or cultivated and open land.

### 1.3.2 Spatiotemporal Analysis of Normalized Difference Vegetation Index and Land Surface Temperature

It can be inferred from the maps in Figures 1.5 and 1.6 that LST and NDVI have an inverse relationship. The LST ranged between 29°C to 43°C during 2005 and 29°C to 44°C during 2014. During 2014, an increasing pattern of LST was found in the built-up areas. The value of NDVI ranged in between −0.9 to 0.1 during 2005 and −0.6 to 0.2 during 2014. By visual inspection, it was inferred that the vast majority of the open grounds and the wetland zones have been converted to agricultural lands because it has been occupied by people. The individuals relocated to create settlements and

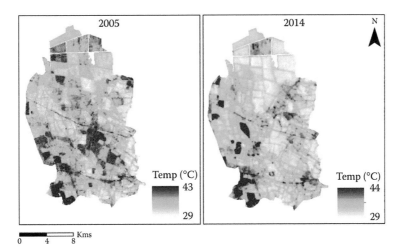

**FIGURE 1.5** Map of LST of Kuttanad for the years 2005 and 2014.

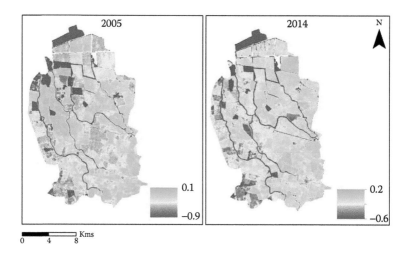

**FIGURE 1.6** Map of NDVI of Kuttanad for the years 2005 and 2014.

adopted agriculture for their livelihood, consequently depleting the water bodies. The regions where the NDVI values were high, the LST was observed to be low. It was additionally noted that the LST of the urbanized zone increased in 2014 compared with that in 2005.

## 1.4 CONCLUSION

Supervised classification was performed to find out the percentage of change in area of water bodies and built up between 2005 and 2014. To validate the classification result for water bodies, NDWI-based image classification was performed. NDVI and LST were also computed to quantify the impact on vegetation spread and temperature change in the study area. It was observed that water bodies occupied 34.60 km$^2$ of the total area in the year 2005 and decreased to 16.56 km$^2$ during 2014. The built-up area was also found to have expanded from 17.48 km$^2$ in 2005 to 38.43 km$^2$ in 2014. Changes showed that the areas that were once covered by vegetative lands became occupied by the built up, and the water bodies were found to be disappearing as a result of encroachment by the farmers converting it into land for cultivation. Some portions of the cultivated and open lands were also found to be occupied by built-up areas. Likewise, the overall vegetation spread was found to be increasing as a result of the different agricultural practices. The overall LST was found to be comparatively low in the year 2014 due to the increase in vegetation cover. Lower temperatures were noted in areas covered by water bodies. However, the local analysis indicated that the LST surrounding the dense built-up area was high as compared to that of less dense built-up area. The result provides an insight into the impact of urbanization on water bodies and surface temperature using RS and geographical information systems as analysis tools.

## REFERENCES

1. Bagan H, Yamagata Y (2012) Landsat analysis of urban growth: How Tokyo became the world's largest megacity during the last 40 years. *Remote Sens. Environ.* 127, 210–222.
2. Raja RAA, Anand V, Kumar AS, Maithani S, Kumar VA (2013) Wavelet based post classification change detection technique for urban growth monitoring. *J. Indian Soc. Remote Sens.* 41, 35–43.
3. Kaliraj S, Meenakshi SM, Malar VK (2012) Application of remote sensing in detection of forest cover changes using geo-statistical change detection matrices—A case study of devanampatti reserve forest, tamilnadu, India. *Nat. Environ. Polluti. Technol.* 11, 261–269.

4. Markogianni V, Dimitriou E, Kalivas DP (2013) Land-use and vegetation change detection in plastira artificial lake catchment (Greece) by using remote-sensing and GIS techniques. *Int. J. Remote Sens.* 34, 1265–1281.
5. Dronova I, Gong P, Wang L (2011) Object-based analysis and change detection of major wetland cover types and their classification uncertainty during the low water period at Poyang Lake, China. *Remote Sens. Environ.* 115, 3220–3236.
6. Zhu X, Cao J, Dai Y (2011) A decision tree model for meteorological disasters grade evaluation of flood. In *Proceedings of 4th International Joint Conference on Computational Sciences and Optimization 2011*, Kunming and Lijiang, Yunnan, China, Institute of Electrical and Electronics Engineers, New York, NY, pp. 916–919.
7. Salmon BP, Kleynhans W, Van Den Bergh F, Olivier JC, Grobler TL, Wessels KJ (2013) Land cover change detection using the internal covariance matrix of the extended Kalman filter over multiple spectral bands. *IEEE J. Sel. Topics Appl. Earth Observations Remote Sens.* 6, 1079–1085.
8. Demir B, Bovolo F, Bruzzone L (2013) Updating land-cover maps by classification of image time series: A novel change-detection-driven transfer learning approach. *IEEE Trans. Geosci. Remote Sens.* 51, 300–312.
9. Ridd MK, Liu JA (1998) A comparison of four algorithms for change detection in an urban environment. *Remote Sens. Environ.* 63, 95–100.
10. Du Z, Linghu B, Ling F, Li W, Tian W, Wang H, Gui Y, Sun B, Zhang X (2012) Estimating surface water area changes using time-series Landsat data in the Gingjiang river basin, China. *J. Appl. Remote Sens.* 6, 063609.
11. Zhaohui Z, Prinet V, Songde MA (2003) Water body extraction from multi-source satellite images, In: *IGARSS 2003. 2003 IEEE International Geoscience and Remote Sensing Symposium. Proceedings (IEEE Cat. No.03CH37477)*, Toulouse, France. doi:10.1109/IGARSS.2003.1295331.
12. Xu H (2006) Modification of normalised difference water index (NDWI) to enhance open water features in remotely sensed imagery. *Int. J. Remote Sens.* 27, 3025–3033.
13. Tang Z, Ou W, Dai Y, Xin Y (2013) Extraction of water body based on Landsat TM5 imagery—A case study in the Yangtze river. *Adv. Inf. Comm. Technol.* 393, 416–420.
14. Li W, Du Z, Ling F, Zhou D, Wang H, Gui Y, Sun B, Zhang X (2013) A comparison of land surface water mapping using the normalized difference water index from TM, ETM+ and ALI. *Remote Sens.* 5, 5530–5549.
15. McFeeters SK (2013) Using the normalized difference water index (NDWI) within a geographic information system to detect swimming pools for mosquito abatement: A practical approach. *Remote Sens.* 5, 3544–3561.
16. McFeeters SK (1996) The use of the normalized difference water index (NDWI) in the delineation of open water features. *Int. J. Remote Sens.* 17, 1425–1432.
17. Sreejith KA (2013) Human impact on Kuttanad wetland ecosystem: An overview. *International Journal of Science, Environment and Technology* 2(4), 679–690.
18. Kikon N, Singh P, Singh SK, Vyas A (2016) Assessment of urban heat islands (UHI) of Noida City, India using multi-temporal satellite data. *Sustain. Cities Soc.* 22, 19–28.
19. Chen D, Huang J, Jackson J (2005) Vegetation water content estimation for corn and soybeans using spectral indices derived from MODIS near- and short-wave infrared bands. *Remote Sens. Environ.* 98, 225–236.
20. Myneni RB, Hoffman S, Knyazikhin Y, Privette JL, Glassy J, Tian Y, Wang Y et al. (2002) Global products of vegetation leaf area and fraction absorbed PAR from year one of MODIS data. *Remote Sens. Environ.* 83, 214–231.
21. Chander G, Markham B (2003) Revised Landsat-5 TM radiometric calibration procedures and post calibration dynamic ranges. *IEEE Trans. Geosci. Remote Sens.* 41(11), 2674–2677.
22. Qin Z, Karnieli A, Berliner P (2001) A mono-window algorithm for retrieving land surface temperature from Landsat TM data and its application to the Israel-Egypt border region. *Int. J. Remote Sens.* 22, 3719–3746.
23. Liu L, Zhang Y (2011) Urban heat analysis using the Landsat TM data and ASTER data: A case study in Hong Kong. *Remote Sens.* 3, 1535–1552.

# 2 Landscape Pattern and Dynamics in a Fast-Growing City, Khamis-Mushyet, Saudi Arabia, Using Geoinformation Technology

*Javed Mallick and Hoang Thi Hang*

## CONTENTS

2.1 Introduction ............................................................................................................. 12
2.2 Study Area and Data Used ...................................................................................... 13
2.3 Methodology ............................................................................................................ 14
2.4 Result ....................................................................................................................... 17
    2.4.1 Land Use and Land Cover Mapping (1990) ................................................ 17
    2.4.2 Land Use and Land Cover Change (1990–2002) ........................................ 20
    2.4.3 Land Use/Land Cover Change Analysis (2002–2014) ................................ 23
    2.4.4 Land Use and Land Cover Change Analysis (1990–2014) ......................... 23
    2.4.5 Urban Expansion of Khamis-Mushyet during 1990–2002–2014 ................ 24
2.5 Conclusions ............................................................................................................. 25
References ....................................................................................................................... 26

**ABSTRACT** Changes in land use and land cover (LU/LC) are among the most important socioeconomic driving forces of global as well as local environmental change. The urban pattern of Khamis-Mushyet city, Aseer province, Kingdom of Saudi Arabia, is an outcome of natural and socioeconomic factors and their utilization by humans in time and space. Land is becoming a scarce resource due to immense economic activities and demographic pressure. Hence, information on LU/LC and possibilities for their optimal use is essential for the selection, planning, and implementation of land use schemes to meet the increasing demands for basic human needs, basically housing. The main objective of this study was to analyze relevant remote sensing data from 1990 and 2014 and identify the locations, types, and trends of the main land cover changes in Khamis-Mushyet during the last 24 years of Khamis-Mushyet city. Therefore, an attempt was made to map the status of LU/LC of Khamis-Mushyet city between 1990, 2002, and 2014 with a view to detect the land transformation and the changes that took place, particularly the built-up land. Land use affects land cover, and changes in land cover affect land use.

From this study it is noticed that there were significant losses in categories of scrublands, and exposed rock and as well as gains in built-up areas. During 1990 to 2014, mainly the scrubland, exposed rock, and scrubland, became built-up areas, especially in the northwest, eastern, and south eastern parts of the study site. The city has witnessed faster decrease in scrublands, exposed rock, and bare soil, between 1990 and 2014. Between the period of 2002 to 2014, there has been more spatial expansion of Khamis-Mushyet city occurred than in comparison to the period between 1990 and 2002. There is a possibility of opportunity of economic base of the city center. This may therefore suggest that the city has reduced in producing functions that attracted migration into the area. The general need is to check it, or else there will be a food crisis apart from the change in micro-climate change over Khamis-Mushyet city.

## 2.1 INTRODUCTION

Studies have shown that there remains only rare landscape on the Earth that is still in its natural state. Due to anthropogenic activities, the land surface is being significantly changed in some manner, and human presence on the Earth and use of land have had a profound effect upon the natural environment, thus resulting in an observable pattern in the land use and land cover (LU/LC) over time.

LU/LC change has become a central component in current strategies for managing natural resources and monitoring environmental changes. To understand how the LU/LC change affects and interacts with the processes of global Earth systems, information is needed on what changes occur, where and when they occur, the rates at which they occur, and the social and physical forces that drive them (Lambin and Ehrlich, 1997). The LU/LC pattern is also an outcome of natural and socioeconomic factors and their utilization by humans in time and space. Land is becoming a scarce resource due to immense agricultural and demographic pressure. Hence, information on LU/LC and the possibilities for their optimal use is essential for the selection, planning, and implementation of land use schemes to meet the increasing demands for basic human needs, basically housing. Spatial transformation in LU/LC is among the most important socioeconomic driving forces of global as well as local environmental change (Vitousek, 1992; Walker and Steffen, 1996). This information also assists in monitoring the dynamics of land use resulting from changing demands of an increasing population. LU/LC change has become a central component in current strategies for managing natural resources and monitoring environmental changes (Lam, 2008).

Viewing the Earth from space is now crucial for understanding the effect of human activities on its natural resource base over time. In the situation of rapid and often unrecorded land use change, observations of the Earth from space provide objective information of human utilization of the landscape. Over the past several years, data from RS satellites have become vital in mapping the Earth's surface features and infrastructures, managing natural resources, and studying environmental changes that are taking place. The combination of new tools—that is, RS and geographic information system (GIS)—is very useful for the study of climate change and ecosystem management. The collection of remotely sensed data facilitates the synoptic analyses of changes in the Earth system's function and pattern, which are taking place at local, regional, and global scales over time; such data also provide an important link between intensive localized ecological research and regional, national, and international conservation and management of biological diversity (Wilkie and Finn, 1996). In regions with rapid urban growth—particularly in areas with few economic resources—remotely sensed data provide an alternative data source to monitor, record, and analyze the changing urban environment (Wentz et al., 2008).

The most general and straightforward change method detected in urban sprawl is the postclassification comparison method. For example, Ji et al. (2006) used multistage Landsat satellite data to monitor urban growth in the Kansas City metropolitan area. Knowledge of LU/LC will certainly contribute to the ongoing and emerging challenges of our world, such as land degradation (Mallick et al., 2014), agricultural disuse, deforestation, air and water pollution, soil carbon emission (Cai, 1996; Fang et al., 2007; Lal, 2002; Liang et al., 2007; Lin et al., 1997), loss of biodiversity (Fu et al., 2003; Lopez-Pujol et al., 2006; Xie et al., 2001), and greater vulnerability to climate change.

The purpose of the study reported here was to analyze relevant RS data from 1990, 2002, and 2014 and to identify the locations, types, and trends of the main land cover changes in that 24-year period in Khamis-Mushyet city, Aseer province, Kingdom of Saudi Arabia. Therefore, an attempt was made to map the status of LU/LC of Khamis-Mushyet to detect the land transformation and the changes that took place, particularly the built-up land, to assess the microclimate change during this period. Land use affects land cover, and changes in land cover affect land use. A change in either, however, is not necessarily the product of the other. Changes in land cover by land use do not necessarily imply degradation of the land. However, many shifting land use patterns driven by a variety of social causes result in land cover changes that affect biodiversity, water, and radiation budgets; trace gas emissions; and other processes that come together to affect the climate and biosphere (Riebsame et al., 1994).

Conventional ground methods of land use mapping are labor intensive and time-consuming and are done relatively infrequently. These maps become outdated with the passage of time, particularly in a rapidly changing environment. According to Olorunfemi (1983), monitoring changes and time series analysis are quite difficult with traditional methods of surveying. In recent years, the use of data and GIS techniques has proved to be of immense value for preparing accurate LU/LC maps and monitoring changes at regular intervals. In inaccessible regions, this technique is perhaps the only method of obtaining the required data on a cost- and time-effective basis.

The objectives of this study are as follows:

- To create a LU/LC classification scheme.
- To integrate information from various sources including RS to examine the distribution of different LU/LC types.
- To determine the trend, nature, rate, location, and magnitude of LU/LC change.

Khamis-Mushyet is an important foothills location that has witnessed remarkable expansion, growth, and developmental activities such as building and road construction, deforestation, and many other anthropogenic activities since its inception in about 1934. This has therefore resulted in increased land utilization and modification and alterations in the status of its LU/LC over time without any detailed and comprehensive attempt (as provided by RS data and GIS) to detect the land change rate. It is therefore necessary for a study such as this to be carried out if Khamis-Mushyet will be able to avoid the associated problems of a growing and expanding city like many others in the world have experienced. The research findings in this study are useful for environmental analysis and urban planning of conservation.

## 2.2 STUDY AREA AND DATA USED

Khamis-Mushyet covers an area of 600 km². The boundary of the study area lies between the latitude of 18°11′55.47″N and 18°22′48.59″N and longitude of 42°37′12.38″E and 42°54′5.3″E (Figure 2.1).

The climatic condition in the study area is semiarid. It averages 355 mm of rainfall annually with the bulk occurring between June and October. The area is sometimes subject to heavy rain; neighboring villages and rural areas can experience flash floods during the winter. The topography of the area is undulating, and the elevation ranges from 1952 to 2257 m mean sea level (Figure 2.2). The site with sedimentary soft, hard silt, and clay rocks surround the mountainous. The area has a severe problem with erosion due to slopes, weak geology, and rain, which affect the productivity of agriculture and forests, and cause erosion and sedimentation. Inhabitants of the mountainous region are dependent on agriculture and a secondary activity for their livelihood. The study area represents a utilized landform, and its socioeconomic activities revolve around utilization of natural resources that need immediate attention in terms of land conservation and development.

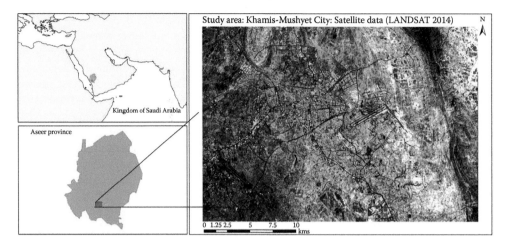

**FIGURE 2.1** Study area.

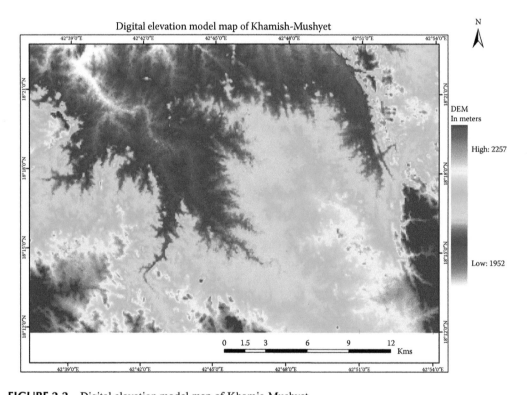

**FIGURE 2.2** Digital elevation model map of Khamis-Mushyet.

Landsat TM, ETM+, and eight datasets were used in this project for preparation of LU/LC maps. Software used were ArcGIS and Microsoft Office in addition to SPSS for statistics computation. Global Positioning System (GPS) was used for field survey measurements.

## 2.3 METHODOLOGY

Landsat datasets are geometrically corrected. For standardization, a topographical sheet was compared with one in Landsat (VNIR) but provided low geometrical accuracy. Also, the GPS readings in the ground during the pre-field campaign did not match the images. Hence, all the images were geometrically rectified to a common Universal Transverse Mercator (UTM-WGS84) coordinate system and were resampled to their respective spatial resolution using the nearest neighbor algorithm.

In keeping with the project objectives, questions and satellite data used the Anderson LU/LC classification description (Table 2.1).

For image classification, a pre-field visit was made to have an idea of the different LU/LC classes existing in our study area. Unsupervised classification was initially performed on Landsat datasets to obtain an idea of the spectral separability of the LU/LC classes. An extensive field survey was done with the prior knowledge using the unsupervised classification to identify sample points in the imagery for different LU/LC classes at various locations using GPS. Based on the collected sample sets for respective LU/LC classes, training sets were selected in the False color composite (FCC) imagery for supervised classification using maximum likelihood classifier (MLC). Supervised classification was done using ground checkpoints and digital topographic maps of the study area. So, training sites were digitized based on the ground truth data (extensive survey using GPS) and higher spatial resolution (Google Earth). Spectral signatures of urban areas and agricultural areas with bare soil were mixed; therefore, their training sites had to be redefined. Texture analysis using Dominance index and kernel window of 5 × 5 pixels also was conducted. Finally, the MLC was run with original bands and the texture image as inputs, producing three final land cover maps of 1990, 2002, and 2014 that were compared. A cross-classification procedure is a fundamental pairwise comparison technique used for two images of qualitative data (Eastman, 1995). An assessment of land cover change and its implication based on cross-classification principles was carried out using image-processing software. For the LU/

### TABLE 2.1
### Classification Description

| Class | Description |
|---|---|
| Built-up area | This category contains dwelling units. These areas are found in the populated urban zones and generally are characterized by impervious surface coverage of 20%–65%. This is represented by the multiple unit structures of densely urban cores and sparsely populated regions surrounded by or adjacent to forested or agricultural lands in the land use class. Residential, commercial, industrial, and occasionally other land uses may be included. |
| Water body | This category comprises areas with surface water, impounded either in the form of ponds, lakes, or reservoirs. These are seen clearly on the satellite image in blue to dark blue or cyan color depending on the depth of water. |
| Agricultural land | This includes the lands primarily used for farming and for production of food, fiber, and other commercial and horticultural crops. It includes land-use crops (irrigated and nonirrigated, fallow, plantation, etc.). |
| Dense vegetation | This includes the areas bearing an association predominantly of trees and other vegetation types (within the notified forest boundaries) capable of producing timber and other forest produce. This category was included in all the areas where the canopy cover/density is more than 40%. |
| Sparse vegetation | This category included all the areas where the canopy cover/density ranges between 10% and 40%. It included the roadside trees, parks, and gulf playground. |
| Scrubland | This area is where the crown density is less than 10% of the canopy cover, generally seen at the fringes of dense vegetation and settlements and that has biotic and abiotic interference. These areas also have exposed soil surface with some vegetation as influenced by human impacts and/or natural causes. It contains scrubs with very low plant cover value as a result of overgrazing, woodcutting, and so on. |
| Bare soil | It includes areas of exposed soil surface as influenced by human impacts and/or natural causes. |
| Exposed rock | This category includes areas of bedrock exposure, volcanic materials, and other accumulation of rock without vegetative cover. |

LC, maps from 1990, 2002, and 2014 were used as input parameters and to identify the locations and magnitude of the major land change, land persistence, and transitions in the study area.

Figure 2.3 shows the flow diagram of methodology for the generation of the LU/LC map for the different time periods, that is, 1990, 2002, and 2014. By using the attribute table classified map, the changes in LU/LC between 1990, 2002, and 2014 were observed. To achieve this information, the first task was to develop a table showing the area in km$^2$ and the percentage of change for each of the two periods measured against each LU/LC category. The trend of change was then calculated by dividing observed change by the sum of changes multiplied by 100 using the following equation:

$$(\text{Trend}) \text{ percentage change} = \frac{\text{observed change}}{\text{sum of change}} \times 100$$

To get the annual rate of change, the percentage change is divided by 100 and multiplied by the number of the study year 1990–2002 (12 years) and 2002–2014 (12 years). The general approach of this project is to map the LU/LC and the detected change. For this medium, resolution satellite datasets were used, namely Landsat (NASA). A general approach for the methodology is shown in Figure 2.3.

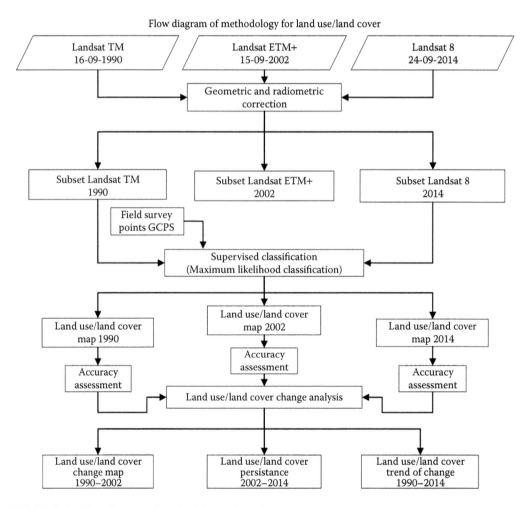

**FIGURE 2.3** Flow diagram of methodology adopted.

## 2.4 RESULT

### 2.4.1 Land Use and Land Cover Mapping (1990)

The LU/LC map was prepared using a supervised classification scheme with an MLC algorithm. The Figure 2.4 shows LU/LC map of 1990; it shows that the high dense and low dense built-up areas are mostly in the central parts of the city and along the roads. Table 2.2 is the information of the LU/LC map of 1990. This table shows the area under different LU/LC classes. The results show that scrublands were the dominant land use in 1990 with 40.89% of the total area followed by exposed rock (31.69%) whereas built-up areas occupied 4.40% of the total area.

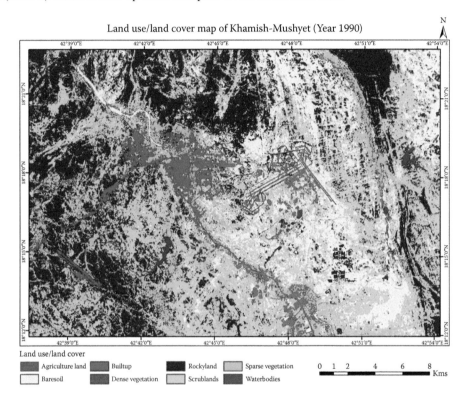

**FIGURE 2.4** LU/LC Map 1990.

**TABLE 2.2**
**Land Use and Land Covers Distribution (1990)**

| S. No. | Land Use and Land Cover Class Name | Area in km$^2$ | Percentage of Total Area |
| --- | --- | --- | --- |
| 1 | Exposed rock | 190.13 | 31.69 |
| 2 | Built-up area | 26.38 | 4.40 |
| 3 | Water bodies | 2.03 | 0.34 |
| 4 | Dense vegetation | 4.61 | 0.77 |
| 5 | Sparse vegetation | 29.35 | 4.89 |
| 6 | Scrublands | 245.36 | 40.89 |
| 7 | Bare soil | 94.17 | 15.70 |
| 8 | Agricultural land | 7.97 | 1.33 |
|   | Total | 600 | 100 |

This situation may be due to the fact that most of the city's population was engaged in primary activities and the highly rugged terrain caused the lack of transportation system almost 24 years before. The agricultural areas distribution showed that the population mostly depended upon farming. This clearly speaks about the presence of fertile agricultural land all around wadies and the low slope area. Of the total area, dense vegetation occupied of only 0.77% and sparse vegetation of 4.89%.

Figure 2.5 (LU/LC in 2002) shows that the built-up area is mainly in the central, southwestern, and eastern parts of the city. The most dominant class is the scrublands (40.13%) followed by exposed rock (30.79%) (Table 2.3). The built-up area occupies an area of just 10.41% of the total area

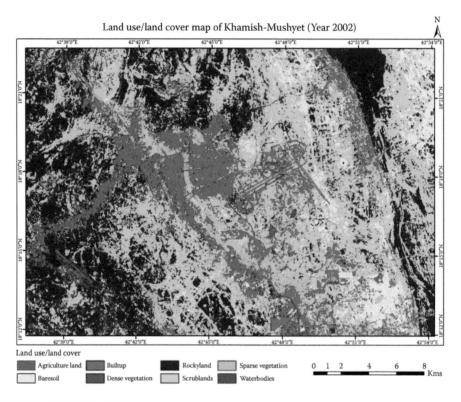

**FIGURE 2.5** LU/LC Map 2002.

**TABLE 2.3**
**Land Use and Land Cover Distribution (2002)**

| S. No. | Land Use and Land Cover Class Name | Area in km² | Percentage of Total Area |
| --- | --- | --- | --- |
| 1 | Exposed rock | 184.73 | 30.79 |
| 2 | Built-up | 62.09 | 14.03 |
| 3 | Water body | 0.27 | 0.05 |
| 4 | Dense vegetation | 13.31 | 2.22 |
| 5 | Sparse vegetation | 39.54 | 6.59 |
| 6 | Scrubland | 240.77 | 40.13 |
| 7 | Bare soil | 39.32 | 6.55 |
| 8 | Agricultural land | 19.61 | 3.27 |
|   | Total | 600 | 100 |

compared to 4.40% in built-up in 1990. This transformation may be connected to the change in the economic base of the city from agriculture to other secondary activities. The area of agricultural land is quite high at 3.27%, which shows that the population increase also increased the demand of agricultural products; hence, the engagement in agricultural practices increased. Apart from this, of the total area, dense vegetation occupied 2.22% and sparse vegetation around 6.59%. There is an increase in the vegetation area in Khamis-Mushyet from 1990 to 2014 because of practice factors.

Figure 2.6 (LU/LC 2014) shows that the built-up area is mainly in the central, northeastern, southwestern, and southeastern parts of the city. The built-up area (Table 2.4) occupies 19.52% of

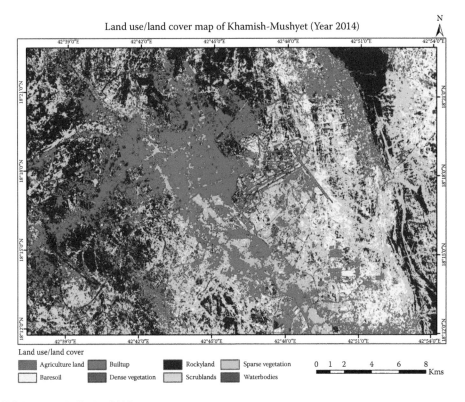

**FIGURE 2.6** LU/LC Map 2014.

**TABLE 2.4**
**Land Use and Land Cover Distribution (2014)**

| S. No. | Land Use and Land Cover Class Name | Area in km² | Percentage of Total Area |
|---|---|---|---|
| 1 | Exposed rock | 183.96 | 30.66 |
| 2 | Built-up area | 117.11 | 19.52 |
| 3 | Water body | 0.30 | 0.05 |
| 4 | Dense vegetation | 6.60 | 1.16 |
| 5 | Sparse vegetation | 28.68 | 4.78 |
| 6 | Scrubland | 197.69 | 32.95 |
| 7 | Bare soil | 42.64 | 7.11 |
| 8 | Agricultural land | 22.67 | 3.78 |
|  | Total | 600 | 100 |

the total area compared to 14.03% in 2002. This transformation may be connected to the change in the economic base of the city from agriculture to other industrial and commercial activities and the concentration of urban expansion in the city center over the prior decade that increased.

The area of agricultural land is 3.78% (Table 2.4), showing that the populations are engaged in commercialized agricultural practices, which is the basis for living. But the agricultural land compared with that shown in the 2002 image data decreased due to urban expansion on agricultural land. Apart from this, of the total area, dense vegetation area occupied 1.16% and sparse vegetation around 4.78%. The decrease in the vegetated area in Khamis-Mushyet from 2002 to 2014 occurred because of urban expansion and climate change.

The accuracy assessment was done for the classified maps of 1990, 2002, and 2014. It is seen that the overall accuracy of classified maps of 1990, 2002, 2014 is 90.14%, 91.76%, and 91.32%, respectively, with kappa statistics of 0.8901, 0.9063, and 0.9026.

### 2.4.2 Land Use and Land Cover Change (1990–2002)

In order to assess the LU/LC change from 1990 to 2002, classified LU/LC maps of the two years were used to run the change detection model in GIS software. The outcome shows where all the land transformation took place and an attribute table (Table 2.5) that shows the quantitative values of land use changes.

Any increase in the area of a particular class from other classes has been termed as a gain, whereas a decrease in the area of a particular class to another class has been termed as a loss, which are shown both in table and figure formats. Table 2.5 represents the net change among the different LU/LC classes between the two different time periods. The result shows that there were significant changes in all LU/LC classes between 1990 and 2002 with the exception of water for which little change was recorded. As far as major changes are concerned, four land categories experienced major transitions. The maximum change of 88.23 $km^2$, that is 35.66%, was recorded in the scrubland category, which was a gain but at the same time it lost 92.46 $km^2$ to other classes (Table 2.5).

Table 2.5 shows gains and losses in the classes from 1990 to 2002. Both scrubland (92.46 $km^2$) and bare soil (62.13 $km^2$) decreased during the 12 years. On the other hand, there is a significant increase in the area of scrubland (88.23 $km^2$), exposed rock (50.12 $km^2$), and built-up area (40.18 $km^2$).

This increase in the built-up area (40.18 $km^2$) took place mainly in the areas of scrubland and exposed rock. This affects the microclimate because this addition of impervious surfaces will lead to an increase in land surface temperature. Apart from this, the area of sparse vegetation, exposed rock, and scrubland decreased in $km^2$ by 8.88, 22.44, and 37.37, respectively, due to urban expansion, most of which transformed the built-up area. Table 2.5 shows the contribution of different LU/LC classes to the net change in the built-up areas. It is seen that a net change of 35.70 $km^2$ was found in build-up area during 1990–2002. This shows that the increase of built-up areas at the expense of agricultural, scrub, and bare soil land will lead to a negative impact on the environment of Khamis-Mushyet.

An important aspect of change detection is to determine what is actually changing to what; that is, which LU/LC class is giving to which other class and where. This information will reveal both the desirable and undesirable changes and the classes that are relatively stable over time. It will also serve as a vital tool in analyzing microclimate change and making environmental management decisions. This process involves a pixel-to-pixel comparison of the classified LU/LC map and the overlaying of one map over the other in GIS software. Table 2.6 shows the category of land transformation that occurred between 1990 and 2002. The major changes that occurred from scrubland to exposed rock area accounted for 43.66 $km^2$ because it might be due to erosion.

## TABLE 2.5
## Gains and Losses and Net Change

| Land Use and Land Cover Classes | Gains and Losses and Net Change between 1990 and 2002 | | | | Gains and Losses and Net Change between 2002 and 2014 | | | | | | Gains and Losses and Net Change between 1990 and 2014 | | | | |
|---|---|---|---|---|---|---|---|---|---|---|---|---|---|---|---|
| | Gain (km²) | Gain (%) | Loss (km²) | Loss (%) | Net Change (km²) | Gain (km²) | Gain (%) | Loss (km²) | Loss (%) | Net Change (km²) | Gain (km²) | Gain (%) | Loss (km²) | Loss (%) | Net Change (km²) |
| Water bodies | 0.00 | 0.00 | −1.78 | 0.72 | −1.78 | 0.00 | 0.00 | −0.17 | 0.07 | −0.17 | 0.00 | 0.00 | −1.74 | 0.59 | −1.74 |
| Exposed rock | 50.12 | 20.26 | −55.52 | 22.44 | −5.4 | 59.44 | 24.28 | −60.21 | 24.57 | −0.77 | 64.49 | 21.94 | −70.66 | 24.03 | −6.17 |
| Scrubland | 88.23 | 35.66 | −92.46 | 37.37 | −4.23 | 58.55 | 23.92 | −101.64 | 41.48 | −43.09 | 74.56 | 25.36 | −121.88 | 41.46 | −47.32 |
| Sparse vegetation | 32.15 | 13.00 | −21.96 | 8.88 | 10.19 | 21.23 | 8.67 | −32.09 | 13.10 | −10.86 | 22.18 | 7.54 | −22.85 | 7.77 | −0.67 |
| Dense vegetation | 11.12 | 4.49 | −2.42 | 0.98 | 8.7 | 3.24 | 1.32 | −9.95 | 4.06 | −6.71 | 5.73 | 1.95 | −3.75 | 1.28 | 1.98 |
| Built-up area | 40.18 | 16.24 | −4.47 | 1.81 | 35.71 | 66.35 | 27.10 | −11.33 | 4.62 | 55.02 | 94.47 | 32.13 | −3.74 | 1.27 | 90.73 |
| Bare soil | 7.28 | 2.94 | −62.13 | 25.11 | −54.85 | 16.79 | 6.86 | −13.47 | 5.50 | 3.32 | 11.24 | 3.82 | −62.77 | 21.35 | −51.53 |
| Agricultural land | 18.32 | 7.41 | −6.67 | 2.70 | 11.65 | 19.21 | 7.85 | −16.15 | 6.59 | 3.06 | 21.32 | 7.25 | −6.61 | 2.25 | 14.71 |

**TABLE 2.6**
**Land Transformation**

| S. No. | Category Changes | Land Transformation between 1990 and 2002 | | Land Transformation between 2002 and 2014 | | Land Transformation between 1990 and 2014 | |
|---|---|---|---|---|---|---|---|
| | | Changes/ Transformation (km²) | Changes (%) | Changes/ Transformation (km²) | Changes (%) | Changes/ Transformation (km²) | Changes (%) |
| 1 | Scrubland to agriculture land | 9.58 | 1.60 | 11.07 | 1.85 | 9.85 | 1.64 |
| 2 | Scrubland to bare soil | 6.48 | 1.08 | 14.80 | 2.47 | 9.24 | 1.54 |
| 3 | Exposed rock to built-up area | 16.24 | 2.71 | 27.46 | 4.58 | 37.41 | 6.24 |
| 4 | Scrubland to built-up area | 15.01 | 2.50 | 20.08 | 3.35 | 35.98 | 6.00 |
| 5 | Scrubland to exposed rock | 43.66 | 7.28 | 42.47 | 7.08 | 53.51 | 8.92 |
| 6 | Bare soil to scrubland | 42.97 | 7.16 | 7.71 | 1.29 | 32.90 | 5.48 |
| 7 | Exposed rock to scrubland | 27.87 | 4.65 | 27.45 | 4.58 | 26.70 | 4.45 |
| 8 | Sparse vegetation to scrubland | 11.23 | 1.87 | 12.31 | 2.05 | 9.94 | 1.66 |
| 9 | Scrubland to sparse vegetation | 14.49 | 2.42 | 12.61 | 2.10 | 11.02 | 1.84 |
| 10 | Change <2 km | 412.47 | 68.75 | 424.04 | 70.67 | 373.45 | 62.24 |
| | Total area | 600 | 100 | 600 | 100 | 600 | 100 |

Other important changes occurred in exposed rock of which 16.24 km² was transformed to built-up land because it may be connected to the change in the economic base (primary activity) of the city. Of the scrubland area, 15.01 km² seems to have been transformed to built-up area; it may be due to the urbanization.

### 2.4.3 LAND USE/LAND COVER CHANGE ANALYSIS (2002–2014)

The result of this comparison shows that there were significant changes in all LU/LC classes between 2002 and 2014 with the exception of water for which little change was recorded. The maximum change 66.35 km² that is, a 27.10% gain, was recorded in the built-up category, but at the same time, it has lost 4.62 km² to some other classes (Table 2.5). This transformation may be connected to the change in the economic base of the city from agriculture to commercial/industrial activity. Any increase in the area of a particular class from other classes has been termed as the gain, and a decrease in the area of a particular class to another class has been termed as the loss, shown both in Table 2.5 and Figure 2.6. Table 2.5 shows the net change among the different LU/LC classes in the two different time periods. Table 2.5 shows gains and losses between the classes from 2002 to 2010. Both scrubland (41.48 km²) and exposed rock (24.57 km²) lost area during the 12 years. On the other hand, there is a very high increase in the built-up area of 66.35 km².

This increase in the built-up area took place mainly in the scrublands areas. Table 2.5 shows 43.09 km² net change. This is a negative change in the terms of environment of an urban area because this is the addition of the impervious surfaces, which will lead to increase in land surface temperature. Contrary, built-up has recorded 55.02 km² net change (increase) in other areas. This is also an indication of environmental change in the city.

Table 2.5 shows the contribution of different LU/LC classes to the net change in the built-up areas: 37.45% from exposed rock and 31.27% from scrubland. This shows that the increase of built-up areas at the expense of the scrub, bare soil, and agricultural land will lead to a negative impact on the environment of Khamis-Mushyet.

Table 2.6 shows the land transformation by category between 2002 and 2014. The major changes to built-up area occurred from exposed rock and scrubland, 27.46 and 20.08 km², respectively, due to a change in economic activity (primary) and urbanization processes. Another major change in the exposed rock area indicates transformation from scrubland (42.47 km²), which may be due to the climate change and land degradation.

### 2.4.4 LAND USE AND LAND COVER CHANGE ANALYSIS (1990–2014)

In order to assess the LU/LC change from 1990 to 2014, classified LU/LC maps of these two time periods were used to run the change detection model in GIS software. The outcome shows where all the land transformation took place and an attribute table (Table 2.5) that shows the quantitative values of land use changes that took place during this period. The result demonstrates that there were significant changes in all LU/LC classes between 1990 and 2014 with the exception of water for which little change was recorded. Figure 2.5 shows the gains and losses of different LU/LC during that 24-year period.

The maximum change of 94.47 km² that is, 32.13%, was recorded in the built-up category, which is the gain, but at the same time it has lost 1.27 km² of the total area to some other classes (Table 2.5), followed by scrubland (loss) of −121.88 km². This change/transformation is due to the concentration of urban population in the city center. Other transformation occurred in exposed rock with a loss of about −70.66 km². this transformation may be due to urban and commercial/industrial activities.

Any increase in the area of a particular class from other classes has been termed as the gain, whereas a decrease in the area of a particular class to another class has been termed as the loss, which is shown in Table 2.5. Another graph was made to represent the net change among the different LU/LC classes between two different time periods. Table 2.5 shows the significant changes during these 2.5 decades. It is seen that both scrublands (−121 km²) and exposed rock (−70.66 km²) decreased. On the other hand, there is a significant increase in the area of built-up (94.47 km²).

This increase in the built-up area took place mainly in the scrubland. This is evident from the facts that 47.32 km² net changes recorded for scrubland in Table 2.5. Similarly, the built-up area had 90.73 km² net change, which is an environmental change in terms of microclimate, including the addition of impervious surfaces, which will lead to an increase in land surface temperature.

Table 2.5 shows the contribution of different LU/LC classes to the net change in the built-up areas. Any increase in the area of a particular class from other classes has been termed as the gain, whereas decrease in the area of a particular class to another class has been termed as the loss, which has been shown both in table and figure. Another graph has been made to represent the net change among the different land use/land cover classes between two different time periods. Table 2.5 shows the significant changes during the past 2.5 decades. It is seen that both scrublands (−121 km²) and exposed rock (−70.66 km²) have been decreased. On the other hand there is a significant increase in the area of built-up 94.47 km². This increase in the built-up area is taking place mainly in the scrubland. This is evident from the facts that 47.32 km² net changes have been recorded in scrublands Table 2.5. Similarly built-up has recorded 90.73 km² net changes in areas, which is an environmental change in terms of micro-climate. This is the addition of the impervious surfaces, which will lead to increase in land surface temperature. Table 2.6 shows the category-wise transformation that has occurred between 1990 and 2002. It is seen that 16.24 km² and 15.01 km² areas under built-up area has increased on exposed rock and scrublands. However the category-wise transformation more occurred during 2002–2014 under built-up area. It has increased on exposed rock and scrublands of 14.80 km² and 27.46 km² respectively. This shows that the increase of built-up areas on scrubland, exposed rock, and bare soil will lead to a negative impact on the environment of Khamis-Mushyet.

An important aspect of change detection is to determine what is actually changing to what, that is, which LU/LC class is changing which other one and where. This information will reveal both the desirable and undesirable changes and the classes that have been relatively stable over time. This information will also serve as a vital tool in microclimate change analysis and environmental management decision-making. This process involves a pixel-to-pixel comparison of the classified LU/LC map and the overlaying of one map over the other in GIS software.

Table 2.6 shows the category-wise of transformation that occurred between 1990 and 2014. The major changes that has been observed from rockyland to built-up land and scrubland to built-up that accounted for 37.41 km² and 35.98 km², which transformed to built-up land. This is due to large-scale migration of people to the physical expansion of the city in the built-up area. Scrublands are transformed to exposed rock due to erosion and anthropogenic activities.

### 2.4.5 Urban Expansion of Khamis-Mushyet during 1990–2002–2014

In 1990, the built-up area occupied 4.40% (26.38 km²) of the total area, whereas it increases just double of 10.41% (62.09 km²) and 19.52% (117.11 km²) in 2002 and 2014 respectively. Figure 2.7 shows a comparison of the expansion of the built-up area from 1990 to 2002 and 2002 to 2014. This transformation may be connected to the change in the economic base of the

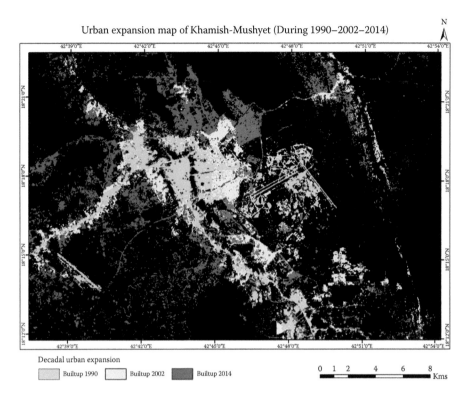

**FIGURE 2.7** Urban Expansion Map during 1990–2002–2014.

city from agriculture to other industrial and commercial activities and other reason over the last three decades.

## 2.5 CONCLUSIONS

In this study, the methodology adopted for mapping LU/LC and its change pattern included RS data and GIS techniques. Classification of multispectral RS satellite data and comparison of land cover maps are essential tools for assessing LU/LC use changes. With the increase in population, Khamis-Mushyet is increasingly crowded day by day. This will lead to an increase in built-up areas for residential and commercial/industrial purposes. Hence, it will contribute to multifaceted problems of the city. It is therefore suggested that emphasis should be on building houses toward the outskirts of the city with special concern on environmental issues.

Application of RS and GIS was found helpful in quantifying past and present LU/LC so that appropriate planning could be made for the future. These methods also have potential for analysis of large (regional or global) areas because the processing times are less compared with a conventional technique. In addition, these two spatial statistics could serve to improve land cover classifications in areas where cover types with contrasting degrees of anthropogenic influences exist. This study investigated the landscape dynamic characteristics of LU/LC variation of Khamis-Mushyet using RS data and GIS technology. The LU/LC transformations were also analyzed according to elevation and slope.

During 1990 to 2014, mainly the scrubland, exposed rock, and bare soil became a built-up area, especially in the northwest, eastern, and southeastern parts of the study site. Khamis-Mushyet witnessed fast decrease in scrubland, exposed rock, and bare soil between 1990 and 2014. Between the period of 2002 and 2014, there was more spatial expansion of the city compared to the period between 1990 and 2002. There is a possible opportunity of an economic base of the city center. This may therefore suggest that the city has reduced its producing functions that attracted migration into the area. The general need is to check it or else there will a food crisis apart from a microclimate change over Khamis-Mushyet.

## REFERENCES

Cai, Z. C. (1996). Effect of land use on organic carbon storage in soils in eastern China. *Water Air and Soil Pollution*, 91, 383–393.

Eastman, J. R. (1995). *IDRIS for Windows, User's Guide*. Clark University, Worcester, MA, 405 p.

Fang, J. Y., Guo, Z. D., Piao, S. L., and Chen, A. P. (2007). Terrestrial vegetation carbon sinks in China, 1981–2000. *Science in China Series D-Earth Sciences*, 50, 1341–1350.

Fu, C. Z., Wu, J. H., Chen, J. K., Qu, Q. H., and Lei, G. C. (2003). Freshwater fish biodiversity in the Yangtze River basin of China: Patterns, threats and conservation. *Biodiversity and Conservation*, 12, 1649–1685.

Ji, W., Ma, J., Twibell, R. W., and Underhill, K. (2006). Characterizing urban sprawl using multi-stage remote sensing images and landscape metrics. *Computers Environment and Urban Systems*, 30, 861–879.

Lal, R. (2002). Soil carbon sequestration in China through agricultural intensification, and restoration of degraded and desertified ecosystems. *Land Degradation & Development*, 13, 469–478.

Lam, N. (2008). Methodologies for mapping land cover/land use and its change. In S. Liang (Ed.), *Advances in Land Remote Sensing: System, Modeling, Inversion and Application* (pp. 341–367). Springer, New York.

Lambin, E. F. and Ehrlich, D. (1997). Identification of tropical deforestation 'hot spots' at broad spatial scales. *International Journal of Remote Sensing*, 18, 3551–3568.

Liang, W., Shi, Y., Zhang, H., Yue, J., and Huang, G. H. (2007). Greenhouse gas emissions from northeast China rice fields in fallow season. *Pedosphere*, 17, 630–638.

Lin, E. D., Liu, Y. F., and Li, Y. (1997). Agricultural C cycle and greenhouse gas emission in China. *Nutrient Cycling in Agroecosystems*, 49, 295–299.

Lopez-Pujol, J., Zhang, F. M., and Ge, S. (2006). Plant biodiversity in China: Richly varied, endangered, and in need of conservation. *Biodiversity and Conservation*, 15, 3983–4026.

Mallick, J., Al-Wadi, H., Rahman, A., and Ahmed, M. (2014). Landscape dynamic characteristics using satellite data for a mountainous watershed of Abha, Kingdom of Saudi Arabia. *Environmental Earth Sciences*, 72, 4973–4984.

Olorunfemi, J. F. (1983). Monitoring urban land – Use in developed countries – An aerial photographic approach. *Environmental International*, 9, 27–32.

Riebsame, W. E., Meyer, W. B., and Turner, B. L. II. (1994). Modeling land-use and cover as part of global environmental change. *Climate Change*, 28, 45.

Vitousek, P. M. (1992). Global environmental change: An introduction. *Annual Review of Ecology and Systematics*, 23(1992), 1–14.

Walker, B. H. and Steffen, W. E. (1996). *Global Change and Terrestrial Ecosystems*. Cambridge University Press, Cambridge, UK.

Wentz, E. A., Nelson, D., Rahman, A., Stefanov, W. L., and Roy, S. S. (2008). Expert system classification of urban land use/cover for Delhi, India. *International Journal of Remote Sensing*, 29(15–16), 4405–4427.

Wilkie, D. S. and Finn, J. T. (1996). *Remote Sensing Imagery for Natural Resources Monitoring*. Columbia University Press, New York, p. 295.

Xie, Y., Li, Z. Y., Gregg, W. P., and Dianmo, L. (2001). Invasive species in China: An overview. *Biodiversity and Conservation*, 10, 1317–1341.

# 3 Understanding the Spatio-Temporal Monitoring of Glaciers
## *Application of Geospatial Technology*

*Shruti Dutta and AL. Ramanathan*

## CONTENTS

3.1 Introduction .................................................................................................................. 28
3.2 Identifying Geomorphological Features from Space ............................................. 28
3.3 Mapping and Inventory of Glaciers .......................................................................... 33
    3.3.1 Manual Delineation of Area/Length .............................................................. 33
    3.3.2 Mapping the Debris-Covered Glaciers through Automated Classification ... 36
3.4 Estimation of Thickness of Glaciers ......................................................................... 38
    3.4.1 Thickness Area Relation Method .................................................................. 39
    3.4.2 Volume–Area Related Thickness Estimations .............................................. 40
    3.4.3 Glacier Flow Mechanics Method .................................................................. 41
    3.4.4 Remote Sensing–Based Methods ................................................................... 41
3.5 Volumetric Estimation ............................................................................................... 42
    3.5.1 Glacier Flow Mechanics Method .................................................................. 42
    3.5.2 Remote Sensing–Based Methods ................................................................... 42
3.6 Extracting Glacier Parameters from Digital Elevation Model ............................... 42
    3.6.1 Deciphering Mass Balance Studies from DEMs .......................................... 43
    3.6.2 Volume–Area Scaling for Mass Balance Studies ......................................... 43
    3.6.3 Accumulation Area Ratio–Equilibrium Line Altitude Method for Mass Balance Estimations ........................................................................................ 44
    3.6.4 The Remote Sensing–Based Geodetic Method ............................................ 46
3.7 Geospatial Technology for Cryosphere and the Way Forward… .......................... 47
References ............................................................................................................................. 47

**ABSTRACT** The glaciers contribute as health indicators to the climate of the region and impacts on the runoff at the lowlands and recharging of the aquifers. However, the monitoring gets severely hindered due to factors such as logistics, accessibility, adverse environmental conditions, and so on. Here, the application of geospatial techniques for spatio-temporal monitoring of glaciers (special reference to Himalayas) has been attempted. Various geomorphological features such as moraines, debris cover, crevasses, and snout can be identified with the help of satellite imageries to assess the glacio-fluvial activity as well as widely employed for the manual delineation of the glacier. A study comprising the mapping of 224 glaciers through manual delineation (1972–2006) in the Beas basin, Himachal Pradesh reveals that the glacier cover reduced from 419 to 371 km², witnessing approximately 11.6% deglaciation (Dutta et al., 2012); whereas the Ravi basin (60 glaciers) gives a high deglaciation rate of about 16.37% during 1972–2006 (Dutta et al., 2013). Few studies have been done with different methods for automated mapping, which become immensely vital when applied over large area and more relevant for clean ice mapping. However, the accuracy of the results gets hindered by the presence of debris cover, proglacial/supra glacial lakes and turbid and frozen lakes. Another important aspect of glacial monitoring is the distribution of ice thickness estimated using surface velocities, slope, and the ice flow law. Digital elevation models (DEMs) are valuable tools for widespread applications in extracting various components of a glaciated environment. A detailed account of various methodologies to estimate the glacier thickness–volume–mass balance have been compiled, since these parameters become utmost crucial to assess the intricacies related to health of the glacier.

## 3.1 INTRODUCTION

The technology used to acquire, manipulate, and store geographic information is referred to as geospatial technology whose various forms are Geographic Information System (GIS), Global Positioning System (GPS), and Remote Sensing (RS). These techniques play a vital and efficient role in the monitoring of glaciers, and its significance is well established now. The glaciers contribute as a natural "thermometer" in the hydrological cycle, being sensitive indicators of any minute changes in the microclimate of the region. They are responsible for providing a resource of water to local areas in the mountains and to the millions of people in the lower reaches (Immerzeel et al., 2010). This resource also has an impact on the runoff at the lowlands and recharging of the aquifers, which are fed by rivers.

However, the monitoring and consistent studies get hampered in the glaciated terrains, which are also difficult in terms of logistics, accessibility, adverse environmental conditions, and inability to conduct field trips round the year. In this scenario, RS techniques are often the most efficient way to monitor a large number of glaciers at the same time. Various characteristics like synoptic view, repetitiveness, and updated information make RS a very powerful tool to study glaciated terrains. A reasonable number of studies focused at the spatio-temporal monitoring of the glaciers have been carried out to accomplish long-term observations and relevant output. The spectral distinction of the features related to glaciers along with accurate resolution (spectral, temporal and spatial) becomes the decisive factor in the studies of glaciers from geospatial data.

Here we analyze the potential and application of geospatial techniques in assessment of spatial and temporal monitoring of alpine glaciers with special emphasis on the Himalayan glaciers. The Indian Himalaya comprises approximately 9575 glaciers (Raina and Srivastava, 2008; Sangewar and Shukla, 2009) covering an area of approximately 37,466 km². Glaciers represent a water resource forming a unique freshwater reservoir that influences runoff downstream. The satellite-aided digital techniques are useful in interpretation of numerous parameters, thus being important in assessment of the glacier health. The following sections deal with the application of geospatial technology in interpretation and information about various parameters related to spatial and temporal variability of glaciers.

## 3.2 IDENTIFYING GEOMORPHOLOGICAL FEATURES FROM SPACE

Geomorphological features can be studied with the help of satellite imageries to assess the glaciofluvial activity at the respective glacier. Glaciofluvial activity is highly controlled by glacier movement and meltwater interaction with the surrounding environment. This section discusses the identification of geomorphological inputs with respect to RS techniques mainly concentrated in the glaciated terrains of Chhota Shigri Glacier, Himachal Pradesh, India (Figure 3.1).

For the assessment of the glacio-fluvial activity with the environment, the glaciated basin can be broadly divided into *inactive* and *active zones* based on their upstream and downstream

# Understanding the Spatio-Temporal Monitoring of Glaciers

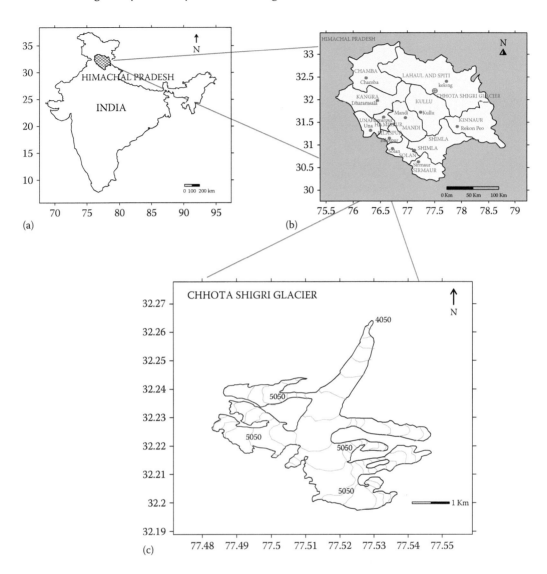

**FIGURE 3.1** Location map of Chhota Shigri Glacier, Himachal Pradesh. (a) Outline of India, (b) route map, and (c) outline map of Chhota Shigri.

characteristics. The area between the snout of the glacier and its terminal moraines forms the inactive zone of the glacier so called because it has lost all its contacts with the present glacier. This zone is primarily dominated by the glacio-morphological features formed by glaciofluvial activity, which in turn reflects a strong indication of the remnants of the past glaciation record. The most prominent and identifiable feature in the inactive zone is the lateral moraines, and the presence of shadow along with these ridges help in their identification (Figure 3.2). The presence of morainic deposition in a distinct pattern gives a strong indication of the various stages of the glaciation encountered in the area. It can be used as the basis for reconstructing the glacier history in elucidating the sequence of glacier episodes before its complete deglaciation from the inactive zone.

The satellite image may also depict some recessional moraines identified in the form of recognizable loops when considering the upward movement of the glacier (Figure 3.3). It also gives an idea about the increase in roughness of the inactive zone from snout to downstream. Sometimes the lower roughness area may exhibit the existence of the hatch-textured surface with an irregular or nonuniform relief in the area. A close observation with the images of Chhota Shigri Glacier reveals that the reflectance of

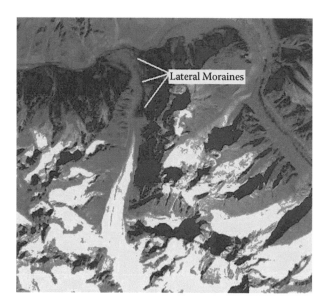

**FIGURE 3.2** Lateral moraines in the inactive zone of the glacier.

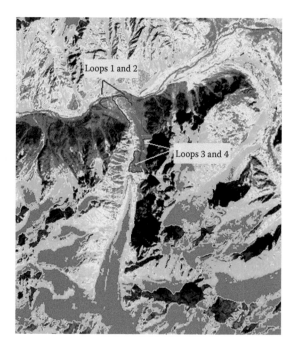

**FIGURE 3.3** Identification of morainic loop.

morainic material changes with distance from the snout toward downstream (Figure 3.4). In the frontal portion of the glacier basin, broad plain shows very low reflectance. This may be related to reworking of the sediment in the region through time, leading into uniform distribution of sediments. Furthermore, the biogenic activities on this surface have also contributed to decreasing the reflectance since this surface was the first to be exposed in deglaciation and got sufficient time for this reworking. But near the present snout, the fresh recessional moraine is in the form of dump material. This is a highly irregular fresh material and gives the high reflectance on imagery.

# Understanding the Spatio-Temporal Monitoring of Glaciers

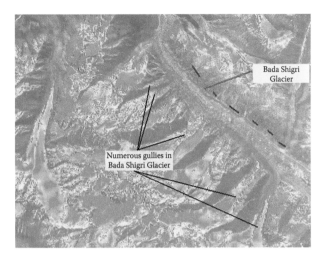

**FIGURE 3.4** Comparative lithological characteristics and variation in reflectance of Chhota Shigri and Bada Shigri glaciers.

The zone comprising the region between the head of the glacier and the terminal moraines and predominated by glacial activity is called active zone. In this region, glacier ice and its movement result in the formation of various glacio-morphological features on glacier surface. The mapping of the glacier ice can be done using various algorithms. The other major glacio-morphological feature identified through RS technique is the debris cover. The generation of the debris is due to high relief around the glacier body and extreme climatic conditions. The debris falls on the surface of the glacier and subsequently, the interaction of these surface moraines with the glacial movement results in the formation of various geomorphological features on the glacier surface. As the glacier moves downward, the moraines travel to the terminal point and concentrate near the lower ablation zone. This results in higher thickness of debris cover in the lower zone and lower thickness in the upper reaches of the glacier. The glacier surface under debris cover is identified using the image (Figure 3.5). The corresponding image clearly depicts the delineation of the glacier boundary with the existing morainic materials.

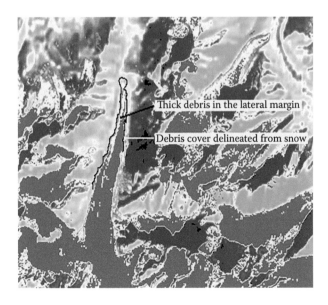

**FIGURE 3.5** Debris cover in the ablation zone of the glacier.

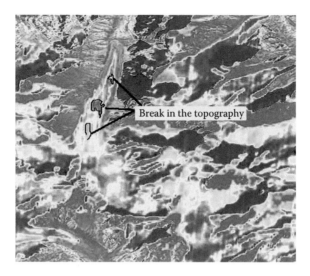

**FIGURE 3.6** Identification of break in the topography, indicating sites for crevasse development.

The crevasses can also be well recognized by careful observation of a step type of a feature (Figure 3.6). They represent a significant break in bedrock topography. At this breaking point, due to bedrock and glacial interaction, it results in formation of crevassed zones. This suggests that these three places probably act as locations for crevasse development. Usually, they are mainly concentrated in the upper zone of the glacier, whereas they are not very clear in the lower zone. The active zone also reflects a strong indication of medial moraines whereas the lateral moraines are well developed near the lower ablation zone and move toward higher reaches. The origin of stream or cave shadow near the glacier can be used as an indication of snout position. It can be easily identified for big glaciers, because the shadow generated is quite big. However, it becomes a little difficult in smaller glaciers with debris. Under these conditions, the collateral features help to identify the snout (Figure 3.7). For identification of the snout, the reflection of the slope and location of the

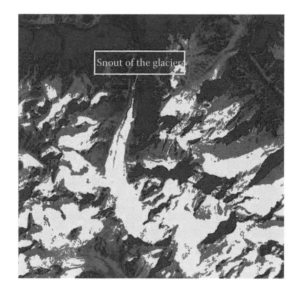

**FIGURE 3.7** Position of the snout.

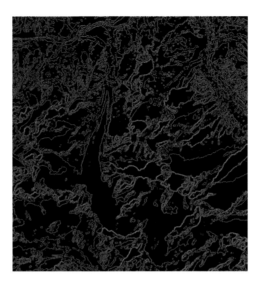

**FIGURE 3.8** Identification of glacier boundary.

lateral moraines with association of recessional moraines and field observations can be of utmost relevance. For identifying the linear feature, the Sobel edge detecting algorithm is used (Figure 3.8).

## 3.3 MAPPING AND INVENTORY OF GLACIERS

Satellite RS in glaciology was used for surveying as early as 1960, when the initial picture taken by first weather satellite TIROS-1 was used to delineate snow cover in Eastern Canada. Since then, a number of orbiting and geostationary satellites with improved spectral, spatial and temporal resolutions have been put into orbit and data sensed by them have been used for weather monitoring and earth resource studies like ice and snow mapping. The potential for operational satellite-based mapping has been enhanced by the development of higher temporal frequency satellites and sensors.

The substantial role that satellite observations can play in glaciology has been persuasively demonstrated in the literatures during the past few decades. Airborne and spaceborne RS has a long history of being used for delineation of glacial facies and for determination of the physical properties of a snowpack. Early work by Ostrem (1975) demonstrated usefulness of RS to glacier mass balance studies through the ability of Landsat Multispectral Sensor (MSS) data to delineate the transient firn line on Norwegian glaciers. The initial studies on glacier inventory has been carried out by manual delineation of boundaries on standard false color composites (FCC) of Landsat MSS and Thematic Mapper (TM) by Williams (1986) and Hall et al. (1992).

### 3.3.1 Manual Delineation of Area/Length

The Survey of India topographic maps (1962) provide the earliest information about the glacial extent prepared by aerial photography and field investigations. It is evident that the topographical mapping using the glacio-morphological features through the plane table and terrestrial photogrammetry is a tedious and herculean task (Bhambri and Bolch, 2009). The RS involves field work but is only limited for obtaining ground control points for validation of the dataset. The multispectral satellite imageries can be utilized to map the changes in glacier area and terminus position, mainly through manual delineation or automated mapping. It becomes a matter of concern in the scenario of algorithms not to be able to differentiate with debris cover or other irregular spectral profiles (Smith et al., 2015). Moreover, this method is subject to the proficiency of the user with respect to the identification and recognition skills of glacio-morphological features on the satellite imageries (Bhambri and Bolch, 2009). Nevertheless, the mapping of Himalayan glaciers has been mostly carried out by manual delineation using the satellite data of variable spatial resolution (Landsat MSS, TM, ETM, LISS III/IV, PAN, AWifs, Resourcesat, Cartosat, etc.).

The manual mapping of the glaciers is preferred because the debris-covered ice and the surrounding bedrock exhibit a similar spectral reflectance in automated classification (Andreassen et al., 2008; Bolch et al., 2010). Various geomorphological features like stream, lateral/medial/terminal moraines, cliffs and snout/terminus along with other parameters such as reflectance of snow/ice and shape of the valley and characteristics like flow lines, texture of the debris, shadow of the steep mountain peaks, presence of vegetated parts of the mountains, and so on play an instrumental role for the delineation of the glacial boundaries (Brahmabhatt et al., 2012). Different algorithms such as addition, subtraction, and band ratioing can be applied to enhance glacial features. FCC of bands and ratioing TM4/TM5 can be used to differentiate debris cover and glacial ice (Bhambri and Bolch, 2009) whereas addition algorithms can help for giving more clarity in identification of peaks.

Using these techniques of manual digitization and application of various algorithms, some landmark studies using RS data have been done to estimate the glacier retreat for Himalayan glaciers. A study conducted of 466 glaciers in Chenab, Parbati, and Baspa basins (Kulkarni et al., 2007) shows an overall deglaciation of 21% during 1962–2001. Another study comprising the mapping of 224 glaciers through manual mapping (Figure 3.9) during the period 1972–2006 in the Beas basin, Himachal Pradesh reveals that the glacier cover reduced from 419 to 371 km², witnessing approximately 11.6% deglaciation in the Beas basin (Dutta et al., 2012). A higher

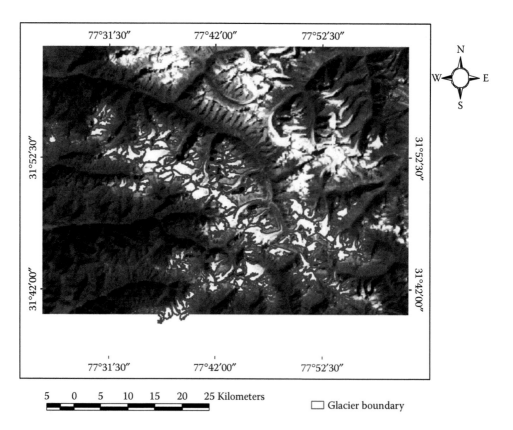

**FIGURE 3.9** Glacier boundary delineation through manual digitization in the Parbati sub-basin, Beas. (From Dutta, S. et al., *J. Earth Syst. Sci.*, 121, 1105–1112, 2012.)

rate of retreat of the glaciers was observed during 1989–2006 as compared to the retreat during 1972–1989 (Figure 3.10). Also, the loss has been more prominent in the glaciers with an areal extent of 2–5 km². The mapping of glaciers through the manual digitization method in the Ravi basin (60 glaciers) gave a high deglaciation rate of about 16.37% (Figure 3.11) in the last four decades (Dutta et al., 2013).

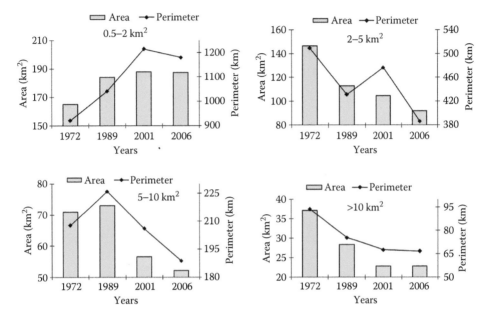

**FIGURE 3.10** Graphical representation of the deglaciation pattern in the Beas basin, 1972–2006. (From Dutta, S. et al., *J. Earth Syst. Sci.*, 121, 1105–1112, 2012.)

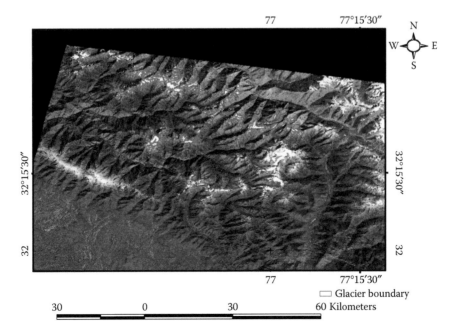

**FIGURE 3.11** Glacier boundary delineation through manual digitization in the Ravi Basin.

### 3.3.2 Mapping the Debris-Covered Glaciers through Automated Classification

Still, the method for manual digitization becomes time consuming and cannot be assumed to be completely error free from the user end. Because of these limitations, automated mapping becomes vital and works on the principle of differential reflectance of snow in the visible and near infrared region (VIS/NIR) with respect to shortwave infrared (SWIR) region of the solar spectrum. Few studies have been done with different methods for automated method as described in Table 3.1.

### TABLE 3.1
### A Tabular Representation of the Literature for Automated Delineation Methods of Debris Covered Glaciers

| | Worker | Methodology | Area of Study |
|---|---|---|---|
| 1 | Lougeay (1974) | Mapping the ice-cored moraine by virtue of its lower radiometric temperature | |
| 2 | Bayr et al. (1994) and Rott (1994) | Proposed thresholds of a ratio image of TM-4 to TM-5 (NIR/SWIR) and TM-3 to TM-5 (RED/SWIR) | Delineate glacier ice area eastern Austrian Alps |
| 3 | Hall et al. (1995) | Proposed the Normalized Difference Snow Index (NDSI) [VIS − SWIR] / [VIS + SWIR]) technique | For identification of snow, United States |
| 4 | Sidjak and Wheate (1999) | Combination of principal components two, three, and four of the masked glacier area, the ratio TM-4/TM-5, and the NDSI | British Columbia, Canada |
| 5 | Bishop et al. (1995, 1999) | Applied artificial neural network (ANN) classifier for estimation of debris cover | Himalayan glaciers |
| 6 | Paul (2001) | Evaluated both ratio image techniques and concluded that the TM-4 to TM-5 ratio technique is the more appropriate for clean-ice glacier mapping | |
| 7 | Philip and Sah (2004) | Used merged IRS ID LISS III and PAN product for the study of glacier landform mapping | Shaune Garang Glacier |
| 8 | Kaushal et al. (2004) | Used the ratio of IRS LISS III band 4 (SWIR band) and band 3 (Red band) for snow mapping | Siachen Glacier |
| 9 | Ranzi et al. (2004) | The results on the use of thermal band to obtain visible and thermal information derived from TERRA–ASTER images for the detection of debris superimposed on ice and pure debris or rocks | Miage Glacier, Mont Bianco/Mont Blanc massif |
| 10 | Paul et al. (2004a); Andreassen et al. (2008) | The ratio RED/SWIR performs better in areas with dark shadow and thin debris cover | Alpine glaciers |
| 11 | Gupta et al. (2005) | NDSI technique on digital IRS LISS-III multispectral data. mapped dry/wet snowcover | Upper Bhagirathi basin (Gangotri Glacier) |
| 12 | Racoviteanu et al. (2008a) | NDSI for glacier mapping | Cordillera Blanca |
| 13 | Racoviteanu et al. (2008b) | NDSI method | Sikkim Himalayas |

*(Continued)*

## TABLE 3.1 (*Continued*)
## A Tabular Representation of the Literature for Automated Delineation Methods of Debris Covered Glaciers

| | Worker | Methodology | Area of Study |
|---|---|---|---|
| 14 | Shukla et al. (2010) | A synergistic multisensor integrating the inputs from thermal and optical RS data, multispectral classification techniques, and the DEM-derived geomorphometric parameters.<br>1. Preprocessing of the RS data and data integration<br>2. Analysis of the temperature differences between the supraglacial and periglacial debris in shaded and illuminated areas<br>3. Multispectral image classification<br>4. Delineation of debris-covered glacier boundaries | Chenab Basin, Himalayas |
| 15 | Bhambri et al. (2011) | A semiautomated mapping based on (ASTER) DEM and thermal data using morphometric parameters. A thermal mask (single band of an ASTER thermal image) was generated along with identification of clean-ice glaciers (band ratio based on ASTER bands 3 and 4) | Garhwal Himalayas |
| 16 | Kamp et al. (2011) | Used the morphometric glacier mapping (MGM) approach to map large debris-covered glaciers through<br>1. Band ratio analysis<br>2. Five thermal bands (ground surface temperature signal)<br>3. Supervised classification<br>4. Cluster analysis of surface morphology | Debris-covered Parkachik and Drang Drung glaciers |
| 17 | Ghosh et al. 2014a | Multisource approach (Paul et al. 2004a; Veettil, 2012; TM4/TM5 band ratio image; slope derived from Cartosat1-DEM) was applied to map the six debris-covered glaciers to classify the supraglacial debris associated only within the six glaciers.<br>Class I: Snow- and ice-covered parts of the glaciers<br>Class II: Debris (supraglacial and periglacial)<br>Class III: Surrounding nonglacierized areas (vegetation, valley rocks, and water) | Greater Himalayan Range, Ladakh |
| 18 | Bhardwaj et al. (2014) | Semiautomated approach combining thermal, optical RS data, and DEM-derived morphometric parameters. A thermal mask (thresholding of surface temperature layer obtained from Landsat TM/ETM + thermal band satellite data). The extent of clean glacier ice was identified by band ratioing and thresholding of TM/ETM+4 and TM/ETM+5 bands to generate final classification maps. | Hamtah and Patsio glaciers |

(*Continued*)

**TABLE 3.1 (*Continued*)**
**A Tabular Representation of the Literature for Automated Delineation Methods of Debris Covered Glaciers**

|  | Worker | Methodology | Area of Study |
|---|---|---|---|
| 19 | Smith et al. (2015) | Rule-based classification algorithm based on spectral, topographic, velocity, and spatial relationships between glacier areas and the surrounding environment. Multiple data sets (velocity fields, USGS HydroSHEDS, SRTM) used to identify glacier classification<br>1. Clean-ice glacier outlines<br>2. "Potential debris areas"<br>3. Low-elevation areas<br>4. Low-velocity areas<br>5. Areas distant from river networks or clean glacier ice<br>6. Areas very distant from clean glacier ice and manual seed points<br>7. The resulting glacier outlines are cleaned with statistical filtering | Central Asia |

The automatic mapping and classification become immensely vital when applied over large areas. This applies especially to NDSI and band ratio classification techniques, being fast, robust (Racoviteanu et al., 2008b), and more relevant for clean-ice mapping. However, the accuracy of the results is hindered because of many factors. The output can be influenced because of the presence of debris cover on the glacier, proglacial/supra glacial lakes as well as proglacial turbid and frozen lakes. Hence, these areas can be designated and wrongly classified as the glaciers. The correct interpretation of the glacial lakes is highly significant since they are associated with the surfacial ablation. If not properly addressed, they can pose a barrier in the mapping of their spatial attributes and viability of the field observations boundaries (Ghosh et al. 2014a; Figure 3.12). Also, the debris flow/mass movement induced postdepositional sedimentation may pose a difficulty in fully automated mapping of debris-covered glaciers in the polygenetic environment of the Himalayas. The mapping of small glaciers with debris cover and adjacent moraines might be restricted due to ASTER stereo data and thermal bands (Bhambri et al., 2011).

## 3.4 ESTIMATION OF THICKNESS OF GLACIERS

The estimation of glacier thickness is one of the key parameters in assessment of the volume and henceforth the mass balance records of any glaciers. These factors become extremely crucial in assessing the intricacies related to the health of the glacier. The ice thickness is required for the transient modeling to further predicting glacier changes. Direct measurement techniques (radio-echo soundings, borehole measurements) are labor intensive and have limited spatial extent (Farinotti et al., 2009). A first estimate of the ice thickness distribution of all glaciers around the globe based on the Randolph Glacier Inventory (RGI) was presented by Huss and Farinotti (2012). Assessment of the glacier thickness is the preliminary step to calculate the volume. A very important method recommended by UNESCO/IASH (1970) is based on the estimation of the average thickness of each glacier using an empirical relation between the average glacier thickness and glacier surface area and is restricted with specific characteristics and climatic zonation of glaciers. A unique area–volume table has been developed for each of three pilot studies used to test the feasibility of the recommended method (UNESCO/IASH, 1970). Subsequent to this, Müller (1970) estimated the mean

# Understanding the Spatio-Temporal Monitoring of Glaciers

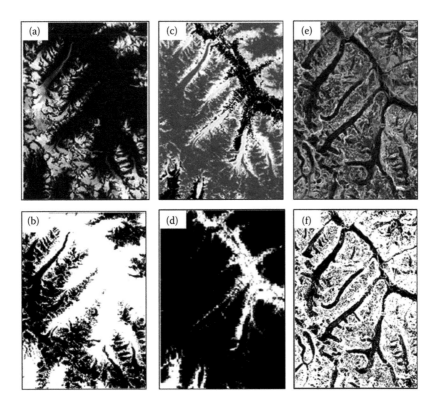

**FIGURE 3.12** Mapping of debris-covered glaciers through satellite images for the glaciers encompassing the Great Himalayan Range. (a) TM4/TM5 ratio image, (b) thresholding applied to TM4/TM5 ratio image showing snow- and ice-covered parts of the glaciers (black), (c) hue component from Intensity Hue Saturation (IHS) image derived after transformation of RGB composite using TM bands 3, 4, and 5, (d) thresholding of hue component of IHS image for demarcation of vegetation-free areas (black), (e) slope image derived from the Cartosat-1 DEM, and (f) slope map showing all slope values of <15 deg (black). (From Ghosh et al. 2014a.)

depth/thickness through statistical method. The same principle has been adopted for Himalayan glaciers as well, although with few departures (Raina and Srivastava, 2008).

Another analysis involving the surface slope data derived from remotely sensed surface elevation measurements has been helpful to provide a useful estimator for ice thickness in absence of other data in South Georgia (Cooper et al., 2007). In addition, the paper shows that useful conclusions can be drawn from regions where it is apparent that the underlying assumptions of the technique break down. A recent study about the estimation of change in the ice volume (Andreassen et al., 2015) in Norway has been done by combining direct ice thickness and total glacier volume observations with different modeling approaches and estimation methods. Power–law relationships for volume/area, volume/length, and volume/area/length have been derived from the plentiful databases available for area and length (Chen and Ohmura, 1990; Bahr et al., 1997; Radic et al., 2008). These can be employed for calculation of thickness and volume (Gantayat et al., 2014). The following methods can be applied for calculation of ice thickness.

## 3.4.1 Thickness Area Relation Method

Bruckl (1970) gave an empirical relationship between the mean thickness and area of the glacier, used by Müller et al. (1976) for the volume estimation of Alpine glaciers. This method has now been adopted for Himalayan glaciers as well.

$$\langle h \rangle = 5.2 + 15.4 x \sqrt{S} \tag{3.1}$$

where:
⟨$h$⟩ is the mean thickness in meters
$S$ is the area of the glacier in km²

### 3.4.2 Volume–Area Related Thickness Estimations

The thickness area relation can be represented as follows,

$$H = cA^\beta \tag{3.2}$$

where:
$H$ is the mean ice thickness
$\beta$ is derived from two scaling parameters, herein defined as $\beta = \gamma - 1$

V–A scaling has been the most frequently used approach for ice volume estimations so far. This method is simple and is applied for volume as a function of area. Glacier volume can be quickly estimated subject to determination of scaling parameters. The general form of V–A scaling relation is

$$V = cA^\gamma \tag{3.3}$$

where:
$V$ represents the glacier volume
$A$ is the glacier area
$c$ and $\gamma$ are two scaling parameters

The volume equation is translated into the thickness–area relation to make it compatible with other methods and ice thickness measurements.

A recent study by Frey et al. (2014) considers three sets of scaling parameter used earlier by Cogley (2011) based on various results on the same area of study. It is, by and large, the only available area-related consideration existing for Karakoram and Indian glaciers, particularly used for all ICIMOD glacier inventories. The details have been illustrated in Table 3.2.

### TABLE 3.2
**Parameters of the Applied V–A Relations. Cogely (2011) Used the Same Relationships for the Glacier Volume Estimations of the Same Region. Volumes Are Calculated in m³ for Input Glacier Areas Measured in m³. $c$ is in m$^{3-2\gamma}$; $\gamma$ is Dimensionless**

| S. No | Source | $c$ | $\gamma$ | Remarks |
|---|---|---|---|---|
| 1 | Chen and Ohmura (1990) | 0.2055 | 1.360 | Used measurement from 63 glaciers to determine the related scaling parameters |
| 2 | Bahr et al. (1997) | 0.191 | 1.375 | Derived the parameters in a theoretical study |
| 3 | LIGG/WECS/NEA (1988) | 0.8433 | 1.3 | Established a thickness-area relation based on ice-thickness measurements on 15 glaciers (Su et al., 1984) |

Used in expression $H = a + cA^{\gamma-1}$
Where $a = -11.32$ m

*Source:* Frey, H. et al., *Cryosphere*, 8, 2313–2333, 2014.

### 3.4.3 GLACIER FLOW MECHANICS METHOD

The thickness can be estimated by using the Nye 1952a theory for the flow of mechanics of an infinitely wide glacier. Accordingly,

$$h = \tau/\rho g \sin \alpha \tag{3.4}$$

where:
- $h$ is the thickness of the ice at a particular slope
- $\rho$ is the density of ice (assumed to be 900 kg m$^{-3}$)
- $g$ is the gravity 9.8 ms$^{-2}$
- $\alpha$ is the slope (from Shuttle Radar Topography Mission [SRTM] digital elevation model [DEM], height corresponding to each $\alpha$ obtained; Cooper et al., (2007)
- $\tau$ is the basal shear stress (between 50 and 150 KPa, ideally taken as 100 KPa) (Hooke, 2005; Cooper et al., 2007).

### 3.4.4 REMOTE SENSING–BASED METHODS

The distribution of ice thickness can be estimated using surface velocities, slope, and the ice flow law (Gantayat et al., 2014). Surface velocities would be calculated using subpixel correlation of the acquired images and the freely available software co-registration of optically sensed images and correlation (COSI-Corr). The best possible correlation would be obtained by an iteratively cross-correlated two images in the phase plane on sliding windows. After performing subpixel correlation, taking a sliding window of 32_32 pixels and a step size of two pixels would give three output images; displacement of north/south image and an east/west displacement image and a signal-to-noise ratio (SNR) describing the quality of correlation. Pixels with displacements >85 m and SNR <0.9 are discarded. Thereafter, a vector field generation from the two displacement images and its proper alignment along the length of the glacier is followed by a calculation of the Eulerian norm of the displacement images. This would result in the magnitude of the resultant displacement. The difference in the time of acquisition between the two images is used to estimate the velocity field.

Ice thickness is estimated using the equation of laminar flow (Cuffey and Paterson, 2010):

$$U_s = U_b + \frac{2A}{n+1} \tau_b^n H \tag{3.5}$$

where:
- $U_s$ and $U_b$ are surface and basal velocities, respectively
- $n$ is Glen's flow law exponent is assumed to be 3
- $H$ is ice thickness
- $A$ is a creep parameter (which depends on temperature, fabric, grain size, and impurity content (= 3.24 × 10$^{-24}$ Pa$^{-3}$ s$^{-1}$ for temperate glaciers); (Cuffey and Paterson, 2010)

The basal stress is modeled as

$$\tau B = f \rho g h \sin \alpha \tag{3.6}$$

where:
- $\rho$ is the ice density (constant value of 900 kgm$^{-3}$; Farinotti et al., 2009)
- $g$ is acceleration due to gravity (9.8 ms$^{-2}$)

*f* is a scale factor, that is, the ratio between the driving stress and basal stress along a glacier and has a range of [0.8, 1] for temperate glaciers. We use *f* = 0.8 (Haeberli and Hoelzle, 1995)

α is the slope, estimated from Advanced Spaceborne Thermal Emission and Reflection Radiometer (ASTER) DEM elevation contours, at 100 m

From Equations 3.5 and 3.6,

$$H = \sqrt{\frac{1.5 U_s}{Af(\rho g \sin\alpha)^3}} \quad (3.7)$$

From this equation, a depth for each area between successive 100 m contours would be calculated and thereafter plotted to provide an ice-thickness distribution for the entire glacier. The ice-thickness values are then smoothed using a 3_3 kernel in order to remove any abrupt changes in spatial ice thickness values. The corresponding thickness is multiplied by the area to obtain the volume of the glacier.

## 3.5 VOLUMETRIC ESTIMATION

Assessment of the ice volume is a critical issue recognized for the well-being of the glacier since the glaciers are potential sources of a huge water reserve. The variability in ice loss across space and time is primarily responsible for sea level rise and global climate change. The traditional ground measurement techniques are not feasible for obvious reasons, RS techniques being an excellent *in situ* complementary method.

### 3.5.1 GLACIER FLOW MECHANICS METHOD

Using Nye (1952b). theory, the ice volume can be calculated for the flow of mechanics of an infinitely wide glacier. This method has been extensively used (Cooper et al., 2007; Li et al., 2012) with an accuracy up to +20% (Driedger and Kennard, 1986). Accordingly, based on Equation 3.4, where the thickness *h* was estimated as

$$h = \frac{\tau}{\rho g \sin\alpha}$$

Average thickness of the glacier thus obtained is multiplied by the area to get the glacier *volume*.

### 3.5.2 REMOTE SENSING–BASED METHODS

Using the value of ice thickness estimated using the equation of laminar flow (Cuffey and Paterson, 2010), from Equation 3.7

$$H = \sqrt{\frac{1.5 U_s}{Af(\rho g \sin\alpha)^3}}$$

The corresponding thickness is multiplied by the area to obtain the *volume* of the glacier.

## 3.6 EXTRACTING GLACIER PARAMETERS FROM DIGITAL ELEVATION MODEL

DEMs are valuable tools for widespread applications in extracting various components of a glaciated environment. The RS-aided delineation coupled with DEMs is useful in assessment of length, termini elevations, median elevations, hypsometry maps, and glacier flow patterns and in deriving ice divides in a semiautomated fashion (Manley, 2007; Racoviteanu et al. (2008b)). Differencing multitemporal DEMs generated from spaceborne or airborne observations is an effective method for monitoring the spatial patterns of glacier thickness and volume changes. However, the RS-based DEMs can be restricted because of factors like spatial resolution of satellite data, exactness of the position on ground as well as verticality of ground control points (GCPs), and contrast and cloud cover in optical satellite images (Bhambri and Bolch, 2011). The GCPs are estimated either through high-precision Differential GPS (DGPS), topographic maps, or satellite data in case of nonavailability of extensive field surveys or maps.

The SRTM-derived DEMs have an advantage of providing elevation data at a global scale. Still, because of slope-induced error and the duration of acquisition of image is in concurrence with the accumulation time in the midlatitudes and the outer tropics, it is rendered unsuitable for studies across a small time scale and smaller glaciers. Subsequently, DEMs can be derived from SPOT5, ASTER, CORONA, or ALOS PRISM with stereo data and three-dimensional monitoring. Considering the best combination of technology and financial implication, Terra ASTER is the most cost-effective optical stereo capability data consisting of 14 spectral bands in three sub system; the visible and near infrared (VNIR; Band 1–3; 15 m), the shortwave infrared (SWIR; Band 4–10; 30 m), and the thermal infrared (TIR; Band 11–14; 90 m). SPOT 5 and ASTER have the highest coverage area (~60 × 60 km) compared with other satellite data.

The following five COTS software are facilitated for end users for processing stereo ASTER data and generation of DEM (Toutin, 2008):

- Geomatica™ OrthoEngine™ of PCI Geomatics (www.pcigeomatics.com)
- OrthoBase Pro module of ERDAS Imagine® and the LPS SP 2 of Leica
- Geosystems Geospatial Imaging (www.gis.leica-geosystems.com)
- ENVI DEM Extraction Module of ENVI 4.2 (www.rsinc.com)
- Desktop Mapping System (DMS)™ of R-WEL (www.rwel.com)
- SilcAst of Sensor Information Laboratory Corp. (www.silc.co.jp)

### 3.6.1 Deciphering Mass Balance Studies from DEMs

The application of DEM coupled with satellite imageries reflecting real delineation is extremely useful to assess the topographical features, namely slope and aspect. Moreover, it has been vital for the derivation of hypsometric maps at a variable time along with estimation of vertical surfacial change at inaccessible locations. These can be considered as indirect evidences of mass balance studies (Berthier et al., 2004; Khalsa et al., 2004). Since the empirical methods are expensive and time consuming and usually are limited to a small number of glaciers, the utilization of RS techniques has opened up a number of opportunities for mass balance observations.

### 3.6.2 Volume–Area Scaling for Mass Balance Studies

This technique utilizes the empirical relationship developed using the scaling between glacier volume and area (Bahr et al., 1997; Bahr and Dyurgerov, 1999). Accordingly,

$$A = L^\alpha \quad \text{and} \quad V = A^\beta$$

where:
  $V$ is the volume
  $A$ is the area
  $L$ is the length
  $\alpha$ and $\beta$ are scaling coefficients derived empirically given a value of 1.36 for valley glaciers and 1.25 for ice sheets based on field observations (Bahr et al., 1997)

Using the parameter of length variation, the change of volume can be represented as follows (Oerlemans et al., 2007):

$$\frac{V}{V_{ref}} = \left(\frac{L}{L_{ref}}\right)^n \tag{3.8}$$

where:

$V_{ref}$ and $L_{ref}$ are reference volume and length, respectively
$h$ is a scaling coefficient (assumed to be 1.4–1.5 for glaciers with no change in width and 2.4–2.5 for ice caps)

An exponent of 2.0–2.1 is recommended keeping in view a wide range of glacier geometry, also in concurrence with Bahr et al., 1997. Based on the scaling relationship, estimated volume is converted to specific mass balance, assuming a constant density of ice as 900 kg/m³ for both the accumulation and ablation zone.

This technique does not address the limitations arising because of the microclimatic effects and presence of debris on the glacier tongue. Moreover, the method also assumes a steady state between climate, ice flow, and ice geometry, which does not hold in actual scenario (Racoviteanu, 2008).

### 3.6.3 Accumulation Area Ratio–Equilibrium Line Altitude Method for Mass Balance Estimations

The accumulation area ratio (AAR)–equilibrium line altitude (ELA) method is one of the direct methods to estimate the mass balance at regional level; it is primarily derived from satellite imageries or field observations. There are three important methods as per the literature:

1. AAR/ELA method (Kulkarni et al. (1992a))
2. Template method (Dyurgerov, 1996)
3. ELA method for mass balance time series (Rabatel et al., 2005, 2008)

The basic assumptions underlying the mentioned methods are as follows:

1. That the accumulation area of the glacier bears a fixed proportion with respect to the total area of the glacier.
2. The transient snowline altitude (SLA) is in perfect coincidence with the yearly equilibrium line altitude (ELA).

The ELA is usually interpreted by using the profound reflectivity variation present in the accumulation zone (fresh zone, high albedo) as compared to the ablation zone (debris cover, less albedo, and less reflective). The marked shift in the zones can be easily outlined with the contour being extracted. The area above the ELA is estimated from satellite images.

1. *The AAR–ELA method* emphasizes on a correlation between AAR and mass balance for a climatic zone wherein the field-based mass balance estimates ($Bn$) and AAR data are compiled with and then plotted for individual glaciers in a climatic zone. Subsequently, by replicating the method for a number of glaciers, a regression line of the form is derived.

$$Bn = a * \text{AAR} + b \tag{3.9}$$

where:
> $Bn$ is the specific mass balance in water equivalent (m)
> AAR is the accumulation area ratio

Using a regression equation between AAR and ELA, a *steady-state AAR* ($AAR_0$), corresponding to the state of equilibrium of the glacier with the climate (zero mass balance) is found. The method has been replicated for a number of glaciers. A steady state AAR value of 0.44 has been concluded for western Himalayas (Kulkarni, 1992). This shows a different value with respect to the reported value of 0.67 for alpine glaciers and of 0.82 for tropical glaciers. Thus, the significance of regional application of this method is evident.

2. *The template method* (Dyurgerov, 1996) takes into account the hypsographic characteristics (area-altitude distribution) of glaciers, which is a reflection of mass balance variation vis-à-vis topography and altitude for the same climatic zone. The glacier area, ELA, and DEM pave the way to assess the hypsographic curve of the glacier so as to relate AAR and ELA as per the following:

$$\text{AAR}(\text{ELA}) = \frac{1}{A_{\text{tot}}} \int_{z_{\text{top}}}^{\text{ELA}} s(z) dz \tag{3.10}$$

where:
> $A_{\text{tot}}$ is the glacier area
> $s(z)dz$ is the change in glacier area as a function of elevation ($z$)

The annual mass balance of the glacier, $Bn$ is represented as

$$Bn = a_1 * \left( \text{AAR}(\text{ELA}) + a_2 \right) \tag{3.11}$$

where the constants $a_1$ and $a_2$ are determined empirically from field measurements.

This method correlates the annual mass balance estimates to ELA and establishes a linear relation between $Bn$ and ELA, which can be further implied to derive the mass balance of the inaccessible glaciers, which have limitations for field verifications, in the same climatic zone. The mass balance estimates for the entire Ak-Shirak range in the Tien Shan has been done through this method (Khalsa et al., 2004).

3. *The ELA method* uses the late summer snowline from satellite imageries and does not use a statistical relationship. Rather this method identifies the snowline at the end of ablation season on the images (presumably the ELA for temperate glaciers) and thereafter extracting the elevation ($ELA_i$) from the DEM. The total volume of ice lost to the glacier is inferred through the geodetic method, and the mean annual mass balance is inferred. The steady state ELA over the time period ($ELA_0$) is calculated using the equation

$$\text{ELA}_0 = \frac{1}{n} \sum_{i=1}^{n} \text{ELA}_i + \frac{\overline{B}}{\partial b / \partial z} \tag{3.12}$$

where:
> $\overline{B}$ is the mean glacierwide mass balance over the time period determined either from field measurements using the glaciologic method or from RS using the geodetic method (described in the next section)
> $\partial b/\partial z$ is the mass balance gradient at the ELA

The annual mass balance $b(t)$ is estimated using the equation:

$$b(t) = \left(\text{ELA}_0 - \text{ELA}_i\right)\frac{\partial b}{\partial z} \tag{3.13}$$

The ELA method is based on the assumption of the proximity and a linear relationship between mass balance gradient ($\partial b/\partial z$) and ELA. The value of $\partial b/\partial z$ has been fixed at 0.78 m per 100 m for Alps glaciers (Rabatel et al., 2005, 2008) and 0.69 m per 100 m for the Chhota Shigri Glacier in the Western Himalaya (Wagnon et al., 2007). The method is more accurate in case of absence of debris cover on the glacier tongue. This method can also be applied for the unsurveyed glaciers with similar climatic signatures. However, in the case of debris cover, the surface energy balance influences the ablation pattern, consequently playing a role in mass balance gradient.

### 3.6.4 THE REMOTE SENSING–BASED GEODETIC METHOD

Since the field surveys are harsh and inaccessible, sometimes the direct measurements may not be present. In these circumstances, the geodetic method is an indirect method that encompasses measurement of elevation changes over a period of time ($\delta h/\delta t$) with the aid of numerous DEMs constructed over the glacier surface. The elevations recorded with the older DEMs may be created from mostly topographical maps are subtracted from the latest DEMs constructed from the RS imageries (SRTM, ASTER, SPOT 5, Cartosat, etc.). This can be done based on a pixel-by-pixel calculation of the glacier or by averaging the elevation change over the entire glacier. The volume estimation is also done accordingly, either by multiplication of the elevation difference by the pixel area to give the volumetric change per pixel ($\delta V/\delta t$) or by multiplying the average elevation with the glacier area to obtain the overall change in volume. This change of volume is then multiplied with the density of glacier (usually assumed a constant density of ice as 900 kg/m$^3$). This method is valid if the elevation remains unchanged due to any nontectonic activity and if the density of the ice mass does not change (Bamber and Rivera, 2007). However, being limited to assessing decadal level changes, this method has certain constraints because of the generation of the following errors:

1. Error because of pixel data saturation, shadow areas, or low contrast during DEM generation from optical stereo data
2. Error due to radar shadow and penetration into snow during DEM generation from radar data
3. Error from old topographical map scanning and digitization of contours
4. Error generation during ellipsoid and datum conversion for subtraction of different DEM derived from various sources (Bhambri and Bolch, 2011)
5. Error generation incurred while accounting for variable resolution DEMs

A study by Berthier et al. (2007) in the Himachal Pradesh (Western Himalayas, India) has been conducted taking into consideration the SRTM and SPOT5 DEM and concluded that the rate of ice loss was two times higher during 1999–2004 than the previous long-term (1979) mass balance records for the Himalayas. Vincent et al. (2013) have deduced the mass balance estimate using the satellite DEM differencing and field measurement, and measured a negative mass balance.

## 3.7 GEOSPATIAL TECHNOLOGY FOR CRYOSPHERE AND THE WAY FORWARD...

The consistent advancement in geospatial technology has paved a way for the development of qualitative as well as quantitative information and understanding of the complex glacier-climate system and its consequent response. There has been significant path breaking progress to monitor the spatial and temporal variability in the cryosphere. The satellite-aided system has been instrumental in providing much needed dataset for inaccessible sites lacking a conventional glaciological method. It is also true that these methods are not completely free from limitations too. Therefore, there is an enormous scope to fill in the gaps and can be achieved by continuous acquisition of high-quality satellite data perfectly in concurrence of the requirements. Consequently, substantial improvements and a wider perspective toward evaluation of various software packages will facilitate our understanding of the complex system and, in turn, will play a decisive role in assessment of future water resources, glacial hazards, and glacier-induced sea level rises.

## REFERENCES

Andreassen, L.M., Huss, M., Melvold, K., Elvehøy, H., Winsvold, S.H. 2015. Ice thickness measurements and volume estimates for glaciers in Norway. *J Glaciol*, 61(228), 763–775.

Andreassen, L.M., Paul, F., Kääb, A., Hausberg, J.E. 2008. Landsat-derived glacier inventory for Jotunheimen, Norway, and deduced glacier changes since the 1930s. *Cryosphere*, 2, 131–145. doi:10.5194/tc-2-131-2008.

Brahmbhatt, R., Bahuguna, I.M., Rathore, B.P., Kulkarni A.V., Nainwal H.C., Shah R.D., Ajai. 2012. A comparative study of deglaciation in two neighbouring basins (Warwan and Bhut) of Western Himalaya. *Curr Sci*, 103(3), 298–304.

Bahr, D.B., Dyurgerov, M. 1999. Characteristic mass-balance scaling with valley glacier size. *J Glaciol*, 45(149), 17–21.

Bahr, D.B., Meier, M.F., Peckham, S.D. 1997. The physical basis of glacier volume–area scaling. *J Geophys Res-Solid*, 102, 20355–20362.

Brückl, E. 1970. Eine Methode zur Volumenbestimmung von Gletschern auf Grund der Plastizitätstheorie. *Arch Meteorol Geophys Bioklimatol* Ser A, 19, 317–328.

Bamber, J.L., Rivera, A. 2007. A review of remote sensing methods for glacier mass balance determination. *Global Planet Change*, 59(1–4), 138–148.

Bayr, J.J., Hall, D.K., Kovalick, W.M. 1994. Observations on glaciers in the eastern Austrian Alps using satellite data. *Int J Rem Sens*, 15(9), 1733–1742.

Berthier, E., Arnaud, Y., Kumar, R., Ahmad, S., Wagnon, P., Chevallier, P. 2007. Remote sensing estimates of glacier mass balances in the Himachal Pradesh (Western Himalaya, India). *Rem Sens Environ*, 108(3), 327–338.

Bhambri, R., Bolch, T. 2009. Glacier mapping: A review with special reference to the Indian Himalayas. *Progr Phys Geogr*, 33(5), 672–704.

Bhambri, R., Bolch, T., Chaujar, R.K. 2011. Mapping of debris-covered glaciers in the Garhwal Himalayas using ASTER DEMs and thermal data. *Int J Rem Sens*, 32(23), 8095–8119.

Bhardwaj, A., Joshi, P.K., Sam, S.L., Singh, M.K., Singh, S., Kumar, R. 2014. Applicability of Landsat 8 data for characterizing glacier facies and supraglacial debris. *Int J Appl Earth Observ Geoinf*, 38, 51–64. doi:10.1016/j.jag.2014.12.011.

Bishop, M.P., Shroder, J.F., Jr., Hickman, B.L. 1999. SPOT panchromatic imagery and neural networks for information extraction in a complex mountain environment. *Geocarto Int*, 14, 19–28.

Bishop, M.P., Shroder, J.F., Jr., Ward, L.J. 1995. SPOT multispectral analysis for producing supraglacial debris-load estimates for Batura Glacier, Pakistan. *Geocarto Int*, 10, 81–90.

Bolch T., Menounos, B., Wheate, R. 2010. Landsat-based inventory of glaciers in western Canada, 1985–2005. *Rem Sens Environ*, 114, 127–137.

Chen, J., Ohmura, A. 1990. Estimation of Alpine glacier water resources and their change since the 1870s, IAHS Publications: Hydrology in mountainous regions, I—Hydrological measurements; the water cycle. In: *Proceedings of Two Lausanne Symposia*, August, Lang, H., Musy, A. (Eds.). IAHS Publications, 193, pp. 127–135.

Cogley, J.G. 2011. Present and future states of Himalaya and Karakoram glaciers. *Ann Glaciol*, 52, 69–73.

Cooper, A.P.R., Tateb, J.W., Cook, A.J. 2007. Estimating ice thickness in South Georgia from SRTM elevation data the international archives of the photogrammetry. *Rem Sens Spatial Inf Sci*, 38(Part II), 592–597.

Cuffey, K.M., Paterson, W.S.B. 2010. *The Physics of Glaciers*, 4th ed. Butterworth-Heinemann, Oxford, UK.

Driedger, P., Kennard, D. 1986. Glacier volume estimation on cascade volcanoes: An analysis and comparison with other methods. *Ann Glaciol*, 8(1), 59–64.

Dutta, S., Ramanathan, AL., Linda, A. 2012. Shrinking glaciers in the Beas basin observed through remote sensing techniques, 1972–2006, Himachal Pradesh, India. *J Earth Syst Sci*, 121(5), 1105–1112.

Dutta, S., Ramanathan, AL., Linda, A. 2013. Estimating glacier changes in the Ravi Basin (1972–2006) through remote sensing techniques. In: *Climate Change and Himalaya-Natural Hazards and Mountain Resources*, Sundaressan, J., Gupta, P., Santosh, K.M., Boojh, R. (Eds.), 262p.

Dyurgerov, M. 1996. Substitution of long-term mass balance data by measurements of one summer. *Zeitschrift für Gletscherkunde und Glazialgeologie*, 32, 177–184.

Farinotti, D., Huss, M., Bauder, A., Funk, M., Truffer, M. 2009. A method to estimate the ice volume and ice-thickness distribution of alpine glaciers. *J Glaciol*, 55, 422–430.

Frey, H., Machguth, H., Huss, M., Huggel, C., Bajracharya, S., Bolch, T., Kulkarni, A., Linsbauer, A., Salzmann, N., Stoffel, M. 2014. Estimating the volume of glaciers in the Himalayan–Karakoram region using different methods. *Cryosphere*, 8, 2313–2333. doi:10.5194/tc-8-2313-2014.

Gantayat, P., Kulkarni, A.V., Srinivasan, J. 2014, Estimation of ice thickness using surface velocities and slope: Case study at Gangotri Glacier, India. *J Glaciol*, 60(220), 277–282.

Ghosh, S., Pandey, A.C., Nathawat, M.S. 2014a. Mapping of debris covered glaciers in parts of the Greater Himalaya Range, Ladakh, western Himalaya, using remote sensing and GIS. *J Appl Rem Sens*, 8(1), 083579. doi:10.1117/1.JRS.8.083579.

Ghosh, S., Pandey, A.C., Nathawat, M.S., Bahuguna, I.M., Ajai. 2014b. Contrasting signals of glacier changes in Zanskar valley, Jammu & Kashmir, India using remote sensing and GIS. *J Indian Soc Rem Sens*, 42(4), 817–827.

Gupta, R.P., Haritashya, U.K., Singh, P. 2005. Mapping dry/wet snow cover using IRS multispectral data in the Himalayas. *Rem Sens Environ*, 97, 458–469.

Hall, D.K., Riggs, G.A., Salomonson, V.V. 1995. Development of methods for mapping global snow cover using moderate resolution imaging spectroradiometer data. *Rem Sens Environ*, 54, 127–140.

Hall, D.K., Williams, R.S. Jr., Bayr, K.J. 1992. Glacier recession in Iceland and Austria. *EOS, Trans Am Geophys Union*, 73, 135–141.

Hooke R.L.B. 2005. *Principles of Glacier Mechanics*, 2nd edition. Cambridge University Press, Cambridge, UK.

Haeberli, W., Hölzle, M. 1995. Application of inventory data for estimating characteristics of and regional climate-change effects on mountain glaciers: A pilot study with the European Alps. *Ann Glaciol*, 21, 206–212.

Huss, M., Farinotti, D. 2012. Distributed ice thickness and volume of all glaciers around the globe. *J Geophys Res*, 117(F4), F04010. doi:10.1029/2012JF002523.

Immerzeel, W.W., Van Beek, L.P.H., Bierkens, M.F.P. 2010. Climate change will affect the Asian water towers. *Science*, 328, 1382–1385.

Kamp, U., Byrne, M., Bolch, T. 2011. Glacier fluctuations between 1975 and 2008 in the Greater Himalaya Range of Zanskar, Southern Ladakh. *J Mt Sci*, 8(3), 374–389.

Kaushal, A., Singh, Y.K., Pal, D.J., Mathur, P. 2004. Snow class stratification and snow line monitoring of a glacier in North Himalayas using advanced remote sensing techniques. In: *International Symposium on Snow Monitoring and Avalanches*, Manali, India, 4 pp.

Khalsa, S.J.S., Dyurgerov, M.B., Khromova, T., Raup, B.H., Barry, R.G. 2004. Space-based mapping of glacier changes using ASTER and GIS tools. *IEEE Trans Geosci Rem Sens*, 42(10), 2177–2183.

Kulkarni, A.V. 1992a. Mass balance of Himalayan glaciers using AAR and ELA methods. *J Glaciol*, 38(128), 101–104.

Kulkarni, A.V., Bahuguna, I.M., Rathore, B.P., Singh, S.K., Randhawa, S.S., Sood, R.K., Dhar, S. 2007. Glacial retreat in Himalaya using Indian remote sensing satellite data. *Curr Sci*, 92(1), 69–74.

Li, H., Ng, F., Li, Z., Qin, D., Cheng, G. 2012. An extended "perfect-plasticity" method for estimating ice thickness along the flow line of mountain glaciers. *J Geophys Res*, 117, F01020. doi:10.1029/2011JF002104.

LIGG/WECS/NEA. 1988. Report on first expedition to glaciers and glacier lakes in the Pumqu (Arun) and Poiqu (Bhote-Sun Kosi) river basins, Xizang (Tibet), China. Science Press, Beijing, China.

Lougeay, R. 1974. Detection of buried glacial and ground ice with thermal infrared remote sensing. In: *Advanced Concepts and Techniques in the Study of Snow and Ice Resources*, Santeford, H.S., Smith, J.L. (Eds.). National Academy of Sciences, Washington, DC, pp. 487–494.

Manley, W.F. 2007. Geospatial inventory and analysis of glaciers: A case study for the eastern Alaska Range. In: *Satellite Image Atlas of Glaciers of the World*, Williams, R.S. Jr., Ferrigno, J.G. (Eds.). USGS Professional Paper 1386-K, U.S. Geological Survey, pp. K424–K439.

Müller, F. 1970 A pilot study for an inventory of glaciers in theEastern Himalayas. In Perennial snow and ice masses: A guidefor compilation and assemblage of data for a world inventory. (UNESCO Technical Papers I Hydrology No. 1) UNESCO, Paris, pp. 47–59.

Müller, F., Caflisch, T., Müller, G. 1976. Firn und Eis der Schweizer Alpen: Gletscherinventar. Z¨urich, Eidgenossische ¨ Technische Hochschule. Geographisches Institut Publ. 57.

Nye, J.F. 1952a. The mechanics of glacier flow. *J Glaciol*, 2(12), 82–93.

Nye, J.F. 1952b. A method of calculating the thickness of ice sheets. *Nature*, 169(4300), 529–530.

Oerlemans, J., Dyurgerov, M., Van de Wal, R.S.W. 2007. Reconstructing the glacier contribution to sea-level rise back to 1850. *Cryosphere Discuss*, 1, 77–97. www.the-cryospherediscuss.net/1/77/2007/.

Ostrem, G. 1975. ERTS data in glaciology: An effort to monitor glacial mass balance from satellite imagery. *J Glaciol*, 15, 403–415.

Paul, F. 2001. Evaluation of different methods for glacier mapping using Landsat TM. In: *Workshop on Land Ice and Snow, Dresden/FRG*, June 16–17, 2000. *EARSeL eProceedings* 1, pp. 239–245.

Paul, F., Huggel, C., Kääb, A. 2004a. Combining satellite multispectral image data and a digital elevation model for mapping debris-covered glaciers. *Rem Sens Environ*, 89(4), 510–518.

Philip, G., Sah, M.P. 2004. Mapping repeated surges and retread of glaciers using IRS: 1C/1D data—A case study of Shaune Garang glacier, northeastern Himalayas. *Int J Appl Earth Observ Geoinf*, 6, 127–141.

Rabatel, A., Dedieu, J.P., Thibert, E., Letreguilly, A., Vincent, C. 2008. Twenty-five years of equilibrium-line altitude and mass balance reconstruction on the Glacier Blanc, French Alps (1981–2005), using remote-sensing method and meteorological data. *J Glaciol*, 54(185), 307–314.

Rabatel, A., Dedieu, J.P., Vincent, C. 2005. Using remote-sensing data to determine equilibrium-line altitude and mass-balance time series: validation on three French glaciers, 1994–2002. *J Glaciol*, 51(175), 539–546.

Racoviteanu, A.E., Arnaud, Y., Williams, M.W., Ordonez, J. 2008a. Decadal changes in glacier parameters in the Cordillera Blanca, Peru, derived from remote sensing. *J Glaciol*, 54, 499–510.

Racoviteanu, A.E., Williams, M.W., Barry, R.G. 2008b. Optical remote sensing of glacier characteristics: A review with focus on the Himalaya. *Sensors*, 8, 3355–3383.

Radic, V., Hock, R., Oerlemans, J. 2008. Analysis of scaling methods in deriving future volume evolutions of valley glaciers. *J Glaciol*, 54(187), 601–612. doi:10.3189/002214308786570809.

Raina, V.K., Srivastava, D. 2008. *Glacier Atlas of India*. Bangalore, India: Geological Society of India, p. 316.

Ranzi, R., Grossi, G., Iacovelli, L., Taschner, T. 2004: Use of multispectral ASTER images for mapping debris-covered glaciers within the GLIMS Project. In: *IEEE International Geoscience and Remote Sensing Symposium* II, pp. 1144–1147.

Rott, H. 1994. Thematic studies in alpine areas by means of polarmetric SAR and optical imagery. *Adv Space Res*, 14, 217–226.

Sangewar, C.V., Shukla, S.P. 2009. Inventory of the Himalayan glaciers. Special Publication No. 34, Geological Survey of India.

Sangewar, C.V., Shukla, S.P. 2009. Inventory of the Himalayan glaciers–A contribution to the International Hydrological Programme (an updated edition) (Spl. Pub., No. 34), 588. Geological Survey of India, Kolkata, India.

Shukla, A., Arora, M.K., Gupta, R.P. 2010 Synergistic approach for mapping debris-covered glaciers using optical-thermal remote sensing data with inputs from geomorphometric parameters. *Remote Sens Environ*, 114(7), 1378–1387.

Sidjak and Wheate (1999) Glacier mapping of the Illecillewaet icefield, British Columbia, Canada, using Landsat TM and digital elevation data. *Int J Rem Sens*, 20(2), 273–284.

Sidjak, R.W., Wheate, R.D. 1999. Glacier mapping of the Illecillewaet icefield, British Columbia, Canada, using, Landsat TM and digital elevation data. *Int J Remote Sens*, 20, 273–284.

Smith, T., Bookhagen, B., Cannon, F. 2015. Improving semi-automated glacier mapping with a multi-method approach: applications in central Asia. *Cryosphere*, 9(5), 1747–1759.

Su, Z., Y. Shi. 2002. Response of monsoonal temperate glaciers to global warming since the Little Ice Age. Quat Int, 97–98, 123–131.

Su, Z., Ding, L., Liu, C. 1984. Glacier thickness and its reserves calculation on Tianshan Mountains. *Xinjiang Geogr*, 7(2), 37–44. [In Chinese]

Toutin, T. 2008. ASTER DEMs for geomatic and geoscientific applications: A review. *Int J Rem Sens*, 29(7), 1855–1875.

United Nations Educational, Scientific, and Cultural Organization/International Association of Scientific Hydrology. 1970. Perennial ice and snow masses: A guide for compilation and assemblage of data for a world inventory. UNESCO/IASH Technical Papers in Hydrology, 1, 59 p.

Vincent C., Ramanathan, AL., Wagnon, P., Dobhal, D.P., Linda, A., Berthier, E., Sharma, P. et al. 2013. Balanced conditions or slight mass gain of glaciers in the Lahaul and Spiti region (northern India, Himalaya) during, the nineties preceded recent mass loss. *Cryosphere*, 7, 569–582. doi:10.5194/tc-7-569-2013.

Veettil. B. K. 2012 A remote sensing approach for monitoring debris-covered glaciers in the high altitude Karakoram Himalayas. *Int J Geomat Geosci*, 2(3), 833–841.

Wagnon, P., Kumar, R., Arnaud, Y., Linda, A., Sharma, P., Vincent, C., Pottakal, J. et al. 2007. Four years of mass balance on Chhota Shigri Glacier, Himachal Pradesh, India, a new benchmark glacier in the western Himalaya. *J Glaciol*, 53(183), 603–611.

Williams, M.W. 1986. Glacier inventories of Iceland: evaluation and use of sources of data. *Ann Glaciol*, 8, 184–191.

# 4 Urban Imprints on City's Environment—Unfolding Four Metro Cities of India

*Richa Sharma and P.K. Joshi*

## CONTENTS

4.1 Introduction .................................................................................................. 51
4.2 Study Area ................................................................................................... 52
4.3 Data .............................................................................................................. 54
    4.3.1 Land Use and Land Cover Mapping ................................................. 55
    4.3.2 Land Surface Temperature Estimation .............................................. 56
4.4 Results .......................................................................................................... 58
    4.4.1 Chennai .............................................................................................. 58
    4.4.2 Delhi .................................................................................................. 60
    4.4.3 Kolkata .............................................................................................. 63
    4.4.4 Mumbai ............................................................................................. 64
4.5 Conclusion ................................................................................................... 68
Acknowledgment .................................................................................................. 69
References ............................................................................................................. 69

**ABSTRACT** The world is urbanizing at the fastest rate ever. Much of the predicted urban growth is occurring in developing countries with a higher rate in Asian countries. This pattern of urban growth is set to continue well into the future. However, knowledge about the patterns of changes is missing. These pattern imprints of urbanization on a city's environment are often unreported or least studied. The lack of studies is due to the unavailability of tools and their applications to understand the phenomena. This chapter emphasizes the application of remote sensing inputs for studying such imprints on a city's environment. By taking examples from the four metro cities (Chennai, Delhi, Kolkata, and Mumbai) of India, the chapter demonstrates changes in the land use land cover and its impact on land surface temperature of the cities. We also demonstrate the usage of satellite remote sensing data over the four decades.

## 4.1 INTRODUCTION

The rampant urbanization across the globe has brought about significant and irreversible land use and land cover (LULC) changes on the earth's surface (Kalnay and Cai, 2003; Foley et al., 2005; Dye, 2008; Taubenböck et al., 2009). The unprecedented growth in population is intensifying the pressure on available resources (Moore et al., 2002), which results in increased urban footprints on a city and surrounding environment. The most prominent of the changes in urban land transformation is the conversion of green cover into impervious surfaces, for instance, conversion of agricultural land into residential built-up areas (Miller and Small, 2003).

Such land transformations severely degrade the quality of the environment by adversely impacting the biodiversity (Savard et al., 2000; McKinney, 2008; Threlfall et al., 2012), soil fertility (Chen, 2007), water quality (Paul and Meyer, 2001; Kaushal et al., 2008), air quality (Atkinson-Palombo et al., 2006), soil nutrients (Groffman et al., 2005), and influences on other natural resources (Huang et al., 2010). A number of studies have analyzed the impact of LULC changes on environmental characteristics such as land surface temperatures (LST) (Fu and Weng, 2016; Chen et al., 2017) and vegetation cover (Zhang et al., 2009; Cui and Shi, 2012).

Urbanization also has huge impacts on the local climate of a city (Tayanc and Toros, 1997; Souch and Grimmond, 2006; Xian and Crane, 2006; Bounoua et al., 2009; Nonomura et al., 2009); the most common manifestation of this impact is in the form of urban heat island (UHI). The UHI indicates a phenomenon of urban areas experiencing higher temperatures in comparison to their surrounding rural environs (Gartland, 2011). In addition to urban–rural temperature variations, UHI causes within-city temperature variations, due to the cities' heterogeneous character. The cities usually have huge intra-urban variation in that nature of urban structure, urban cover, urban fabric use, and urban metabolism (Yan et al., 2014). Land conversion from green to urban use affects the thermal properties of land resulting in an UHI (Rinner and Hussain, 2011; Ryu and Baik, 2012). Other factors contributing in to the development of UHI are urban structures (Unger, 2004), fabric or material use (Chudnovsky et al., 2004), and anthropogenic waste (Quah and Roth, 2012) that results in entrapping more heat and reducing wind movement. UHI can have implications for a city other than just heating the local environment. This warming of the local climate results in increased energy demands for cooling (Kolokotroni et al., 2012), undesirable health impacts, especially during heat waves (Conti et al., 2005; Tan et al., 2010), disturbed hydrology (Lin et al., 2011); poor air quality (Lai and Cheng, 2009); and disrupted urban biodiversity (Beaubien, 2013; Lai and Cheng, 2009). Remote sensing of UHI is typically studied using LST derived from the various thermal satellite imageries such as the Advanced Very High Resolution Radiometer (AVHRR) of the National Oceanic and Atmospheric Administration (NOAA) (Dousset, 1989; Gallo et al., 1993; Lee, 1993), moderate-resolution imaging spectroradiometer (MODIS) on board Aqua/Terra (Tran et al., 2006; Imhoff et al., 2010; Schwarz et al., 2011), and relatively high-resolution Landsat (Kim, 1992; Aniello et al., 1995; Nichol, 1996; Stathopoulou and Cartalis, 2007; Yuan and Bauer, 2007; Guo et al., 2015) and Advanced Spaceborne Thermal Emission and Reflection Radiometer (ASTER) (Nichol, 2005; Nichol et al., 2009; Buyantuyev and Wu, 2010; Liu and Zhang, 2011).

India was a rapidly urbanizing country that was expected to have seven megacities by 2010 (UN 2012). The megacities in India have witnessed up to a five-time increase in urban population in the past 50 years (Taubenböck et al., 2009). Hence, the current research focuses on studying the urban expansion and its impact on environment in the four metro cities of India. This work attempts to quantify and analyze the environmental impacts of spatiotemporal changes in the cities' land use.

## 4.2 STUDY AREA

This chapter presents studies of four metro cities of India: (1) Chennai, (2) Delhi, (3) Kolkata, and (4) Mumbai (Figure 4.1). All the four cities have grown tremendously over the past few decades. Unbridled and unplanned urbanization makes them ideal examples of the impacts of environmental degradation and manifestations of changing urban climate such as UHI. A brief description of the four cities is mentioned in Table 4.1.

*Chennai*, the capital of the state of Tamil Nadu, is located on the southeastern coast of India. Due to its proximity to the equator, the city experiences hot and humid weather through all the seasons of the year. Late May to June is the hottest part of the year with maximum temperatures around 45°C. During monsoons from mid-June to September, the city receives large amounts of rainfall. The city experiences short winters (November to February) with moderate temperature around 15°C–22°C.

# Urban Imprints on City's Environment—Unfolding Four Metro Cities of India

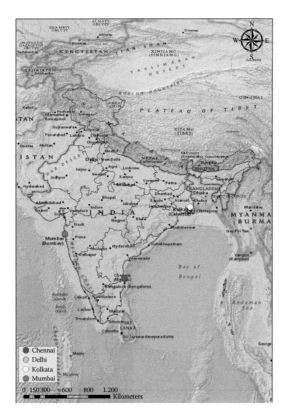

**FIGURE 4.1** The location of the four cities on a map of India.

**TABLE 4.1**
**Basic Information on Demography and Climate of the Cities**

| City | Location | Population in Millions (2011) City Limits | UA | Area (km²) | Climate (Koppen Classification) |
|---|---|---|---|---|---|
| Chennai | 13°5′ N and 80°16′ E | 7.08 | 8.6 | 426 | Tropical wet and dry climate without extreme variation in seasonal temperature |
| Delhi | 28°36′ N and 77°12′ E | 11 | 16.3 | 1484 | Overlap between monsoon-influenced humid subtropical and semiarid areas |
| Kolkata | 22°34′ N and 88°24′ E | 4.5 | 14.1 | 185 | Tropical wet-and-dry climate with dry winters |
| Mumbai | 18°55′ N and 72°54′ E | 12.5 | 21 | 603 | Tropical wet and dry/savanna climate with a pronounced dry season in the low-sun months |

*Delhi*, the capital city of India, is situated in the northern part of India. It is bordered by the states of Haryana on three sides and by Uttar Pradesh on the east. The city is the most populous city in India after Mumbai. It has a rich historical past, and it has served as the capital to various kingdoms and empires since sixth century BC.

*Kolkata* is the capital of the eastern state of West Bengal in India. The city is linearly spread north–south along the east bank of river Hooghly. The city has monthly mean temperatures ranging from 19°C to 30°C. Summers are extremely hot and humid with maximum temperatures rising to 40°C and; winters last for only few months with seasonal lows dipping to 9°C. The region is also prone to cyclones and has an annual rainfall of 1582 mm (Census of India, 2011). The city is considered as the commercial, financial, and cultural center of east India and the northeastern states.

*Mumbai*, the capital city of the state of Maharashtra (in the western part of India) and the financial capital of India, is one of the most populous cities in the world. It has the highest gross domestic product (GDP) and is the wealthiest city of the country. The city thus experiences very high rates of migration. It is bounded by the Indian Ocean on three sides: west, south, and southeast. On the north of the city is Sanjay Gandhi National Park. Some of the coastal parts of the city are covered by mangroves.

## 4.3 DATA

UHI is often studied using the thermal infrared information retrieved from remote sensing data (Yuan and Bauer, 2007). Satellite data are increasingly gaining importance for UHI studies because of its advantage of synoptic coverage and ease of access (Chen et al., 2006). Land surface temperature (LST) has often been acknowledged as an important parameter for urban environmental analysis (Xiao and Weng, 2007) because it closely relates to other biophysical parameters such as the normalized difference vegetation index (NDVI), normalized difference built-up index (NDBI), and normalized difference water index (NDWI) and is influenced by them (Xu et al., 2010). All these parameters could be easily calculated from remotely sensed satellite data. For the current study, Landsat satellite images were used. Landsat data are the most commonly used satellite data for environmental studies across the world (Sobrino et al., 2004). Landsat data consist of seven bands of which the first three are visible bands (red, green, and blue), the fourth band is near-infrared, the fifth and seventh bands fall in shortwave infrared regions, and the sixth band is a long-wave infrared thermal band (Table 4.2). The enhancements to ETM+ from TM essentially include the addition of a panchromatic band and increased spatial resolution of thermal band (band 6).

### TABLE 4.2
### Details and Characteristics of the Satellite Data Used

| Band | Spatial Resolution (m) | | | Spectral Resolution (µm) | | |
|---|---|---|---|---|---|---|
| | TM | ETM+ | OLI | TM | ETM+ | OLI |
| Ultra-blue (coastal/aerosol) | – | – | 30 | – | – | 0.43–0.45 |
| Blue | 30 | 30 | 30 | 0.45–0.52 | 0.45–0.52 | 0.45–0.51 |
| Green | 30 | 30 | 30 | 0.52–0.60 | 0.52–0.60 | 0.53–0.59 |
| Red | 30 | 30 | 30 | 0.63–0.69 | 0.63–0.69 | 0.64–0.67 |
| Near infrared (NIR) | 30 | 30 | 30 | 0.76–0.90 | 0.76–0.90 | 0.85–0.88 |
| Shortwave infrared (SWIR1) | 30 | 30 | 30 | 1.55–1.75 | 1.55–1.75 | 1.57–1.65 |
| Thermal infrared (TIRS1) | 120 | 60 | 100 | 10.4–12.50 | 10.4–12.50 | 10.60–11.19 |
| Thermal infrared (TIRS2) | – | – | 100 | – | – | 11.50–12.51 |
| Shortwave infrared (SWIR2) | 30 | 30 | 30 | 2.08–2.35 | 2.08–2.35 | 2.11–2.29 |
| Panchromatic | – | 15 | 15 | – | 0.52–0.90 | 0.50–0.68 |

## 4.3.1 LAND USE AND LAND COVER MAPPING

Land use relates to the manner in which the surface on earth is used by human anthropogenic activities while land cover relates to the surface observed on earth without the influence of any other activity (Jansen and Di Gregorio, 2003). Understanding LULC changes is very crucial for interpreting any environmental change over different time periods. The satellite images from Landsat used in this study were procured from the United States Geological Survey (USGS) website (www.earthexplorer.usgs.gov). The dates were chosen carefully so that the images for a city over the different years should cover the same season and be relatively cloud free (Table 4.3). The images procured were in Geographic Tagged Image File Format (GeoTIFF) and were projected in the Universal Transverse Mercator (UTM) conformal projection with the World Geodetic System 1984 (WGS84) datum.

LULC mapping was performed using the supervised classification technique. The maximum likelihood classifier, the most commonly used decision rule, was applied. This makes use of a discriminant function to assign pixel to the class with the highest likelihood (Chen and Stow, 2002). Mahalanobis metric was used to classify pixels in which each pixel was treated as being statistically independent (Sharma et al., 2015). The LULC categories were identified based on visual image interpretation keys. For each class, five training sites were given.

For the city of *Chennai*, six LULC classes were identified: (1) *agriculture*, (2) *built-up area*, (3) *marshland*, (4) *vegetation*, (5) *open area*, and (6) *water*. For mapping the LULC in *Delhi*, seven classes were identified: (1) *agriculture*, (2) *dense built-up area*, (3) *forest*, (4) *open area*, (5) *scrub*, (6) *sparse built-up area*, and (7) *water*. Six broad categories were identified for the LULC mapping of *Mumbai*: (1) *forest*, (2) *mangroves*, (3) *vegetation*, (4) *open area*, (5) *built-up area*, and (6) *water*. The eight LULC classes identified for mapping *Kolkata* city included (1) *river*, (2) *wetland water*, (3) *wetland vegetation*, (4) *fallow*, (5) *vegetation*, (6) *open area*, (7) *sparse settlement*, and, (8) *dense settlement*.

Most of the LULC classes mapped for the four cities were common but some classes were unique to one of the cities or sometimes the same class varied slightly from city to city. *Vegetation* was the LULC class that was common for all four cities. It includes urban vegetation and other canopy. However, in case of Mumbai, two other categories of vegetation were considered: *mangrove* and *forest*. *Forest* class covered land under forestation, especially the Sanjay Gandhi National Park. *Mangrove* was another green cover category that specifically was composed of the mangrove forest along the seacoast. It has a unique spectral signature due to the presence of a particular species of trees and salt water. Similarly, the vegetation from wetlands was also distinguished as a separate category in Kolkata. Another category common to all cities was *water* that included various water

**TABLE 4.3**
**List of Landsat Datasets Used for LULC Classification**

| City | Path/Row | Date of Acquisition | Sensor |
| --- | --- | --- | --- |
| Chennai | 142/51 | September 26, 1991 | Landsat 5 TM |
|  |  | October 28, 2000 | Landsat 7 ETM |
|  |  | September 9, 2014 | Landsat 8 OLI TIRS |
| Delhi | 146/40 and 147/40 | March 1, 1998 | Landsat 5 TM |
|  |  | March 7, 2003 | Landsat 7 ETM |
|  |  | March 4, 2011 | Landsat 7 ETM |
| Kolkata | 138/44 | January 19, 1989 | Landsat 5 TM |
|  |  | January 26, 2006 | Landsat 7 ETM |
|  |  | January 21, 2010 | Landsat 7 ETM |
| Mumbai | 148/47 | October 25, 1998 | Landsat 5 TM |
|  |  | October 23, 2009 | Landsat 5 TM |

bodies, big (lake in the national park in Mumbai) and small (creeks, etc.) and seawater. However, in case of Kolkata, *water* was further differentiated as *river* and *wetland water* because wetland water was ecologically unique and fragile and was highly threatened by the scenario of urbanization taking place in the city during the research. *Open-area* comprised lands that lay exposed without any vegetation cover and sandy and silt deposition structures. *Built-up* category covered the urban features and impervious structures like asphalt and concrete roads; however, in the case of Kolkata, this category was further divided into *dense* and *sparse settlements*. The cities of Chennai and Kolkata also have land covered with various agricultural practices. Hence, Chennai has one LULC class called *agriculture* whereas in Kolkata, the land was left fallow during the season under consideration. Hence, Kolkata has a *fallow* land use category instead of agriculture. The city of Chennai also has a unique land use category, the *marshland*. This class covers the freshwater marsh of the Pallikarnai wetland. A class unique to the city of Delhi was *scrub*, which consists of arid vegetation such as *Prosopis* species. Also it is important to note that in Delhi, the *forest* class consists of the ridge forest as well as the urban vegetation such as parks and roadside plantations. It should be mentioned that the ridge forest comprises arid and semiarid vegetation as well.

### 4.3.2 Land Surface Temperature Estimation

Qin's mono-window algorithm (Qin et al., 2001; Sun et al., 2010) was used to retrieve LST from Landsat thermal bands. LST retrieval was done in a three-step process as proposed in the mono-window algorithm (Qin et al., 2001): (1) digital number (DN) to at-sensor brightness temperature, (2) estimated emissivity, and (3) emissivity correction to convert at-sensor brightness temperature into LST. Three parameters required to achieve this were emissivity, transmissivity, and effective mean atmospheric temperature. Brightness temperature was retrieved from the thermal band (band 6) of the data using calibration parameters accessed from the Landsat handbook provided by NASA. The data were radiometrically corrected using the following calibration equations provided by NASA; hence, DN was converted into radiance (Chander et al., 2009):

$$\text{Radiance} = \frac{L_{max} - L_{min}}{(QCal_{max} - QCal_{min})} \times (QCal - QCal_{min}) + L_{min}$$

where:
$L_{max}$ and $L_{min}$ are spectral radiances for thermal band
QCal represents the DN values
$QCal_{min}$ and $QCal_{max}$ are digital numbers 1 and 255

The radiance thus retrieved was converted to at-sensor brightness temperature employing modified Planck's equation.

$$T_b = \frac{K_2}{\ln\left(\frac{K_1}{L} + 1\right)}$$

where:
$T_b$ is the at-sensor brightness temperature (in K)
$L$ is the spectral radiance (in W/(m²/sr-mm)) calculated previously
$K_1$ (607.76 W/(m²/sr-mm)) and $K_2$ (1260.56 K) are prelaunch satellite calibration constants

Since at-sensor brightness temperature assumes all objects to be black bodies, they have an emissivity of 1. However, black body is only a theoretical concept; thus, there is the need to estimate emissivity for different objects to retrieve their actual surface temperatures. NDVI is often used to estimate the emissivity. The brightness temperature ($T_b$) thus obtained was corrected for emissivity using NDVI (Valor and Caselles, 1996; Zhang et al., 2006). For estimating NDVI values, radiance values of band 3 (red) and band 4 (NIR) were calculated using

$$\text{Reflectance} = \frac{\pi \times d^2 \times \text{Radiance}}{\text{ESUN} \times \cos\theta}$$

where:
  $d$ represents earth-sun distance in astronomical units
  ESUN is the mean solar exoatmospheric irradiance
  $\theta$ is the solar zenith angle

The reflectance computed for band 3 and band 4 was used to derive NDVI, which was then used for determining emissivity images. These images were then employed to correct previously obtained at-sensor brightness temperature. Emissivity for the NDVI range of 0.157–0.727 could be expressed as follows:

$$\varepsilon = 1.0094 + 0.047 \times \ln(\text{NDVI})$$

For NDVI values out of this range, different emissivity values were assigned to three different ranges. For NDVI values of less than −0.18, emissivity of 0.985 was given. Emissivity of 0.955 was assigned to NDVI ranging between −0.18 and 0.157. For NDVI values beyond 0.727, emissivity of 0.99 was assigned.

$$T_s = \left\{ a(1-C-D) + \left[ b(1-C-D) + C + D \right] \times T_B - D \times T_a \right\} \times C$$

$$C = \varepsilon \times \tau$$

$$D = (1-\tau)(1+(1-\varepsilon)) \times \tau$$

where:
  $T_s$ is LST
  $a$ and $b$ are constants of values 67.355351 and 0.458606, respectively
  $T_b$ is the at-sensor brightness temperature
  $T_a$ is the effective mean atmospheric temperature
  $C$ and $D$ are the parameters derived from emissivity and transmissivity as mentioned in previous equations following the mono-window algorithm (Qin et al., 2001)

For Landsat TM/ETM+, band 6 has been extensively exploited to study thermal dynamics of various earth surface features (Qin et al., 2001). LST for Landsat TM/ETM+ was thus derived using thermal infrared band—band 6—while for Landsat OLI band 10 (Adnan et al., 2015) was utilized. LST is strongly influenced by the amount of incoming solar radiations; thus, while studying the influence of land use transformations on surface temperature, it is important to keep other climatic factors unchanged. Hence, in such a case, it is recommended to compare data from the same season to minimize the influences of external factors on the energy budget. For example, comparing a 1991 summer image with a 2015 winter image might give spurious results with a sharp decline in LST values. However, this drop in LST was not indicative of UHI mitigation but was due to decreased incoming solar radiation in winter months as compared to summer months. Considering this fact, the data for LST calculation were decided based on an overlap of seasons through all the years in consideration for each city. For instance, in the case of Chennai, images from winter months (January and November) were available for 1991 and 2000 but were not available for 2014. Similarly, spring data (March) could be retrieved for 2000 and 2014 but were not available for 1991. For summer season (May), images could be found for the years 2000 and 2014 but not for 1991. The rainy season (September and October) was the only season for which data were available for all the years; hence, September 26, 1991; October 28, 2000; and September 9, 2014, images were used for estimating the LST. A similar procedure was followed while selecting images for all the four cities.

## 4.4 RESULTS

### 4.4.1 Chennai

A total of 100 stratified random points were used to assess the accuracy of classified maps. Overall accuracy for 1991 was 0.89, for 2000 was 0.89, and for 2014 was 0.81, and kappa was 0.87, 0.87, and 0.76, respectively.

Over the years, *open area* and *marshland* were found to decrease consistently while *built-up area* was observed to show consistent increase (Figure 4.2). *Vegetation* first exhibited an initial fall from 1991 to 2000 (by 27.5%) and registered a slight increase of 5% in the second phase. *Agriculture* first increased slightly, by 12 km$^2$, between 1991 and 2000 and then came down to 13 km$^2$ from 53 km$^2$ in 2014. It is to be noted that the increase in *agriculture* in the initial time period was from *marshland* area, this area eventually losing to *built-up* class. Amid all the land use dynamics, it is important to note that *marshland*, an ecologically critical LULC class, was observed to have steadily declined from 68 km$^2$ in 1991 to 44 km$^2$ in 2000 and finally to 23 km$^2$ in 2014. Thus, overall from 1991 to 2014, the *marshland* within the administrative boundaries shrank by 66%. The *built-up* class, which initially was more confined to the parts of the study area extending from centre to north and east with total coverage of 91 km$^2$ in 1991, spread over the entire domain in next two decades: 96 km$^2$ in 2000 and 139 km$^2$ in 2014.

Figure 4.3 represents the spatial distribution of LST zones over the city for the years 1991, 2000, and 2014 and the quantitative distribution of LST zones by each land use class. It was observed that high LST zones were initially more confined to the north and northwestern parts of the city while over the years, the newly built-up areas tended to be more in higher LST zones (4–6). Also, the open areas (such as beaches) and drier parts of marshland appeared to be hotter than the rest. This was further reinforced with statistics from the radar plots. LST zones 1 (coldest) and 6 (hottest) appeared to be negligibly small in terms of area covered. Looking at LST zone 5 (very hot), it was observed

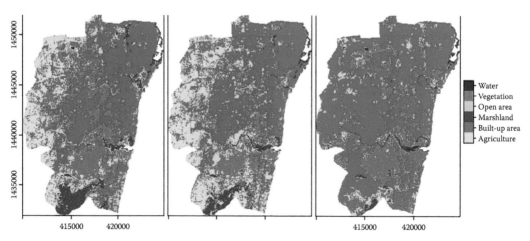

**FIGURE 4.2** LULC maps for Chennai.

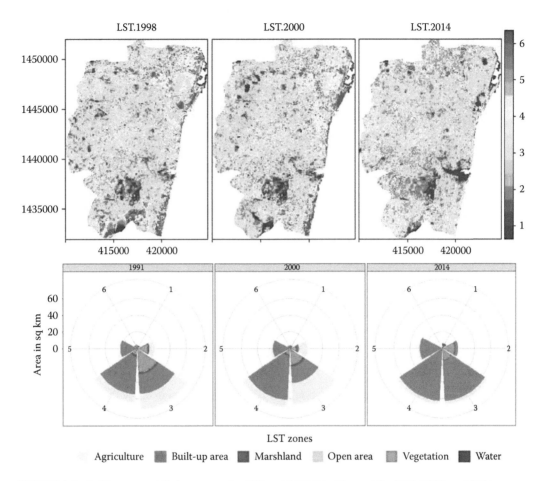

**FIGURE 4.3** LST zones and their areas under different LULC in Chennai for 1991, 2000, and 2014.

that there had been an overall increase in area covered by this category (from 23.76 km² in 1991 to 30 km² in 2014) and specifically the contribution from *built-up area* to this category increased over the years from 19 km² in 1991 to 27 km² in 2014. LST zone 3 (medium hot), which was predominantly contributed by *built-up area*, *vegetation*, and *agriculture*, decreased from 74 km² in 1991 to 67.6 km² in 2000 to 67.3 km² in 2014. *Vegetation* LULC class had relatively higher coverage for LST zones 2 and 3 (cold and medium zones); this was found to be decreasing over the years, particularly for the medium zone. The overall area under LST zones 1 (coldest) and 2 (cold) did not change much over the years. The cooling effect of *water* class was evident from the fact that it fell only under the coldest to medium LST zones with maximum area (84%) of this LULC class being under LST zone 1.

Figure 4.4 shows a boxplot for LST values for various LULC categories over the years 1991, 2000, and 2014 for Chennai. Over the years, *water* had a minimum mean LST (30.26°C in 1991, 31.57°C in 2000, and 33.19°C in 2014) while *open area* had the highest mean LST throughout the years (34.62°C in 1991, 37.23°C in 2000, and 36.79°C in 2014). *Open-area* class in the Chennai LULC map mainly included the sandy beaches, which have high reflectance and no vegetation cover and did not retain moisture, which shot up the radiant temperature of such sandy land covers. It is interesting to note that much fluctuation was observed for all the LST LULC categories except for marshland, which recorded negligible changes in mean LST (ranging from 31.24°C to 31.49°C). *Vegetation* and *agriculture* classes showed LST higher than *water* but was always lower than built-up area. In 2014, however, agriculture had a mean LST (35.64°C), quite close to that of

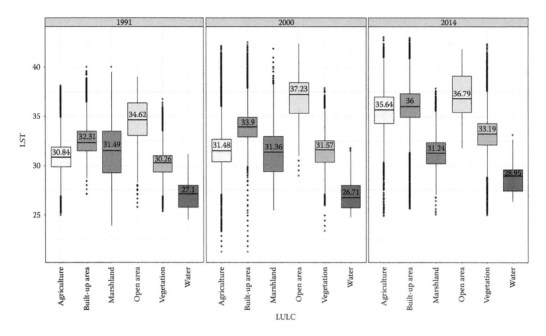

**FIGURE 4.4** LST statistics for various LULC classes over the years for the city of Chennai.

built-up area (36°C), which could be due to two factors: first, there was an increasing encroachment in agricultural lands by the built-up area that could have caused mixing of the thermal signature of the two land use classes; second, there could have been patches of fallow land within the cropped agricultural lands that could have resulted in increased mean LST.

### 4.4.2 Delhi

A total of 100 stratified random points were used to assess the accuracy of classified maps. Overall accuracy for 1998, 2003, and 2011 were 83%, 79%, and 82%, with kappa of 0.74, 0.69, and 0.73, respectively.

Delhi witnessed a loss in *agricultural* land from 1998 to 2003 (Figure 4.5), mainly in the extreme north and southwest. The land was lost to *sparse* and eventually a *dense built-up area* in industrial areas of Narela and Bawana (north) and to *open area* eventually converting to *sparse* and *dense built-up area* in the Indira Gandhi Airport area and the Dwarka subcity (southwest). The green or vegetation classes had much dynamics in *agriculture*, *scrub*, and *forest*. Agriculture declined

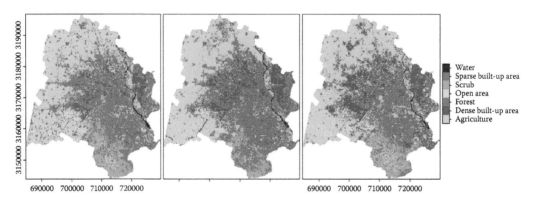

**FIGURE 4.5** LULC maps for Delhi.

initially (from 510.6 km² in 1998 to 457.5 km² in 2003) but had a slight increase in the second time period (from 457.5 km² in 2003 to 512 km² in 2011), but on detailed analysis of spatial patterns, it was observed that the increase in agriculture was confined more along the river banks. Forest exhibited a fall of 10 km² from 1998 to 2003, but an increase of 20 km² in 2003 to 2011 phase. *Forest* exhibited a more to-and-fro the transformative relation with agriculture and sparse built-up area. This was because both *agriculture* and *forest* are vegetation classes and hence have similar spectral signatures, making it hard to accurately distinguish between the two. This could be easily observed in southern parts of the city where intermixing of *agriculture* and *forest* was prominent. Because *forest* class includes the vegetation within the city including the vegetation present in *sparse built-up* area, there were high chances of intermixing of these classes where heterogeneity was high. This ambiguity was clearly noticeable in the central parts of the study domain. The dense built-up area in the city rose over the decades, increasing from a land cover of 4.8% in 1998 to 9.3% in 2003 and then to 12.8% in 2011, spreading rapidly along the highways, evidently in the north, toward the west and southwest. No absolute change in *sparse built-up* cover was observed for the first phase; however, it did display a change in spatial pattern as it gained more land from the *open area* and *agriculture* in the north, west, and southwest and simultaneously lost out to *dense built-up area* in the southwest, moving from central to northwest. However, during the time period from 2003 to 2011, sparse built-up area experienced a loss in land cover to dense built-up area predominantly in East Delhi, Dwarka, the southeastern tip of the city, and the north industrial zone. *Open area* had a dynamic relationship with other classes such as *scrub* and *sparse built-up area*. *Open area* has been grabbing land from agriculture for conversion to the built-up area; however, there was a spectral mixing of these two classes often resulting in a misclassification displaying false conversions of *sparse built-up area* into *open area*. The *scrub* and *open area*, however, did exhibit a factual to-and-fro conversion because scrub consisted of arid vegetation, which during dry periods makes the land devoid of any vegetation, thus exhibiting spectral properties of an *open area*. Figure 4.5 also exhibits the loss of *scrub* land (which was mainly confined to the Asola Bhati Wildlife Sanctuary in the south) to encroachment for *built-up area* and *agriculture*. The *water* class did not change considerably over the years.

The spatiotemporal distribution of LST zones over the city for the years 1998, 2003, and 2011 is depicted in Figure 4.6. The areas that prominently and consistently appeared to be in relatively hotter zones were the industrial belts of Mayapuri, Wazirpur, Azadpur, and Jahangirpuri and the industrial zones of Narela and Bawana along with the airport area in the southwest of the city. Spots of higher LST zones in the west and southwest could be observed in 1998 but then disappeared in later years; these were actually the *open area* that became covered with *sparse built-up area* interspersed with vegetation. In contrast to hot zones, all the agricultural rural land in the city peripheries consistently remained in colder LST zones (very cold and cold). The radar plot (Figure 4.6) indicates that very cold and cold zones of LST were dominated by *agriculture*. LST zone 3 (medium) was primarily agriculture in 1998 but was eventually replaced with *open area* and *scrub*. The major portion of the area of the city fell under LST zone 4 (hot) all through the years. The zone was primarily composed of *sparse built-up area* in 1998, which continually decreased gradually with simultaneous increase in contribution from *dense built-up area*, *open area*, and *agriculture*. High LST in *agriculture* could be due to seasonal fallow land. The zone also exhibited decreasing *forest* land cover.

The LST statistics for different land use classes evidently show a cooling effect of water and vegetation (Figure 4.7). The water and agriculture classes had the least average LST over the years with median LST for agriculture being 26.8°C, 25.6°C, and 27.6°C for 1998, 2003, and 2011 and for water being 27.9°C, 26.3°C, and 28°C, respectively. Ideally, *water* should have lowest LST; however, the *water* class in Delhi comprised of the river water of Yamuna was highly polluted and turbid, thus containing loads of impurities and increasing its LST. These two classes were followed by *forest*, which had an average LST of 31.3°C, 30.9°C, and 31.2°C in 1998, 2003, and 2011, respectively. The land uses with highest LST were *dense built-up area* (32.1°C in 1998, 32.5°C in 2003, and 32.8°C in 2011),

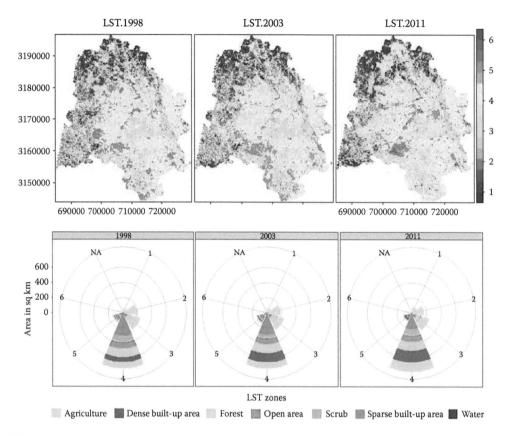

**FIGURE 4.6** LST zones and their areas under different LULC in Delhi for 1998, 2003, and 2011.

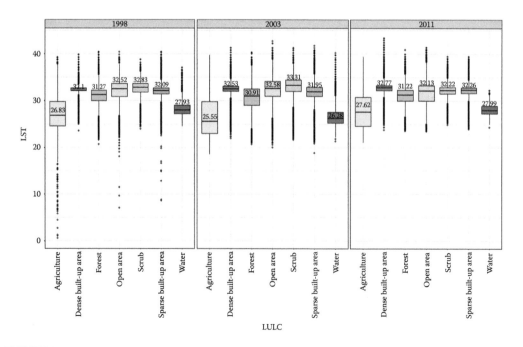

**FIGURE 4.7** LST statistics for various LULC classes over the years for the city of Delhi.

sparse built-up area (32.1°C in 1998, 32°C in 2003, and 32.3°C in 2011), open area (32.5°C in 1998, 32.6°C in 2003, and 32.1°C in 2011), and scrub (32.8°C in 1998, 33.3°C in 2003, and 32.2°C in 2011).

### 4.4.3 Kolkata

The overall accuracy of LULC maps for 1989, 2006, and 2010 was 90%, 88%, and 86%, respectively. The kappa statistics for 1989, 2006, and 2010 were 0.8726, 0.8455, and 0.8212, respectively. Considerable amount of cloud cover in the satellite images resulted in lower accuracy for fallow land and open area in comparison to other LULC classes. However, the LULC maps evidently had reasonable overall accuracy and were sufficient for detecting the increasing built-up growth in the city across the different time spans.

Figure 4.8 clearly shows considerable changes in LULC over time. Although overall increase in the built-up area was around 4 km², intra-built-up variations are important to be considered here. Of *sparse built-up* land, 27% was converted to *dense built-up* land from 1989 to 2006 and from 2006 to 2010, 62% of total *sparse built-up area* was transformed to *dense built-up area*. *Sparse built-up area* in turn engulfed 1.4 km² and 2.2 km² land from the *vegetation* land use class. Vegetation also contributed 3 km² to dense built-up area from 2006 to 2010. Due to higher spectral similarity between *built-up* and *fallow* land, there was a to-and-fro land transformation detected for these two land use classes. For instance, during 1989 to 2006, 6.7 km² of *fallow* land was converted to *built-up area* while 12.7 km² was observed to revert back. And in the 2006–2010 time period, *fallow* to *built-up* conversion was 10.3 km², and reverse transformation of land use took place for 3.8 km². The *built-up area* to *fallow* transitions, however, was due to less accurate distinction between *built-up area* and *fallow* land pixel due to similar spectral signatures of the two. The lowest producer's accuracy for *fallow* class was 42% (in 2010) while the lowest user's accuracy was 64% (in 2006). *Sparse built-up area* also had relatively low user's accuracies (67% in 2010). Wetland area including both *wetland water* and *wetland vegetation* was almost halved over the time and so was *vegetation* class.

On analyzing the spatial pattern of LST zones for the city over the years, it was observed that high LST zones were initially more concentrated in the west and some parts of the north; however, over the years, these zones also spread to southeastern parts of the city (Figure 4.9). LST zones 1 (coldest) and 6 (hottest) appear to be negligibly small in terms of area covered. LST zone 3 (medium) was observed to have slightly increased over the years by 1–2 km². However, a notable increase was observed for LST zone 4 (hot), mainly contributed by *dense built-up* land. Dense built-up land under hot LST zone increased from 20 km² to 30 km². Similarly, contribution from *dense built-up area* increased the area under LST zone 4. LST zone 5 (very hot) showed a fall in contribution from *dense built-up area* and an increase in *fallow*, which as pointed out previously was due to misclassification between *built-up*, *open area*, and *fallow* land. *Vegetation* LULC class had relatively higher coverage for LST zones 2 and 3 (cold and medium zones) in 1989, which had almost disappeared by 2010, particularly in zone 3. Overall area under LST zones 1 (coldest) and 2 (cold) did not change

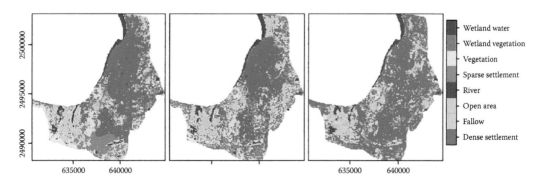

**FIGURE 4.8** LULC maps for Kolkata.

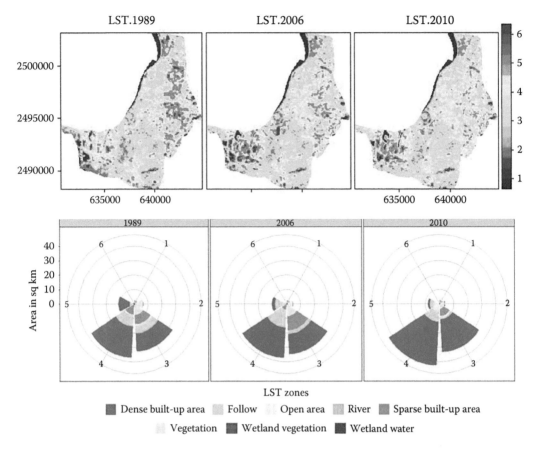

**FIGURE 4.9** LST zones and their areas under different LULC in Kolkata for 1989, 2006, and 2010.

much over the years. The *river* was found to constitute only zone 1 while *wetland* classes spread over zones 1 and 2.

Figure 4.10 helps us understand the relationship between LST and the altering different LULC over the years 1989, 2006, and 2010. As observed for Chennai over the years, Kolkata also had a minimum mean LST for water land use represented by *river*. This was followed by vegetation, which over the years had the second coldest land use class with a mean LST of 27.3° in 1989, 30.3° in 2006, and 32.4° in 2010. The *wetland water and vegetation* classes had a mean LST higher than *vegetation* (but sometimes lower) and *river* but always lower than all other land use categories. The classes that appear as the hottest ones were *dense built-up area* (29.8°C in 1989, 32°C in 2006, and 34.7°C in 2010), *fallow* land (29.6°C in 1989, 33°C in 2006, and 36.4°C in 2010), *open area* (29.5°C in 1989, 34.6°C in 2006, and 36.8°C in 2010), and *sparse built-up area* (29.3°C in 1989, 31°C in 2006, and 33.5°C in 2010). These four classes have specific features that result in higher radiant temperatures. *Open-areas* and *fallow* land, for instance, were devoid of vegetation cover, and in some instances (like Khidripur on the southeast that serves as a port) have slums, small industrial units, and other buildings with tin roofs that exhibit low thermal emittance, increasing their LST.

### 4.4.4 MUMBAI

The overall accuracy for the 1998 LULC map of Mumbai was 87% and for 2009 was 91%. The two years had overall kappa values of 0.842 and 0.8885, respectively. The producer's accuracy for *mangroves* in both years was relatively low (62.5% and 75% in 1998 and 2009, respectively). This was

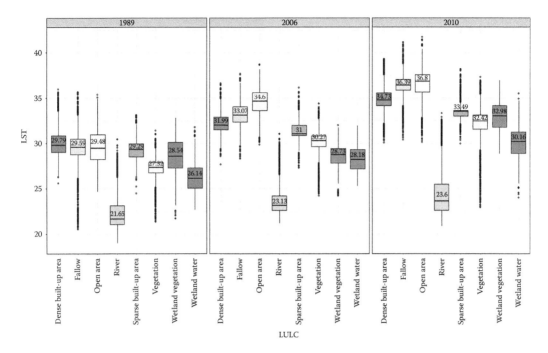

**FIGURE 4.10** LST statistics for various LULC classes over the years for the city of Kolkata.

mainly in areas where mangroves were sparse. Such land was interspersed with open areas, and thus the spectral values of these two classes often are mixed up, resulting in lower producer's accuracy of *mangroves* and lower user's accuracy of *open area* (64.3% in 1998 and 80% in 2009). *Forest* was the only class that had 100% user's accuracy in both the years. But it has relatively lower producer's accuracy, mostly due to peripheral areas of forest where the spectral signature of *forest* and *vegetation* class is mixed up, thus bringing down the user's accuracy of *vegetation* class.

The LULC maps for Mumbai (Figure 4.11) show a decrease in *forest* from 1998 to 2009 along the peripheral areas of the forest patch in the north. However, some improvement in *forest* was observed in the southern patch. *Mangroves* showed a huge loss in the northwestern and northern parts of the city. *Vegetation* class exhibited an improvement at the cost of forest in Sanjay Gandhi National Park, the forest in northern parts of Mumbai. However, *vegetation* received a blow in the central parts of the city where *built-up area* intensified during the study period. *Open-area* claimed land from *mangrove* cover. *Water* appears to have remained more or less the same.

The LULC statistics and analysis of their change reveal a two-way relationship between *forest* and *vegetation* and *vegetation* and *open area* in which each class converted to the other and a part of the latter class reverted to the former. For instance (Table 4.1), 9.6 km² was converted to the *vegetation* class, and at the same time, 7.1 km² of *vegetation* was transformed into forest. The most noteworthy transformation in LULC statistics was the conversion of *vegetation* (22.9 km²) into *built-up area*. The highly critical ecosystem of *mangrove* lost a land area of 11 km² to *open area*. *Forest* and *vegetation* lost 4.9 km² and 16.7 km² from 1998 to 2009, respectively. Another important observation is the increase of 30 km² (14%) of *built-up area* in 2009 as compared to 1998.

LST mapping was performed for 1998 and 2009, ranging from 30°C to 55°C. The LST zoned maps (Figure 4.12) show increasingly hotter LST zones from 1998 to 2009 in the central parts of the city. The *water* class and the *forest* showed lowest LST zones (zones 1 and 2) during both years. LST zones in *mangroves* area were higher than these but lower than other land use classes. LST zones of built-up category were generally 3 (medium), 4 (hot), and 5 (very hot) in both the years. LST zone 1

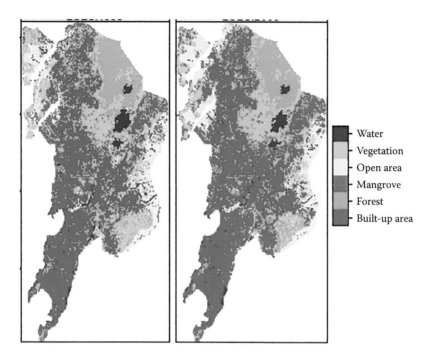

**FIGURE 4.11** LULC maps of Mumbai for the years 1998 and 2009.

(coldest) covered a negligible area of the study domain and was mainly composed of *forest*. LST zone 2 (cold) showed a slight decrease from 1998 to 2009. This category mainly covers the land under *forest*, *vegetation*, and *water*, which was supported by LULC statistics indicating a fall in green cover classes such as *forest* and *vegetation*. LST zone 3 registered an increase of 6 km$^2$ over the years. The zone majorly consists of *vegetation*, and *open area*, *built-up area*, and some fraction claimed from *mangroves* land. The increase in this zone was mainly from the increase in *built-up area* (10 km$^2$) and *open area* (21 km$^2$). *Vegetation* and *mangrove* in this zone, however, registered a decrease of 9 km$^2$ and 7 km$^2$, respectively. The hot LST zone (4) had a decline of 10 km$^2$ from 1998 to 2009. The zone registered an increase from built-up area (10 km$^2$) and a decrease from open area (16 km$^2$) and vegetation (4 km$^2$). Zone 5 (very hot) had *built-up area* as the sole contributor and an increase of 10 km$^2$. The hottest zone of LST was composed of only *built-up area* and showed a slight fall of 1.5 km$^2$.

Due to increasing encroachment (as shown by LULC map), which was engulfing the land along the peripheries of the Sanjay Gandhi National Park, the temperature pattern along the park fringes were altered. Areas of nonurban land that were converted to built-up area during this period appear to be getting warmer. Potential UHI spots in the central parts of the city are Shivaji Nagar, Nirankar Nagar, New Gautam Nagar, Vaibhav Nagar, and Ramabai Ambedkar Nagar in the southeast; Jogeshwari East near Sanjay Gandhi National Park; and Asalfa, Mairwadi, Mohili, Lokmanya Tilak Nagar, and Dharavi.

The fragmentation in forest was well reflected by an increase in LST zone from 2 to 3 and higher zones passing through the forest in the south and southeast. *Built-up* class primarily fell in medium and hot to hottest zones of LST, zones 3, 4, 5, and 6. It can also be observed from the radar plot in Figure 4.12 that built-up area's fraction in LST zones 3, 4, and 5 increased over the time period. It should be noted that *built-up* and *open areas* engulfed land from *vegetation*, *forest*, and *mangrove* in zones 3 and 4. *Forest* class was mainly dominated by cold LST (zone 2), which shrank slightly from 1998 to 2009. *Mangroves* cover relatively less area overall and fall more into zones 3 and 2 and have lost major portions of land to *open area* and *built-up area* in zone 3. *Water* also was dominated by zones 2 and 3 but formed a very small part of the area under consideration. Zones 3 and 4

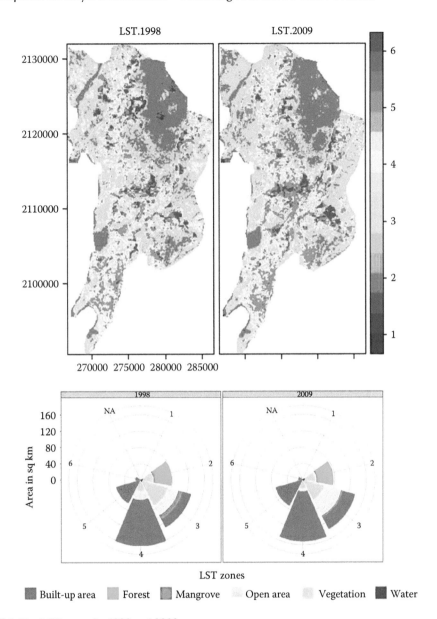

**FIGURE 4.12** LST maps for 1998 and 2009.

dominated the major area of *open area* land use class, and that class seems to have gained under zone 3 but lost to *built-up area* in zone 4. Dominant LST zones for *vegetation* area zones 2, 3, and 4 with the majority of this class falling under the medium zone (zone 3) (Figure 4.13).

In contrast to the observation for Chennai and Kolkata, the minimum average temperature in Mumbai was not for *water* but for *forest* class. This could be because the water in the image was confined to seashores and the lake in the forest. Since the shore water has higher salts, making its appearance turbid, and the quality of lake water was not good, the water in the given image lost impurities, making it warmer than it should have been. *Mangroves* and *water* stand second in terms of median LST followed by *open area* and *vegetation*. *Built-up* always was the hottest of all the classes. The average LST for all the land use classes in 2009 imagery was lower than that for 1998 with *forest*, *mangroves*, *open area*, *vegetation*, and *water* cooling by 3°C–4°C, whereas built-up area cooled down only by 0.8°C, indicating higher differences between LST for *built-up* and other classes.

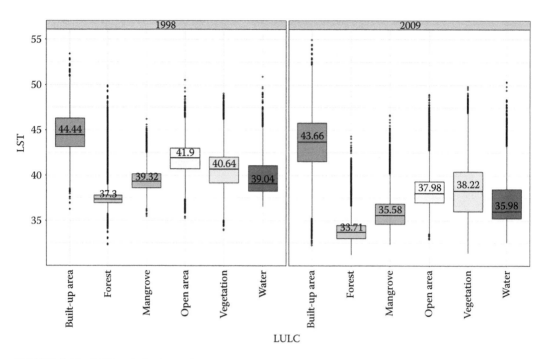

**FIGURE 4.13** LST statistics for various LULC classes over the years for the city of Mumbai.

## 4.5 CONCLUSION

This study aimed to map the LULC transitions taking place in the four metropolitan cities of India over three decades and provide information for understanding the implications of these changes for the thermal environment of the respective urban centers. It was observed that all cities were experiencing depletion in green cover with simultaneous rise in built-up land covers. This was resulting in loss of pervious soil covers to highly impervious urban materials. These urban areas also have the tendency to retain heat for longer time into the night as well. The cities were hence increasingly becoming hotter by decades as shown by the increasing area in higher LST zones, including 4 (hot), 5 (very hot), and 6 (hottest).

In addition to analyzing the surface warming in the urbanizing landscape of the four Indian metro cities, this research also highlights the thermal signatures of different land use categories. This analysis illustrates the cooling effect of green land use classes such as forest, agriculture, urban plantations along areas such as roads or parks and the mangroves (as in case of Mumbai), and water. Water was observed to be consistently cooler than all other land use classes in all the cities for all the years. Thus, in addition to discussing the problems of urbanization in terms of its impact on surface temperatures, this work also presents the probable solution areas for it. Another important aspect emphasized by this research is the impact of urbanization on ecologically sensitive areas such as wetlands and marshlands. In the case of Kolkata, the research also illustrates how wetlands contribute to maintaining the surface temperatures, indicating their ecological sensitivity and the cooling effects that they offer.

Such studies greatly contribute to monitoring the land use changes taking place in rapidly and rampantly urbanizing cities in developing countries such as India. There is a huge opportunity to build further research on this work, such as investigating the impact of changes in thermal environment on the public health (Tomlinson et al., 2011; Araujo et al., 2015), and understanding the more social linkages of these changes (Buyantuyev and Wu, 2010). In addition to providing information for academic audiences, such research helps urban managers to understand the problem areas in their cities.

## ACKNOWLEDGMENT

Authors are grateful to the Department of Science and Technology (DST), Ministry of Science and Technology, Government of India for a project grant. PKJ thanks DST Promotion of University Research and Scientific Excellence (PURSE) for financial support. RS thanks VITO and BELSPO for postdoctoral funding.

## REFERENCES

Adnan NA, Noralam NFF, Salleh SA, Latif ZA (2015) Utilizing Landsat imageries for land surface temperature (LST) analysis of the Penang Island. *IEEE*, pp. 193–198.
Aniello C, Morgan K, Busbey A, Newland L (1995) Mapping micro-urban heat islands using Landsat TM and a GIS. *Comput Geosci* 21:965–969.
Araujo RV, Albertini MR, Costa-da-Silva AL, et al. (2015) São Paulo urban heat islands have a higher incidence of dengue than other urban areas. *Braz J Infect Dis* 19:146–155.
Atkinson-Palombo CM, Miller JA, Balling RC (2006) Quantifying the ozone "weekend effect" at various locations in Phoenix, Arizona. *Atmos Environ* 40:7644–7658.
Beaubien EG (2013) Spring flowering trends in Alberta, Canada: Response to climate change, urban heat island effects, and an evaluation of a citizen science network. PhD thesis, University of Alberta.
Bounoua L, Safia A, Masek J, et al. (2009) Impact of urban growth on surface climate: A case study in Oran, Algeria. *J Appl Meteorol Climatol* 48:217–231.
Buyantuyev A, Wu J (2010) Urban heat islands and landscape heterogeneity: Linking spatiotemporal variations in surface temperatures to land-cover and socioeconomic patterns. *Landsc Ecol* 25:17–33.
Chandramouli C, General R (2011) Census of India 2011. Provisional population totals. Government of India, New Delhi, India.
Chander G, Markham BL, Helder DL (2009) Summary of current radiometric calibration coefficients for Landsat MSS, TM, ETM+, and EO-1 ALI sensors. *Rem Sens Environ* 113:893–903.
Chen D, Stow D (2002) The effect of training strategies on supervised classification at different spatial resolutions. *Photogramm Eng Remote Sens* 68:1155–1162.
Chen J (2007) Rapid urbanization in China: A real challenge to soil protection and food security. *Catena* 69(1):1–15.
Chen X-L, Zhao H-M, Li P-X, Yin Z-Y (2006) Remote sensing image-based analysis of the relationship between urban heat island and land use/cover changes. *Rem Sens Environ* 104:133–146.
Chen Y-C, Chiu H-W, Su Y-F, et al. (2017) Does urbanization increase diurnal land surface temperature variation? Evidence and implications. *L.andsc Urban Plan* 157:247–258.
Chudnovsky A, Ben-Dor E, Saaroni H (2004) Diurnal thermal behavior of selected urban objects using remote sensing measurements. *Energy Build* 36:1063–1074.
Conti S, Meli P, Minelli G, et al. (2005) Epidemiologic study of mortality during the summer 2003 heat wave in Italy. *Environ Res* 98:390–399. doi:10.1016/j.envres.2004.10.009.
Cui L, Shi J (2012) Urbanization and its environmental effects in Shanghai, China. *Urban Clim* 2:1–15.
Desa UN (2012) World urbanization prospects, the 2011 revision. Final Report with Annex Tables. United Nations Department of Economic and Social Affairs, New York.
Dousset B (1989) AVHRR-derived cloudiness and surface temperature patterns over the Los Angeles area and their relationships to land use. *IEEE*, pp. 2132–2137.
Dye C (2008) Health and urban living. *Science* 319:766–769.
Foley JA, DeFries R, Asner GP, et al. (2005) Global consequences of land use. *Science* 309:570–574.
Fu P, Weng Q (2016) A time series analysis of urbanization induced land use and land cover change and its impact on land surface temperature with Landsat imagery. *Rem Sens Environ* 175:205–214.
Gallo K, McNab A, Karl T, et al. (1993) The use of NOAA AVHRR data for assessment of the urban heat island effect. Center for Advanced Land Management Information Technologies–Publication 1.
Gartland L (2011) *Heat Islands: Understanding and Mitigating Heat in Urban Areas*. Routledge, London, UK.
Groffman PM, Dorsey AM, Mayer PM (2005) N processing within geomorphic structures in urban streams. *J North Am Benthol Soc* 24:613–625.
Guo G, Wu Z, Xiao R, et al. (2015) Impacts of urban biophysical composition on land surface temperature in urban heat island clusters. *Landsc Urban Plan* 135:1–10.
Huang J, Wang R, Shi Y (2010) Urban climate change: A comprehensive ecological analysis of the thermo-effects of major Chinese cities. *Ecol Complex* 7:188–197. doi:10.1016/j.ecocom.2009.11.001.

Imhoff ML, Zhang P, Wolfe RE, Bounoua L (2010) Remote sensing of the urban heat island effect across biomes in the continental USA. *Remote Sens Environ* 114:504–513.

Jansen LJ, Di Gregorio A (2003) Land-use data collection using the "land cover classification system": Results from a case study in Kenya. *Land Use Policy* 20:131–148.

Kalnay E, Cai M (2003) Impact of urbanization and land-use change on climate. *Nature* 423:528–531.

Kaushal SS, Groffman PM, Mayer PM, et al. (2008) Effects of stream restoration on denitrification in an urbanizing watershed. *Ecol Appl* 18:789–804.

Kim HH (1992) Urban heat island. *Int J Rem Sens* 13:2319–2336.

Kolokotroni M, Ren X, Davies M, Mavrogianni A (2012) London's urban heat island: Impact on current and future energy consumption in office buildings. *Energy Build* 47:302–311.

Lai L-W, Cheng W-L (2009) Air quality influenced by urban heat island coupled with synoptic weather patterns. *Sci Total Environ* 407:2724–2733.

Lee H-Y (1993) An application of NOAA AVHRR thermal data to the study of urban heat islands. *Atmos Environ Part B Urban Atmos* 27:1–13.

Lin C-Y, Chen W-C, Chang P-L, Sheng Y-F (2011) Impact of the urban heat island effect on precipitation over a complex geographic environment in northern Taiwan. *J Appl Meteorol Climatol* 50:339–353.

Liu L, Zhang Y (2011) Urban heat island analysis using the Landsat TM data and ASTER data: A case study in Hong Kong. *Rem Sens* 3:1535–1552.

McKinney ML (2008) Effects of urbanization on species richness: A review of plants and animals. *Urban Ecosyst* 11:161–176.

Miller RB, Small C (2003) Cities from space: Potential applications of remote sensing in urban environmental research and policy. *Environ Sci Policy* 6:129–137.

Moore M, Gould P, Keary BS (2002) Global urbanization and impact on health. *Int J Hyg Environ Health* 206:269–278.

Nichol J (2005) Remote sensing of urban heat islands by day and night. *Photogramm Eng Rem Sens* 71:613–621.

Nichol JE (1996) High-resolution surface temperature patterns related to urban morphology in a tropical city: A satellite-based study. *J Appl Meteorol* 35:135–146.

Nichol JE, Fung WY, Lam K, Wong MS (2009) Urban heat island diagnosis using ASTER satellite images and "in situ" air temperature. *Atmos Res* 94:276–284.

Nonomura A, Kitahara M, Masuda T (2009) Impact of land use and land cover changes on the ambient temperature in a middle scale city, Takamatsu, in Southwest Japan. *J Environ Manage* 90:3297–3304.

Paul MJ, Meyer JL (2001) Streams in the urban landscape. *Ann Rev Ecol Syst* 32:333–365.

Qin Z, Karnieli A, Berliner P (2001) A mono-window algorithm for retrieving land surface temperature from Landsat TM data and its application to the Israel-Egypt border region. *Int J Rem Sens* 22:3719–3746.

Quah AK, Roth M (2012) Diurnal and weekly variation of anthropogenic heat emissions in a tropical city, Singapore. *Atmos Environ* 46:92–103.

Rinner C, Hussain M (2011) Toronto's urban heat island—Exploring the relationship between land use and surface temperature. *Rem Sens* 3:1251–1265.

Ryu Y-H, Baik J-J (2012) Quantitative analysis of factors contributing to urban heat island intensity. *J Appl Meteorol Climatol* 51:842–854.

Savard J-PL, Clergeau P, Mennechez G (2000) Biodiversity concepts and urban ecosystems. *Landsc Urban Plan* 48:131–142.

Schwarz N, Lautenbach S, Seppelt R (2011) Exploring indicators for quantifying surface urban heat islands of European cities with MODIS land surface temperatures. *Rem Sens Environ* 115:3175–3186.

Sharma R, Chakraborty A, Joshi PK (2015) Geospatial quantification and analysis of environmental changes in urbanizing city of Kolkata (India). *Environ Monit Assess* 187:1–12.

Sobrino JA, Jiménez-Muñoz JC, Paolini L (2004) Land surface temperature retrieval from LANDSAT TM 5. *Rem Sens Environ* 90:434–440.

Souch C, Grimmond S (2006) Applied climatology: Urban climate. *Prog Phys Geogr* 30:270–279.

Stathopoulou M, Cartalis C (2007) Daytime urban heat islands from Landsat ETM+ and Corine land cover data: An application to major cities in Greece. *Sol Energy* 81:358–368.

Sun Q, Tan J, Xu Y (2010) An ERDAS image processing method for retrieving LST and describing urban heat evolution: A case study in the Pearl River Delta Region in South China. *Environ Earth Sci* 59:1047–1055.

Tan J, Zheng Y, Tang X, et al. (2010) The urban heat island and its impact on heat waves and human health in Shanghai. *Int J Biometeorol* 54:75–84.

Taubenböck H, Wegmann M, Roth A, et al. (2009) Urbanization in India—Spatiotemporal analysis using remote sensing data. *Comput Environ Urban Syst* 33:179–188.

Tayanc M, Toros H (1997) Urbanization effects on regional climate change in the case of four large cities of Turkey. *Clim Change* 35:501–524.

Threlfall CG, Law B, Banks PB (2012) Sensitivity of insectivorous bats to urbanization: Implications for suburban conservation planning. *Biol Conserv* 146:41–52.

Tomlinson CJ, Chapman L, Thornes JE, Baker CJ (2011) Including the urban heat island in spatial heat health risk assessment strategies: A case study for Birmingham, UK. *Int J Health Geogr* 10:42.

Tran H, Uchihama D, Ochi S, Yasuoka Y (2006) Assessment with satellite data of the urban heat island effects in Asian mega cities. *Int J Appl Earth Obs Geoinfor* 8:34–48.

Unger J (2004) Intra-urban relationship between surface geometry and urban heat island: Review and new approach. *Clim Res* 27:253–264.

Valor E, Caselles V (1996) Mapping land surface emissivity from NDVI: Application to European, African, and South American areas. *Rem Sens Environ* 57:167–184.

Xian G, Crane M (2006) An analysis of urban thermal characteristics and associated land cover in Tampa Bay and Las Vegas using Landsat satellite data. *Therm Rem Sens Urban Areas* 104:147–156. doi:10.1016/j.rse.2005.09.023.

Xiao H, Weng Q (2007) The impact of land use and land cover changes on land surface temperature in a karst area of China. *J Environ Manage* 85:245–257.

Xu Y, Qin Z, Wan H (2010) Spatial and temporal dynamics of urban heat island and their relationship with land cover changes in urbanization process: A case study in Suzhou, China. *J Indian Soc Remote Sens* 38:654–663.

Yan H, Fan S, Guo C, et al. (2014) Quantifying the impact of land cover composition on intra-urban air temperature variations at a mid-latitude city. *PLoS ONE* 9:e102124. doi:10.1371/journal.pone.0102124.

Yuan F, Bauer ME (2007) Comparison of impervious surface area and normalized difference vegetation index as indicators of surface urban heat island effects in Landsat imagery. *Rem Sens Environ* 106:375–386.

Zhang J, Wang Y, Li Y (2006) A C++ program for retrieving land surface temperature from the data of Landsat TM/ETM+ band6. *Comput Geosci* 32:1796–1805.

Zhang Y, Odeh IO, Han C (2009) Bi-temporal characterization of land surface temperature in relation to impervious surface area, NDVI and NDBI, using a sub-pixel image analysis. *Int J Appl Earth Obs Geoinfor* 11:256–264.

# 5 Predictive Modeling of a Metropolitan City in India Using a Land Change Modeling Approach

*Akanksha Balha and Chander Kumar Singh*

## CONTENTS

5.1 Introduction .................................................................................................................... 74
5.2 Land Change Modeling Approaches ............................................................................... 74
    5.2.1 Statistical Regression Models ............................................................................. 75
    5.2.2 Artificial Neural Networks .................................................................................. 75
    5.2.3 Markov Chain Models ......................................................................................... 75
    5.2.4 Cellular Automata Models .................................................................................. 79
    5.2.5 Agent-Based Models ........................................................................................... 79
    5.2.6 Economic Models ............................................................................................... 79
5.3 Land Change Modeling for Present and Future Environmental Dynamics .................... 79
5.4 Urban Land Change Modeling: Development and Integration with Remote Sensing and Geographical Information System ........................................................................... 80
5.5 Case Study of Lucknow ................................................................................................... 80
    5.5.1 Study Area ........................................................................................................... 80
    5.5.2 Data and Software Used ...................................................................................... 81
    5.5.3 Methodology ........................................................................................................ 81
        5.5.3.1 Preparation of Land Use and Land Cover of Lucknow District ............ 81
        5.5.3.2 Prediction of Future Land Use and Land Cover of the Lucknow District .... 81
    5.5.4 Analysis and Results ........................................................................................... 83
    5.5.5 Conclusion ........................................................................................................... 84
References ................................................................................................................................ 84

**ABSTRACT** Land change modeling is a phenomenon to simulate the land-use changes occurred due to the interactions between humans and environment. The coupled human-environment interaction is the basis of land change modeling. A wide variety of land change models are present in the literature, a brief review of which is mentioned in this chapter. The progress in the development of land change modeling approaches over the past few decades can be understood from the wide availability of models at different scales and contexts. Land change modeling has proven to be a leading tool in advancing our understanding of land-use dynamics and in predicting future land-use change scenarios. Coupling of land change models with other modeling techniques has paved the way to interdisciplinary study to analyze different environmental problems. With the mention of the need and significance of land change modeling, the chapter discusses the case study of Lucknow in India, which shows more urbanization in the study area in a future scenario. The case study reveals that such studies will be useful at decision-making levels in studies related to prediction and planning. The chapter also touches upon the development of urban land change models, in specific over a number of past decades and their advancement with the emergence of remote sensing and GIS. The chapter is an attempt to provide a synoptic view of land change modeling.

## 5.1 INTRODUCTION

Land use and land cover (LULC) refers to manmade and physical features on the earth's surface (Rawat and Kumar, 2015). It represents how the earth's surface is being covered with forests, water bodies, settlements, and so on in a specific area. The changes in LULC occur due to interactions between humans and the environment, thus forming a complex and dynamic phenomenon (Koomen and Stillwell, 2007). It is important to study LULC change because it reveals the changes that have occurred in space and time (Liu and Yang, 2015). That is why LULC changes are modeled to understand and analyze the phenomenon for say, urban sprawl, to provide an informative tool about different future scenarios for decision-makers (Koomen and Stillwell, 2007). LULC change models are thus referred to as instruments to understand and examine the causes and the consequences of LULC change.

By definition, land use and land cover are different terms; however, for the purpose of ease, in place of land use land cover, we use the term *land use* primarily in this chapter. This term is used considering its definition that states that a land-use system is referred to as a land-use type with interlinked determining factors having functional roles connected with each other (Verburg et al., 2006).

Considering the complex and dynamic interlinked human-environmental relationships, various theoretical and methodological challenges exist in modeling the changes in land-use systems (Liu and Yang, 2015). Hence, to deal with them, various modeling theories and approaches exist in literature. In this chapter, we offer a brief review of various modeling approaches. In the next sections, the capability of land change modeling to analyze different environmental issues in present and future scenarios and the development of urban land change modeling are covered. At the end of the chapter, a separate section presents the case study of the Lucknow district in India, in which we mention the prediction of future LULC and analysis of temporal changes in the LULC of the Lucknow district for the time period 2000–2031 using land change modeling.

## 5.2 LAND CHANGE MODELING APPROACHES

The past few decades have witnessed the development of a wide variety of land change models. A brief review of studies conducted by different authors on various kinds of land change models is included in Table 5.1.

### TABLE 5.1
### A Brief Description of the Review Studies on Various Land Change Models by Different Authors

| S. No | Type of Land Change Models | References |
|---|---|---|
| 1 | Deforestation models | Lambin (1997) |
| 2 | Integrated urban models | Miller et al. (1999); U.S. EPA (2000) |
| 3 | Agricultural intensification models | Lambin et al. (2000) |
| 4 | Land use models | Briassoulis (2000); Verburg et al. (2004) |
| 5 | Framework of 19 models based on spatial, temporal, and human-choice complexity | Agarwal et al. (2001) |
| 6 | Multiagent modeling techniques | Parker et al. (2003); Bousquet and Le Page (2004) |
| 7 | Recent advancement in modeling approaches | Veldkamp and Lambin (2001); Veldkamp and Verburg (2004); Verburg and Veldkamp (2005) |

Diversity of models can be understood by the specific application and research question that the model addresses, different spatial scales of the model, dominant land-use change processed by the model, simulation method, and underlying theory of the model (Verburg et al., 2006). Such large diversity of modeling approaches has made it difficult for the researchers to categorize different approaches into a concrete classification system. Before reviewing the different land change models, we discuss the broad characteristics defined by Verburg et al. (2006) to classify these models in Table 5.2.

The various widely used land-use models reviewed in this chapter are mentioned next.

### 5.2.1 Statistical Regression Models

Statistical regression models are based on empirical relationships between dependent variables (e.g., land-use changes) and explanatory variables (i.e., environmental and socioeconomic factors) (Liu and Yang, 2015). The established relationship between the variables generates transition potential maps to predict possible future land changes. Widely used statistical methods for land change modeling include Conversion of Land Use and its effects at Small regional extent (CLUE-S) (Verburg et al., 2002) and logistic regression (Hu and Lo, 2007).

Less computational resource demand and easy operability cause statistical regression models to be used widely in land change modeling (Liu and Yang, 2015). Although they give important information about key factors of land-use changes, they are deterministic in nature in comparison to more advanced models. Hence, this can explain complex interactions and temporal dynamics and complexity within coupled human environmental system only to an extent.

### 5.2.2 Artificial Neural Networks

Artificial neural networks (ANN) are based on machine-learning algorithms (Li and Yeh, 2002). ANN is similar to a regression model in terms of function; both try to link land change and its potential driver factors (Liu and Yang, 2015). The distinguishing feature of ANN lies in its capability to identify nonlinear relationships with the help of hidden layers incorporated in the model network. The flexible and nonlinear behavior of neural network defines the robustness of the model (Lesschen et al., 2005) and helps in predicting future changes.

As the relationship between the variables remains invisible due to hidden layers, the model is not very interpretable (Liu and Yang, 2015). That is why, it is often criticized as a "black box." ANN functions on an assumption of stationarity, that is, the trends from the past and the present will continue to behave in the same way in the future. ANN predicts future land-use changes on the basis of knowledge and information gained from the patterns and behaviors observed from historical data. This oversimplifies the complexity of temporal dynamics of land change phenomenon.

### 5.2.3 Markov Chain Models

Markov chain (MC) models use discrete stochastic methodology to identify the transition potential of changes in land use (Liu and Yang, 2015). The observed land-use change data lead to the generation of transition probabilities, which show the probability that one land-use type will convert into another land-use type within a given time frame in the future.

Like ANN, Markov models also function on the assumption of stationarity. A change in transition probabilities in regular patterns over a period of time can allow Markov models to behave dynamically (Howard et al., 1995). Due to its ability to automatically generate land transition probability with time series data, MC models are coupled with more complex models, for example, cellular automata (CA) and agent-based models (Liu and Yang, 2015) that are explained later in the section.

## TABLE 5.2
## Description of Broad Characteristics (Defined by Verburg et al., 2006) to Classify Different Land Change Models

### Characteristic Type I
**On the basis of spatial nature**

**Spatial Model**

It is used when spatial variation in land use needs to be represented taking into account the spatial variation in the social and biophysical environment. Spatial details are represented either at the level of individual pixels in a raster format data or in the form of other spatial entities, for example, administrative units.

For example, Conversion of Land Use and its Effects (CLUE) model, the SLEUTH model and GEOMOD (Pontius et al., 2001; Verburg et al., 2002; Goldstein et al., 2004).

**Nonspatial Model**

It is used to model the rate and magnitude of land-use change without considering the spatial variation in the system.

For example, a parcel-level nonspatial model for deforestation in a part of the Amazon region by Evans et al. (2001) and a model for Sahel region (SALU) by Stéphenne and Lambin (2001).

### Characteristic Type II
**On the basis of temporal nature**

**Static Model**

It is a regression model that explains the spatial distribution and variation of land-use changes on the basis of hypothesized drivers (Nelson and Hellerstein, 1997; Chomitz and Thomas, 2003).

It can forecast future LULC change but it is not responsible for feedbacks and path dependencies in the system.

It evaluates our understanding and knowledge about drivers of land-use change.

**Dynamic Model**

It analyzes the trajectories of land-use changes and project future land-use change.

Most multiagent models and spatial models such as GEOMOD, CLUE and SLEUTH are included in dynamic models.

### Characteristic Type III
**On the basis of focus**

**Descriptive Model**

Based on actual land use and the dominant land-use changing phenomena, it simulates the land-use functioning to explore future land-use patterns.

Such models are used to make land-use change projections for scenario analysis.

**Prescriptive Model**

It is an optimization model; it models an area on the basis of a given set of objectives in a way that produces a best fit land-use model according to the defined objectives.

Such models are relevant for policy analysis (van Ittersum et al., 2004).

Limitations:
(1) providing no insights into actual land-use change trajectories and the intermediate conditions; or
(2) assuming optimal economic behavior by the land change actors or problem in considering nonoptimal behavior of people into the model and so on. (Rabin, 1998),
(3) does not hold much significance at an aggregate level but it has relevance for fine scale land-use change modeling and for evaluating different actors (Lambin et al., 2000).

*(Continued)*

**TABLE 5.2 (*Continued*)**
**Description of Broad Characteristics (Defined by Verburg et al., 2006) to Classify Different Land Change Models**

| | | |
|---|---|---|
| **Characteristic Type IV** | **Inductive** | **Deductive** |
| On the basis of theory | Most land-use models are based on inductive models. In this, statistical correlations between land change and set of variables define the model. Mostly, studies are not entirely inductive. Theoretical knowledge and understanding of decision-making process guide in choosing the determining factors for modeling. Inductive approach plays its role in quantifying the relationships between different variables. For example, different models exist where relations between different variables are expressed by statistical measures (Verburg and Chen, 2000) or by a set of rules (Parker et al., 2003). | These depend on theoretical concepts that predicts pattern on the basis of processes. Only processes that provide explanation for observed land-use pattern can be pursued by pattern-driven studies or deductive models (Laney, 2004). For example, Walker and Solecki's (2004) and Walker's (2004) theoretical models for deforestation and wetland conversion. |
| **Characteristic Type V** | **Pixel Based** | **Agent Based** |
| On the basis of simulated objects | Basic unit of analysis is either a polygon, line, or a pixel (in raster-formatted data). At the most local level, the unit of analysis and agents of land-use change match each other with high rate of agreement whereas in higher hierarchical level, the agents of change cannot be represented spatially and hence model simulations and decision-making units do not come in agreement with each other. | Uses individual agents as unit of simulations, which are characterized by their autonomous nature and their ability to make decisions that link behavior to the environment. An agent can refer to any level of hierarchy (a herd, a cohort, a village, etc.); it need not necessarily be an individual (Parker et al., 2003; Bousquet and Le Page, 2004). Such multiagent systems give importance to decision-making process of agents and to the environment in which agents are linked. Demerits: (1) Linking agent behavior to the actual land-use system is a difficult task (Rindfuss et al. 2002, 2004). (2) Illustrating the spatial nature of agents in the system is another challenging task in agent-based modeling. |

(*Continued*)

**TABLE 5.2 (Continued)**
**Description of Broad Characteristics (Defined by Verburg et al., 2006) to Classify Different Land Change Models**

| Characteristic Type VI | Regional | Global |
|---|---|---|
| On the basis of scale/level of model | For regional applications, many land-use models have been developed that vary in extent ranging from a few km² to country level with varying resolutions. | In context of global scale land-use models, not much development has been seen.<br>Moreover, the developed global scale land-use models are not used specifically for LULC studies.<br>Instead, they are used for studies aiming at analysis of climate change, biodiversity loss, food production, and so on, and LULC plays an important part in such analyses.<br>Limitation: Their coarser spatial resolution doesn't allow the regional details to be illustrated precisely and therefore impacts the decision-making process.<br>Spatial models of large world regions such as EURURALIS (Klijn et al., 2005) are expected to overcome this limitation and bridge the gap between global and local scale models. |

## 5.2.4 CELLULAR AUTOMATA MODELS

CA models are developed on the basis of a static and cell-based framework (Liu and Yang, 2015). Every cell in the model exhibits a state on the basis of its present status and interactions with its neighborhood by using a set of transition rules that can be transferred to other states. The Slope, Land use, Exclusion, Urban extent, Transportation, Hillshade (SLEUTH) model by Clarke et al. (1997) for simulating urban growth and development is a widely used CA model.

These models have achieved higher demand than all other modeling approaches due to their dynamic modeling capabilities (Liu and Yang, 2015). In the context of the functioning of CA models, it is important to note that these models do not take into account the drivers of change and human decision-making abilities. This is so because cells in the models cannot move and the transitions represent physical processes of land-use changes.

## 5.2.5 AGENT-BASED MODELS

Agent-based models (ABMs) are based on the assumption that considers "agent" as the main driver of a system (Parker et al., 2003; Xie et al., 2007). Agents interact with each other as well as with environment at multiple scales (Liu and Yang, 2015). This leads the agent to gain rationality and information-processing ability for making decisions, thus impacting the behavior of the system (Miller and Page, 2007). These models are believed to play a leading role in land change modeling because they simulate and represent human-environmental interactions and feedbacks in land-use systems (Liu and Yang, 2015). However, due to its complexity in design and implementation, the potential of ABM to simulate real-world phenomenon are yet to be explored completely Also, the lack of data to represent and validate human decision making and agents' interaction at the microlevel may make the improvements in ABMs somewhat difficult and challenging.

## 5.2.6 ECONOMIC MODELS

Economic models incorporate human choices and decisions and economic behaviors and thus involve the human dimension of land-use changes (Liu and Yang, 2015). In these models, microeconomic behavior evaluates the demand and supply relations and generates land-use patterns. The idea behind economic models of land-use changes is market equilibrium. The models operate at either an aggregate scale (e.g., sector-based models) or a disaggregate scale (e.g., spatially disaggregate models). The sector-based model represents the global economy and explains land allocation as per demand and supply (Sohngen et al., 1999), and the spatially disaggregate model represents the individual decision making at the microlevel and explains land-use decisions driven by profits, cost maximization, or utility maximization (Bockstael, 1996; Wu et al., 2004). Economic models help in simulating and predicting nonmarginal land changes (Liu and Yang, 2015). But the complexity of human choices and the lack of data make it difficult for economic models to lay the foundation of underlying assumptions.

A variety of land-use models gives researchers an opportunity to choose the best fit model to simulate land-use processes. Despite the wide availability of models, no single modeling approach is superior to the other (Verburg et al., 2006). The objective of land-change studies determines the choice of model to be used. The dominant land-use change and the spatial and temporal scales of the study also determine the suitability of the specific model. Sometimes the issue of unavailable data may also become one of the restrictions in choosing a suitable model.

## 5.3 LAND CHANGE MODELING FOR PRESENT AND FUTURE ENVIRONMENTAL DYNAMICS

A complete understanding of land-use systems can be gained by the application and analysis of integrated models (Liu and Yang, 2015). Because integrated models incorporate the feedback of the land-use system as well as other environmental systems (Koomen and Stillwell, 2007), integration of land change models with other environmental models help in investigating the land-use system and coupled human–environment interactions in a comprehensive manner (Liu and Yang, 2015). An integrated model helps in understanding the interactions between various biotic and abiotic components of ecosystem, for example, human behavior with hydrological phenomenon, climate change, biodiversity, soil degradation, and so on. Such integrated studies expand the boundaries of land change modeling beyond the examination of only land change dynamics. Studies conducted by Kundu et al. (2017), Reddy et al. (2017), and Zare et al. (2017) affirm the applicability of land change models integrated with other models in analyzing environmental problems in current as well as future scenarios. A motivation for performing integrated land-use change studies is that these studies aid in the decision-making processes at the policy, planning, and management levels (Liu and Yang, 2015). It can receive further impetus due to the capability of land change models to predict future scenarios.

## 5.4 URBAN LAND CHANGE MODELING: DEVELOPMENT AND INTEGRATION WITH REMOTE SENSING AND GEOGRAPHICAL INFORMATION SYSTEM

With reference to Webster's and Collins dictionaries, the term *model* can be understood as a simplified structure or representation of a realistic system (Liu, 2008). Hence, modeling is an attempt to represent the various phenomena of a realistic system in a simplified manner to understand and to predict the nature of the system. This capability of modeling has made its use widespread in urban planning, predictive scenario analysis, and many other studies.

The origin of the application of models in the urban domain can be traced to von Thünen's classical model of agricultural location (von Thünen, 1826). Weber's (1909) industrial location, Burgess' (1925) concentric zone model, Christaller's (1933) central place theory, and other urban models developed in the same era. These models are static in nature and possess little or no representation of the dynamic nature of urban growth and development (Liu, 2008). As a consequence of the computerized technical support and practical need during the quantitative revolution from late 1950s to 1960s, the application of urban models increased. Since then, advancements have been carried out in urban models and related studies.

Development of a geographical information system (GIS) and its integration with urban modeling has provided a rich source of spatial data (Liu, 2008). It has brought an altogether new wave of development in urban modeling with advanced spatial techniques. Advancement in spatial sciences along with dynamic, flexible self-organizing modeling systems that can make local decisions to transform into global patterns has led to the implementation of CA models in urban modeling (Tobler, 1979; Wu and Webster, 1998). Along with GIS technology, advancement in remote sensing has also added a new dimension to urban modeling (The State of Land Change Modeling, 2014; Kamusoko, 2017) by allowing it to be merged with human, geographical, and socioeconomic data to study urban land-use dynamics (Verburg et al., 2006). In recent times, many research studies (Jjumba and Dragićević, 2012; Megahed et al., 2015; Toure et al., 2016; Feng, 2017) have been conducted in which different modeling techniques have been merged with GIS, satellite data, and urban models to examine urban dynamics and urban growth patterns, quantify urban sprawl, and predict future urban scenarios. In addition to this, urban models have also been used to assess the impacts of future development (Al-shalabi et al., 2013). A novel approach utilizing spatial metrics and remote sensing data has been carried out to improve the urban land-use change analysis and modeling (Herold et al., 2005).

## 5.5 CASE STUDY OF LUCKNOW

### 5.5.1 Study Area

The study area for the present study is Lucknow district (Figure 5.1) situated in the Uttar Pradesh state of India. Geographically, the district is located at 26.78° N and 80.89° E at 123 m above mean sea level and covers an area of 2528 km². It is situated in the middle of the Indus–Gangetic plain with the Gomti River as its chief geographical feature.

# Predictive Modeling of a Metropolitan City in India

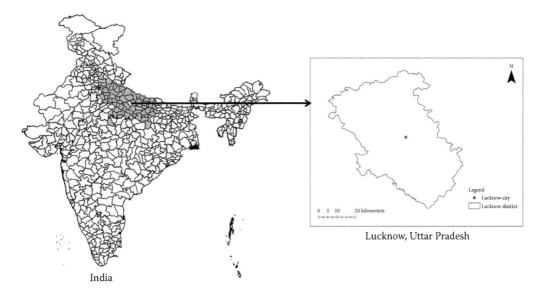

**FIGURE 5.1** Study area.

The river flows through the district dividing it into trans-Gomti and cis-Gomti regions. In the middle of the district lies the city of Lucknow, one of the fastest developing cities and capital of the fourth largest and the most populated state of the country, Uttar Pradesh. Lucknow is the eleventh most populous city and twelfth most populous urban agglomeration of the country. Lucknow possesses a humid subtropical climate with cool, dry winters and dry, hot summers. It receives an average rainfall of 896.2 mm.

### 5.5.2 Data and Software Used

In the present study, high-resolution satellite images of LISS IV and LISS III merged for 2017 and 2007, respectively, and Landsat 7 ETM+ (resampled at 5 m) for 2000 were used. ERDAS IMAGINE 2014 was used for preprocessing the satellite images and for preparing LULC of Lucknow district for different years. IDRISI Selva v. 17.0 was used to predict future LULC of Lucknow.

### 5.5.3 Methodology

#### 5.5.3.1 Preparation of Land Use and Land Cover of Lucknow District

The satellite images of previously mentioned specifications were preprocessed and used to prepare LULC of Lucknow for 2000, 2007, and 2017. A maximum likelihood algorithm of supervised classification was used to prepare the LULC. Seven LULC classes were identified: water, built up, dense vegetation, sparse vegetation, cropland, fallow land, and open land (Table 5.3).

Figure 5.2 presents LULC maps of the Lucknow district for different years. Accuracy assessments of all the LULC images have been done with accuracy of more than 85% in all the images achieved.

#### 5.5.3.2 Prediction of Future Land Use and Land Cover of the Lucknow District

Land Change Modeler (LCM) in IDRISI Selva 17.0 was used to predict LULC of Lucknow for 2031. Land-use images of years 2007 and 2017 set to similar spatial extent, projection, and legend were used as input in the model. Driver variables were chosen considering the dynamics of the LULC of the study area. Drivers chosen were distance from built up, from roads, from cropland, from fallow land, from open land, and from rail lines. Layers for roads and rail lines were procured

**TABLE 5.3**
**Description of the Land Use and Land Cover Classes Identified in the Study Area, Mentioning the Different Features**

| S. No | LULC Classes | Description |
|---|---|---|
| 1 | Water | River, canals, ponds, and other natural or manmade water bodies |
| 2 | Built up | Residential, commercial, and industrial area, public facilities zone, transportation area, and other impervious land |
| 3 | Dense vegetation | Forest area and densely placed trees areas |
| 4 | Sparse vegetation | Urban tree canopy, green spaces (e.g., parks, lawns, golf course); includes plantations |
| 5 | Cropland | Agricultural land having crops and other growing items |
| 6 | Fallow land | Harvested agricultural land |
| 7 | Open land | Open nongreen area, barren land; includes unusable land |

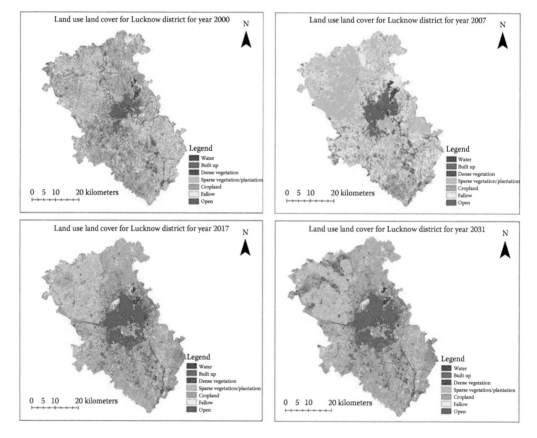

**FIGURE 5.2** Land use and land cover (LULC) maps for Lucknow district for time period 2000–2031, depicting the changing land use on a temporal scale.

from OpenStreetMap (OSM) (https://www.openstreetmap.org/). Drivers work depending upon their relation with the land-use dynamics of the area. For example, distance to roads and distance to built up are significant in urban growth studies because it is believed that in cities, when an urban area starts expanding over time, the built-up area first grows near to the roads or to the already built up area. Considering the land-use development of the study area, relevant transitions were chosen in the model to create transition potentials using Multi-layer perceptron (MLP) neural network. Thereafter, future LULC of the study area was generated.

### 5.5.4 Analysis and Results

The MLP neural network–based land change modeler was used to predict the land-use change. Transitions and driver variables chosen while taking into consideration the land-use dynamics of the area and the changes in LULC in the Lucknow district from 2000 to 2031 are presented in Figure 5.2 and in terms of area ($km^2$) in Table 5.4.

The change in land use was estimated on the basis of model calculations. The temporal analysis and prediction of land-use change was done at a gap of at least seven years to find significant growth and changes in the land use. The results shown in Table 5.4 indicate that built up increased from 304.94 $km^2$ in 2000 to 326.07 $km^2$ and 552.19 $km^2$ in 2007 and 2017, respectively. The trend for the built-up class continues in the predicted scenario in which the built-up area will expand tremendously to 814.28 $km^2$ by 2031. The area covered by water increased from 2000 to 2007 due to the rise in the number of water bodies in the study area but in subsequent years, it remained the same. The area covered by dense vegetation was predicted to decrease from 114.93 $km^2$ in 2000 to 72.84 $km^2$ in 2031. The sparse vegetation showed a decreasing trend except in 2007. In-depth analysis of the overall areal statistics and land-use maps of the study area determined that the large availability of open land in Lucknow might have led to the planting of trees or setting up of green areas. This explains the huge increase in the sparse vegetation and the decrease in open land area in 2007. Besides sparse vegetation and other land-use conversion, the increase in built up in 2007 may also have played a role in decreased open land in 2007 as indicated in Figure 5.2. Cropland and fallow land together constitute the agricultural area in the study area. The image for 2007 belongs to summer months (May–July), and the images for 2000 and 2017 are of winter months (October–March); land use of 2007 had relatively more area under fallow land than cropland as compared to the land use of other years for varying seasons. However, the total area covered by agricultural land decreased (1.6% during 2000–2007 and 1.5% during 2007–2017) over the total period of time with a significant decrease of more than 116 $km^2$ (9.5% during the predicted period of 2017–2031). The study area has a large amount of land covered with open land. With reference to the reasons already mentioned, the open land probably converted to sparse vegetation and built-up area and therefore showed a significant decline in 2007. In subsequent years also, the open land showed a decreasing trend only.

### TABLE 5.4
### Area (in $km^2$) Covered by Different Land Use and Land Cover Classes in Lucknow District over the Period of 2000–2031

| S. No | LULC Classes | 2000 | 2007 | 2017 | 2031 |
|---|---|---|---|---|---|
| 1 | Water | 11.86 | 15.12 | 15.13 | 15.14 |
| 2 | Built up | 304.94 | 326.07 | 552.19 | 814.28 |
| 3 | Dense vegetation | 114.93 | 107.53 | 97.64 | 72.84 |
| 4 | Sparse vegetation | 452.43 | 610.43 | 423.83 | 323.83 |
| 5 | Cropland | 1123.10 | 445.56 | 986.5 | 920 |
| 6 | Fallow land | 261.35 | 916.50 | 354.71 | 305 |
| 7 | Open land | 259.39 | 106.79 | 98.00 | 76.91 |

## 5.5.5 Conclusion

This study has generated a future land-use scenario for the Lucknow district for 2031 and has analyzed how the land use in the district will have changed from 2000 to 2031. The study performed on high-resolution satellite images found the major land-use types in the area and the important drivers responsible for bringing the change in land-use over the time span. The drivers and the land-use transitions confirm that combinations of human–environment interactions are the cause of land change modeling. Land change modeler in IDRISI Selva v. 17.0 has been a promising tool in predicting future scenarios and for understanding land-use dynamics. The study can help experts and decision-makers to analyze the changing landscape and prospective changes in the study area to suggest relevant inputs at the planning and management level. It will benefit urban planning and help to resolve many urban environmental issues.

Hence, the chapter concludes that land change modeling plays a pivotal role in examining the problems related to land-use dynamics and various environmental issues. In the future, we expect that with further improvements in modeling framework and techniques, land change models will explore more dimensions in solving the same problems, in both present and future scenarios.

## REFERENCES

Agarwal, C., Green, G.M., Grove, J.M., Evans, T.P., and Schweik, C.M. 2001. *A Review and Assessment of Land Use Change Models: Dynamics of Space, Time, and Human Choice.* Bloomington, IN: Center for the Study of Institutions, Population, and Environmental Change, Indiana University and USDA Forest Service.

Al-shalabi, M., Billa, L., Pradhan, B., Mansor, S., and Al-Sharif, A.A. 2013. Modelling urban growth evolution and land-use changes using GIS based cellular automata and SLEUTH models: The case of Sana'a metropolitan city, Yemen. *Environmental Earth Sciences,* 70(1): 425–437.

Bockstael, N.E. 1996. Modeling economics and ecology: The importance of a spatial perspective. *American Journal of Agricultural Economics,* 78(5): 1168–1180.

Bousquet, F., and Le Page, C. 2004. Multi-agent simulations and ecosystem management: A review. *Ecological Modelling,* 176(3): 313–332.

Briassoulis, H. 2000. *Analysis of Land Use Change: Theoretical and Modeling Approaches.* Morgantown, WV: Regional Research Institute, West Virginia University.

Burgess, E.W. 1925. The growth of city: An introduction to a research project. In *The City,* R.E. Park, E.W. Burgess, and R.D. McKenzie (Eds.), pp. 47–62. Chicago, IL: The University of Chicago Press.

Chomitz, K.M., and Thomas, T.S. 2003. Determinants of land use in Amazonia: A fine-scale spatial analysis. *American Journal of Agricultural Economics,* 85(4): 1016–1028.

Christaller, W. 1933 (translation 1966). *Central Places in Southern Germany.* Englewood Cliffs, NJ: Prentice Hall.

Clarke, K.C., Hoppen, S., and Gaydos, L. 1997. A self-modifying cellular automaton model of historical urbanization in the San Francisco Bay area. *Environment and Planning B: Planning and Design,* 24(2): 247–261.

Evans, T.P., Manire, A., de Castro, F., Brondizio, E., and McCracken, S. 2001. A dynamic model of household decision-making and parcel level land cover change in the eastern Amazon. *Ecological Modelling,* 143(1): 95–113.

Feng, Y. 2017. Modeling dynamic urban land-use change with geographical cellular automata and generalized pattern search-optimized rules. *International Journal of Geographical Information Science,* 31(6): 1198–1219.

Goldstein, N.C., Candau, J.T., and Clarke, K.C. 2004. Approaches to simulating the "March of Bricks and Mortar." *Computers, Environment and Urban Systems,* 28(1): 125–147.

Herold, M., Couclelis, H., and Clarke, K.C. 2005. The role of spatial metrics in the analysis and modeling of urban land use change. *Computers, Environment and Urban Systems,* 29(4): 369–399.

Howard, D.M., Howard, P.J.A., and Howard, D.C. 1995. A Markov model projection of soil organic carbon stores following land use changes. *Journal of Environmental Management,* 45(3): 287–302.

Hu, Z., and Lo, C.P. 2007. Modeling urban growth in Atlanta using logistic regression. *Computers, Environment and Urban Systems,* 31(6): 667–688.

Jjumba, A., and Dragićević, S. 2012. High resolution urban land-use change modeling: Agent iCity approach. *Applied Spatial Analysis and Policy*, 5(4): 291–315.

Kamusoko, C. 2017. Importance of remote sensing and land change modeling for urbanization studies. In *Urban Development in Asia and Africa*, The Urban Book Series, Y. Murayama, C. Kamusoko, A. Yamashita, and R. Estoque (Eds.). Singapore: Springer.

Klijn, J.A., Vullings, LAE, van de Berg, M., van Meijl, H., van Lammeren, R., van Rheenen, T., Eickhout, B., Veldkamp, A., Verburg, P.H., and Westhoek, H. 2005. EURURALIS 1.0: A scenario study on Europe's rural areas to support policy discussion. Background document. Wageningen University and Research Centre/Environmental Assessment Agency (RIVM).

Koomen, E., and Stillwell, J. 2007. Modelling land-use change. In *Modelling Land-Use Change, The GeoJournal Library*, E. Koomen, J. Stillwell, A. Bakema, and H.J. Scholten (Eds.), vol 90. Dordrecht, the Netherlands: Springer.

Kundu, S., Khare, D., and Mondal, A. 2017. Individual and combined impacts of future climate and land use changes on the water balance. *Ecological Engineering*, 105: 42–57.

Lambin, E.F. 1997. Modelling and monitoring land-cover change processes in tropical regions. *Progress in Physical Geography*, 21(3): 375–393.

Lambin, E.F., Rounsevell, M.D.A., and Geist, H.J. 2000. Are agricultural land-use models able to predict changes in land-use intensity? *Agriculture, Ecosystems & Environment*, 82(1): 321–331.

Laney, R. M. 2004. A process-led approach to modeling land change in agricultural landscapes: A case study from Madagascar. *Agriculture, Ecosystems & Environment*, 101(2): 135–153.

Lesschen, J.P., Verburg, P.H., and Staal, S.J. 2005. *Statistical Methods for Analysing the Spatial Dimension of Changes in Land Use and Farming Systems*. Nairobi, Kenya: International Livestock Research Institute.

Li, X., and Yeh, A.G.O. 2002. Neural-network-based cellular automata for simulating multiple land use changes using GIS. *International Journal of Geographical Information Science*, 16(4): 323–343.

Liu, T., and Yang, X. 2015. Land change modeling: Status and challenges. In *Monitoring and Modeling of Global Changes: A Geomatics Perspective*, J. Li and X. Yang (Eds.), pp. 3–16. Dordrecht, the Netherlands: Springer.

Liu, Y. 2008. *Modelling Urban Development with Geographical Information Systems and Cellular Automata*. Boca Raton, FL: CRC Press.

Megahed, Y., Cabral, P., Silva, J., and Caetano, M. 2015. Land cover mapping analysis and urban growth modelling using remote sensing techniques in Greater Cairo Region—Egypt. *ISPRS International Journal of Geo-Information*, 4(3): 1750–1769.

Miller, E.J., Kriger, D.S., and Hunt, J.D. 1999. TCRP web document 9: Integrated urban models for simulation of transit and land-use policies. Final report. University of Toronto Joint Program in Transportation and DELCAN Corporation, Toronto, Canada.

Miller, J.H., and Page, S.E. 2007. *Complex Adaptive Systems: An Introduction to Computational Models of Social Life*. Princeton, NJ: Princeton University Press.

Nelson, G.C., and Hellerstein, D. 1997. Do roads cause deforestation? Using satellite images in econometric analysis of land use. *American Journal of Agricultural Economics*, 79(1): 80–88.

Parker, D.C., Manson, S.M., Janssen, M.A., Hoffmann, M.J., and Deadman, P. 2003. Multi-agent systems for the simulation of land-use and land-cover change: A review. *Annals of the Association of American Geographers*, 93(2): 314–337.

Pontius, R.G., Cornell, J.D., and Hall, C.A. 2001. Modeling the spatial pattern of land-use change with GEOMOD2: Application and validation for Costa Rica. *Agriculture, Ecosystems & Environment*, 85(1): 191–203.

Rabin, M. 1998. Psychology and economics. *Journal of Economic Literature*, 36: 11–46.

Rawat, J.S., and Kumar, M. 2015. Monitoring land use/cover change using remote sensing and GIS techniques: A case study of Hawalbagh block, district Almora, Uttarakhand, India. *The Egyptian Journal of Remote Sensing and Space Science*, 18(1): 77–84.

Reddy, C.S., Singh, S., Dadhwal, V.K., Jha, C.S., Rao, N.R., and Diwakar, P.G. 2017. Predictive modelling of the spatial pattern of past and future forest cover changes in India. *Journal of Earth System Science*, 126(1): 8.

Rindfuss, R.R., Walsh, S.J., Mishra, V., Fox, J., and Dolcemascolo, G.P. 2002. Linking household and remotely sensed data, methodological and practical problems. In *People and the environment. Approaches for linking houeshold and community surveys to remote sensing and GIS*, J. Fox, R.R. Rindfuss, S.J. Walsh and V. Mishra (Eds.) pp. 1–31. Kluwer Academic, Dordrecht Boston London.

Rindfuss, R.R., Walsh, S.J., Turner, B.L. II, Fox, J., and Mishra, V. 2004. Developing a science of land change: Challenges and methodological issues. *Proceedings of the National Academy of Sciences of the United States of America*, 101(39): 13976–13981.

Sohngen, B., Mendelsohn, R., and Sedjo, R. 1999. Forest management, conservation, and global timber markets. *American Journal of Agricultural Economics*, 81(1): 1–13.

Stéphenne, N., and Lambin, E.F. 2001. A dynamic simulation model of land-use changes in Sudano-sahelian countries of Africa (SALU). *Agriculture Ecosystems & Environment*, 85: 145–161.

The State of Land Change Modeling. 2014. *Advancing Land Change Modeling: Opportunities and Research Requirements*. Washington, DC: The National Academies Press.

Tobler, W.R. 1979. Cellular Geography. In *Philosophy in Geography, S. Gale and G. Olsson* (Eds) pp. 379–386. Reidel, Dordrecht.

Toure, S., Stow, D., Shih, H.C., Coulter, L., Weeks, J., Engstrom, R., and Sandborn, A. 2016. An object-based temporal inversion approach to urban land use change analysis. *Remote Sensing Letters*, 7(5): 503–512.

U.S. EPA (Environmental Protection Agency). 2000. *Projecting Land-Use Change: A Summary of Models for Assessing the Effects of Community Growth and Change on Land-Use Patterns*. Cincinnati, OH: U.S. Environmental Protection Agency, Office of Research and Development.

van Ittersum, M.K., Roetter, R.P., van Keulen, H., de Ridder, N., Hoanh, C.T., Laborte, A.G., Aggarwal, P.K., Ismail, A.B., and Tawang, A. 2004. A systems network (SysNet) approach for interactively evaluating strategic land use options at sub-national scale in South and South-east Asia. *Land Use Policy*, 21(2): 101–113.

Veldkamp, A., and Lambin, E.F. 2001. Predicting land-use change. *Agriculture, Ecosystems & Environment*, 85(1–3): 1–6.

Veldkamp, A., and Verburg, P. H. 2004. Modelling land use change and environmental impact. *Journal of Environmental Management*, 72(1): 1–3.

Verburg, P. H., and Chen, Y. 2000. Multi-scale characterization of land-use patterns in China. *Ecosystems*, 3(4): 369-385.

Verburg, P.H., Kok, K., Pontius Jr, R.G., and Veldkamp, A. 2006. Modeling land-use and land-cover change. In *Land-use and Land-cover Change*, E.F. Lambin and H. Geist (Eds.), pp. 117–135. Berlin, Germany: Springer.

Verburg, P.H., Schot, P.P., Dijst, M.J., and Veldkamp, A. 2004. Land use change modelling: Current practice and research priorities. *GeoJournal*, 61(4): 309–324.

Verburg, P.H., Soepboer, W., Limpiada, R., Espaldon, M.V.O., Sharifa, M., and Veldkamp, A. 2002. Land use change modelling at the regional scale: The CLUE-S model. *Environmental Management*, 30: 391–405.

Verburg, P.H., and Veldkamp, A. 2005. Introduction to the special issue on spatial modeling to explore land use dynamics. *International Journal of Geographical Information Science*, 19: 99–102.

von Thünen, J.H. 1826. Der Isolierte Staat (English translation by C. M. Wartenberg, edited by P. Hall, von Thünen's Isolated State, 1966. Oxford, UK: Pergamon Press).

Walker, R. 2004. Theorizing land-cover and land-use change: The case of tropical deforestation. *International Regional Science Review*, 27(3): 247–270.

Walker, R., and Solecki, W.D. 2004. Theorizing land-cover and land-use change: The case of the Florida Everglades and its degradation. *Annals of the American Association of Geographers*, 94: 311–238.

Weber, A. 1909. *Über den Standort der Industrien*. Tübingen, Germany: Reine Theorie des Standort.

Wu, F., and Webster, C.J. 1998. Simulation of land development through the integration of cellular automata and multicriteria evaluation. *Environment and Planning B: Planning and Design*, 25(1): 103–126.

Wu, J., Adams, R.M., Kling, C.L., and Tanaka, K. 2004. From microlevel decisions to landscape changes: An assessment of agricultural conservation policies. *American Journal of Agricultural Economics*, 86(1): 26–41.

Xie, Y., Batty, M., and Zhao, K. 2007. Simulating emergent urban form using agent-based modeling: Desakota in the Suzhou-Wuxian region in China. *Annals of the Association of American Geographers*, 97(3): 477–495.

Zare, M., Mohammady, M., and Pradhan, B. 2017. Modeling the effect of land use and climate change scenarios on future soil loss rate in Kasilian watershed of northern Iran. *Environmental Earth Sciences*, 76(8): 305.

# 6 Performance Analysis of Different Predictive Algorithms for the Land Features Modeling

*Pradeep Kumar, Rajendra Prasad,
Arti Choudhary, and Sudhir Kumar Singh*

## CONTENTS

| | |
|---|---|
| 6.1 Introduction | 88 |
| 6.2 Description of Study Area and Materials | 89 |
| 6.3 Methodology | 90 |
|     6.3.1 Image Processing of Satellite Data | 90 |
|         6.3.1.1 Image Preprocessing | 90 |
|     6.3.2 Separability Analysis | 91 |
|         6.3.2.1 Transformed Divergence Method | 92 |
|         6.3.2.2 Jefferies Matusita Distance Method | 92 |
|     6.3.3 Image Classification | 92 |
|         6.3.3.1 Support Vector Machines-Based Classification | 92 |
|         6.3.3.2 Artificial Neural Network-Based Classification | 93 |
|         6.3.3.3 Random Forest-Based Classification | 94 |
|     6.3.4 Selected Measures | 94 |
|         6.3.4.1 Marginal Rates | 94 |
|         6.3.4.2 F-Measure | 95 |
|         6.3.4.3 Jaccard Coefficient | 95 |
|         6.3.4.4 Classification Success Index | 95 |
|     6.3.5 Statistical Significance of Classification Accuracy by Z-Test | 95 |
| 6.4 Results and Discussion | 96 |
|     6.4.1 Support Vector Machines-Based Summary of Classification Accuracy | 99 |
|     6.4.2 Artificial Neural Network-Based Summary of Classification Accuracy | 101 |
|     6.4.3 Random Forest-Based Summary of Classification Accuracy | 102 |
|     6.4.4 Postprocessing Summary of Classification Accuracy | 103 |
|     6.4.5 Analyses of Statistical Significance in the Classification Accuracy of Two Algorithms | 104 |
| 6.5 Conclusion | 105 |
| Acknowledgment | 105 |
| References | 105 |

**ABSTRACT** Agricultural crop classification is obligatory to realize the physical and climatic requirements of diverse crops. Resourcesat -2 satellite is a key data source for the regional to global cropland other land feature classification studies. Indian Remote Sensing satellite sensor, linear imaging self-scanning (LISS-IV) data acquired on April 6, 2013, in Rabi season-derived information, was assessed for the diverse crop and other land features classification in Varanasi district, Uttar Pradesh, India. The most popular algorithms—support vector machines (SVMs), artificial neural network (ANN) and random forest (RF)—were analyzed for the land features classification. The spectral separability analysis was completed before the classification to check the separation between 16 different classes. The separability analysis using the transformed divergence (TD) method has shown high separation between the classes compared with the Jefferies Matusita (JM) distance method. The classification results derived from the kernel-based SVMs were found better than the RF and ANN algorithms except for SVMs with sigmoid kernel. The classification accuracy results were enhanced after postprocessing using different filters by pixel window size of 3 × 3 and compared. The performance analysis was also done on the basis of marginal rates, F-measure, Jaccard's coefficient of community (JCC) and classification success index (CSI).The statistical significance in the classification accuracy results between two algorithms was analyzed using Z-test.

## 6.1 INTRODUCTION

Crops are the primary and essential necessities for the livelihood and for any growing country of the world. Timely and precise information about diverse crop provides a vital and imperative role in the local area crop management during different growing seasons (Yang et al., 2007). This information is very necessary for production estimation, timely transportation of crop products, and accurate crop price determination. However, there may be some difficulties in separating these crop species due to variations in soil properties, planting dates, fertilization, irrigation, intercropping, pest conditions, and tillage practices (Ryerson et al., 1997). Crop classification is associated with the global climate change, agricultural environment, and urban development. Therefore, it has attracted widespread interests in the community of geography, ecology, hydrology, GIS, remote sensing, environment, and so on (Srivastava et al., 2012).

Availability of different satellite imagery, image-processing algorithms, and advancement in digital image processing increased the potential to find accurate crop information such as crop type, crop condition, and growth of diverse crops in agriculture (Turker and Arikan, 2005; Akbari et al., 2006). The spectral knowledge of diverse crops in satellite imagery is obligatory to be used as training data for the discrimination of crops in the same or another area (Nidamanuri and Zbell, 2012). High spatial resolution, multispectral satellite Resourcesat-2, sensor linear imaging, and self-scanning (LISS-IV) data in optical and near infrared bands emerged as a possible approach for monitoring crop types and other land features (Kumar et al., 2015). Foremost restrictions on discriminating agricultural crop types by satellite imagery relate to the resemblance of diverse crop and field-to-field inconsistency of plant reflectance of the same crops. The particular combinations of the crops in a specified region, the array of specific crop phenologies, and spatial and spectral variability within fields are also major limitations (Wheeler and Misra, 1980; Buechel et al., 1989). The LISS-IV sensor data have the potential to capture and somewhat remove these limitations. The usefulness of multispectral satellite imageries has been recognized for the discrimination and mapping of diverse crop and other land features by several researchers (WitDe and Clevers, 2004; Conrad et al., 2010; Turker and Ozdarici, 2011; Kumar et al., 2015).

Crop classification maps are essential for the assessment of the amount and type of crops harvested in a certain area and for the management and assessment of agricultural disaster compensation; however, those algorithms have not been recognized till now (Sonobe et al., 2014). Nonparametric algorithms such as support vector machines (SVMs), back propagation artificial neural network (ANN), and random forest (RF) have been found more robust than the conventional statistical algorithms and create no assumptions regarding the statistical nature of the data. These are good additions to the existing catalogue of image classification algorithms that may allow accurate classification. The SVMs algorithm is more and more used algorithm in the remote-sensing community due to its ability to produce better classification results than the other algorithms even with limited and spectral-mixed training data (Foody and Mathur, 2006; Mountrakis et al., 2011; Shao and Lunetta, 2012). The output of SVMs algorithm depends on the input pixels and it pointing out that the training is potentially a significant stage for augmenting classification accuracy (Pouteau and Collin, 2013). However, major concerns regarding the design of SVMs classifier such as the choice of kernel-specific parameters, suitable kernel function, regularization parameter, and strategies for multiclass classification can affect the accuracy of crop discrimination results (Huang et al., 2002; Pal, 2009; Mountrakis et al., 2011).

ANN algorithm has been well proven as a useful tool in remote sensing and in the other applications (Mas and Flores, 2008; Kavzoglu, 2009; Gupta et al., 2015, 2016; Mishra and Rai, 2016). The ANN process is mainly parallel disseminated, which is made from the simple processing units. ANN has a regular tendency to store experiential information and to make it available in use with an unseen data set (Dastorani et al., 2010; Islam et al., 2012). Other related studies (e.g., Paola and Schowengerdt, 1995; Mas and Flores, 2008) have demonstrated the superiority of ANNs because they do not assume the training samples to be normally distributed. For example, a back propagation ANN classification process involves a repeated feed forward and back propagation process to minimize the root mean square error (RMSE) between the models' predicted and target values. However, a few studies have reported some problems regarding use of back propagation ANN for classifying crop and other land cover features (Foody and Arora, 1997; Kavzoglu and Mather, 2003). ANN algorithms run slowly during the training phase due to the involvement of several setting parameters and nature of input data (Arora et al., 2000). The classification accuracy depends on many factors and may be affected by variation in the dimensionality of satellite imagery and training and testing data sets (Foody and Arora, 1997). RF algorithm is a nonparametric rule-based algorithm that can be trained rapidly. This algorithm is very effective for the classification through complex and nonlinear landscape patterns. Furthermore, the RF algorithm runs efficiently on large data sets and does not require normally distributed model training data (Rodriguez-Galiano et al., 2012). SVM needs several user-defined parameters, whereas RF algorithm requires only two parameters to be set. Classification results by RF were found to equal those of SVM (Pal, 2005).

At the advent of a number of satellites, no insufficiency of data since remotely sensed sources; however, appropriate classification accuracy has remained a big challenge. Despite many research papers published in the field of remote sensing, there is still a lack of comparative studies on different supervised classification algorithms using LISS-IV data. The present study, demanding the assessment of SVMs with respect to linear, radial basis, polynomial of degree 2, sigmoid function kernels, ANN, and RF, endeavored to determine the crop type and other land feature discrimination and accuracy enhancement by postprocessing. The spectral separability along with different bands was analyzed for better separation between the crops/noncrops by transformed divergence (TD) and Jeffries–Matusita (JM) distance methods. The classification accuracy of crop/noncrop results was investigated statistically by Z-test and results were compared. This study offers significant information regarding the presence of a diverse crop in an area dominated by agricultural practices. Such information is vital for the successful management and monitoring of diversity and crop productivity and so on.

## 6.2 DESCRIPTION OF STUDY AREA AND MATERIALS

The Varanasi or Banaras district is situated on the banks of the Ganges River in Uttar Pradesh, India. Some part of the Varanasi district, an area about 12,576 ha including Banaras Hindu University (BHU) agriculture farm house, was selected as the research area. Wheat is the most significant crop

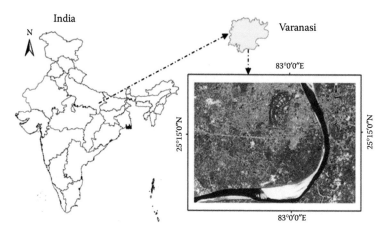

**FIGURE 6.1** Location map of the study area shown by LISS-IV imagery.

of Rabi season in Varanasi. The widespread ground information for different classes was collected using GPS on April 6, 2013. The area is shown in Figure 6.1.

Today it is necessary to utilize imagery having high spatial resolution for better discrimination of diverse crop and noncrop. So, LISS-IV satellite data with 5.8-m spatial and 10 bits radiometric resolutions, containing bands B2 (green, 0.52–0.59 μm), B3 (red, 0.62–0.68 μm), and B4 (NIR, 0.77–0.86 μm), were acquired on April 6, 2013. Single-date LISS-IV multispectral imagery can be utilized for the crop classification if the imagery is acquired during the optimum period for crop discrimination for a given region. Bearing in mind the cost of imagery and weather constraints, it is more effective to use single-date imagery for crop discrimination studies if the majority of the crops can be covered on a single date in a region; otherwise, multidate images may be necessary (Yang et al., 2011).

## 6.3 METHODOLOGY

### 6.3.1 IMAGE PROCESSING OF SATELLITE DATA

Figure 6.2 displays the flowchart of the adopted methodology to discriminate between diverse crop and noncrop.

#### 6.3.1.1 Image Preprocessing

In remote-sensing applications such as image classification, atmospheric correction is not required on the same calendar date. This is due to the fact that atmospheric correction for single-date imagery is regularly comparable to subtracting a constant from all sample pixels in a spectral band (Song et al., 2001). The three different band data were layer stacked to get the FCC image. The image was geometrically corrected. The required area was clipped from the original image, which was found helpful to diminish the dimension of the image file to focus only on the region of interest. This procedure not only eliminates the unnecessary data in the file but also speeds up the processing time, which is essential for the classification algorithms. The regions of interest (ROI) files were generated after the random collection of ground samples. The spectral separability before the classification was performed to check the separation between the classes. One ROI file was used to train the classification algorithms and the other ROI file was used to test the classifiers. The training/testing pixels used for the discrimination are given in Table 6.1.

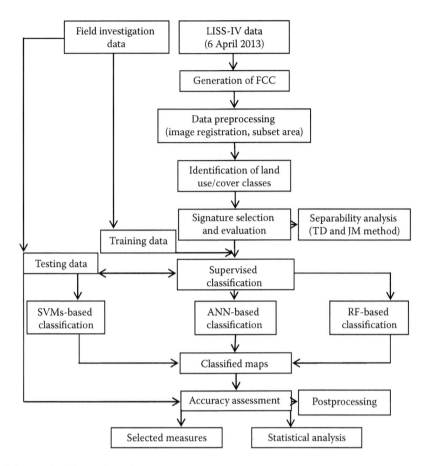

**FIGURE 6.2** Methodology adopted.

**TABLE 6.1**
**Training and Testing Pixels Used in the Classification**

| Class Name | Training Pixels | Testing Pixels | Class Name | Training Pixels | Testing Pixels |
|---|---|---|---|---|---|
| Barley | 379 | 126 | Sugarcane | 316 | 108 |
| Wheat | 604 | 200 | Other crops | 734 | 247 |
| Lentil | 381 | 125 | Water | 625 | 209 |
| Mustard | 368 | 122 | Sand | 663 | 220 |
| Pigeon pea | 352 | 111 | Built up | 462 | 154 |
| Linseed | 337 | 114 | Fallow land | 427 | 142 |
| Corn | 491 | 164 | Sparse vegetation | 780 | 269 |
| Pea | 385 | 128 | Dense vegetation | 739 | 244 |

### 6.3.2 Separability Analysis

To define the spectral separability between classes, the TD (Swain and Davis, 1978) and JM (Richards, 1999) distance methods were analyzed.

### 6.3.2.1 Transformed Divergence Method

The TD method was used for the class separability analysis between two different classes the classes before the image classification. TD is a statistical distance measurement between the classes, calculated from means and covariance matrices of each class. The range of TD values is from 0.0–2.0 and shows how well the chosen training samples were statistically separated. The TD values larger than 1.9 show that the classes have good separability and were obtained using the equation set by Swain and Davis (1978) as follows:

$$TD_{ij} = 2\left(1 - \exp\left(\frac{-D_{ij}}{8}\right)\right) \quad (6.1)$$

$D_{ij}$ is the divergence among two signatures and can be obtained using the following equation:

$$D_{ij} = \frac{1}{2}tr\left((C_i - C_j)(C_i^{-1} - C_j^{-1})\right) + \frac{1}{2}tr\left((C_i^{-1} - C_j^{-1})(\mu_i - \mu_j)(\mu_i - \mu_j)^T\right) \quad (6.2)$$

where:
 $i$ and $j$ are the two classes or signatures being compared
 $C_i$ and $C_j$ are the covariance matrices of signatures $i$ and $j$
 $\mu_i$ and $\mu_j$ are the mean vectors of signatures $i$ and $j$
 $tr$ indicates the trace function that computes the sum of the components on the chief diagonal
 $T$ is the transpose of the matrix used

### 6.3.2.2 Jefferies Matusita Distance Method

JM distance indicates how well a selected spectral class pair is statistically separated. The measurement is based on Bhattacharya distance. JM distance for two classes $a$ and $b$ was computed by the following equation:

$$JM_{ab} = \sqrt{2(1 - \exp(-\alpha))} \quad (6.3)$$

$$\alpha = \frac{1}{8}(\mu_a - \mu_b)^T \left(\frac{C_a + C_b}{2}\right)^{-1} (\mu_a - \mu_b) + \frac{1}{2}\ln\left[\frac{\frac{1}{2}|C_a + C_b|}{\sqrt{|C_a||C_b|}}\right] \quad (6.4)$$

where:
 $C_a$ and $C_b$ are the covariance matrices used for the categories $a$ and $b$
 $\mu_a$ and $\mu_b$ are the mean values for the categories $a$ and $b$
 $T$ signifies the transposition of a vector

The JM distance method delivers a catalog between 0.0 to 2.0, values >1.7 demonstrate that the classes are well separated (ITT Industries Inc., 2006). A JM value <1.0 indicates poor separability between the pair of crops used in the classification.

### 6.3.3 IMAGE CLASSIFICATION

### 6.3.3.1 Support Vector Machines-Based Classification

The SVMs algorithm is a nonparametric algorithm associated with the statistical learning theory. This algorithm was designed in the late 1970s, but its use in remote sensing began to increase about a decade ago (Vapnik, 1998; Mountrakis et al., 2011). SVMs employ a user-defined kernel function to design a set of nonlinear verdict boundaries in the original data set into linear boundaries of a higher dimension (Han et al., 2007). This algorithm classifies the sample training data set into an upper dimensional space and finds the best hyperplanes that separate the classes with the lowest number of classification errors. An optimal hyperplane is firmed using the training data set, and its simplification ability is verified using validation data. The best hyperplanes are located using training samples that lie at the boundaries of class distribution in a feature space. The data sets to train the algorithm that defines the hyperplane of maximum margin are called *support vectors* (Vapnik, 1998; Huang et al., 2002). The remaining training samples can be discarded which are making any involvement to guess hyperplane locations (Brown et al., 2000). An appropriate choice of kernel permits the data to become mostly independent in the feature space in spite of being nonseparable in the original input space (Srivastava et al., 2012). The kernel functions—namely, linear, polynomial of order 2, radial basis, and sigmoidal—were used. The gamma parameter (i.e., 0.25) and penalty parameter were set to their value (i.e., 1000), causing all the pixels in the training data to unite in a class. The zero identification probability threshold was used to limit image pixels to acquire exactly one class label and no unclassified pixels remain left (Petropoulos et al., 2011).

### 6.3.3.2 Artificial Neural Network-Based Classification

The ANN algorithm is a layered feed-forward model in the Environment for Visualizing Images (ENVI) version 5.1 for the supervised learning. In the satellite image classification, every neuron in the input layer signifies one input feature as one satellite image band. In case of an output layer, each neuron resembles one of the classes to be classified. The backpropagation ANN algorithm is one of the most frequently used form of neural computing in remote sensing (Srivastava et al., 2012). ANN is able to process massive amounts of complex and noisy data. The main importance of the ANN algorithm is its proficient adaptability to simulate nonlinear and difficult patterns with suitable topological structures (Atkinson and Tatnall, 1997). The learning rate parameter was set to 0.01, and the momentum value was set to 0.99 for a single hidden layer, and the stopping criterion was fixed to 0.001.Weights in the ANN were initialized using a uniform distribution, and the iteration process was stopped when the RMSE reached at the optimum level. The structure of the ANN is shown in Figure 6.3.

The network has three layers; hidden and output layers contain processing elements at each node. The input layer nodes are simply interfaces to the input data and do not do any processing. The input arrays are the features associated in the classification or are the multispectral trajectories of the training pixels, one band per node (Paola and Schowengerdt, 1995). According to Schowengerdt (2006), within each processing node, there is a summation and transformation. At each hidden layer node $j$, the operation performed on the input pattern $p_i$ produces the output $h_j$,

$$\text{hidden layer: } S_j = \sum_j w_{ji} p_i \tag{6.5}$$

$$h_j = f(S_j) \tag{6.6}$$

which is directed to each output layer node $k$, where the output $o_k$ is calculated as

$$\text{output layer: } S_k = \sum_j w_{kj} h_j \tag{6.7}$$

$$O_k = f(S_k) \tag{6.8}$$

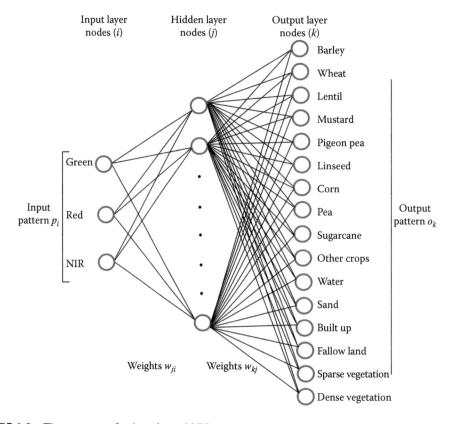

**FIGURE 6.3** The structure of a three-layer ANN.

### 6.3.3.3 Random Forest-Based Classification

The RF classification algorithm is a robust and ensemble learning algorithm. The classification analysis using this algorithm was made via statistical software R. The RF classification algorithm builds many decision trees based on a random bootstrapped training sample data set. Individually, a tree is made by using a different subset from the available training data set, and the nodes in the tree are split using the best split variable from a group of randomly selected variables. This approach delivers robustness to overfitting and can process thousands of dependent/independent variables without deleting variables. To split the nodes, user-defined parameters such as the number of trees and the number of variables are used. The simplification error always converges on increasing the number of the trees. Therefore, selecting the larger numbers of decision trees is recommended for the RF classification (Breiman, 2001). Some samples that are contained within another subset named *out-of-bag* (OOB) do not exist in the training subset. These remaining OOB elements can also be classified for evaluating the performance of the algorithm by the tree (Rodriguez-Galiano et al., 2012).

### 6.3.4 Selected Measures

#### 6.3.4.1 Marginal Rates

A number of asymmetric class measures such as true positive rate (TPR) and true negative rate (TNR) are reference oriented. They consider the columns (true classes) of the error matrix and are estimated as follows:

$$TPR_i = n_{TP}/(n_{TP} + n_{FN}) \tag{6.9}$$

$$TNR_i = n_{TN}/(n_{TN} + n_{FP}) \qquad (6.10)$$

The correspondent estimation oriented measures such as positive predictive value (PPV) and negative predictive value (NPV) based on error matrix rows (estimated classes) (Witten and Frank, 2005) are assessed as follows:

$$PPV_i = n_{TP}/(n_{TP} + n_{FP}) \qquad (6.11)$$

$$NPV_i = n_{TN}/(n_{TN} + n_{FN}) \qquad (6.12)$$

where:
$n_{TP}$ (number of true positives) = $n_{ii}$ (number of correctly classified pixels in a row $i$)
$n_{FP}$ (number of false positives) = $n_{i+}$ (sum of pixels in the error matrix over row $i$)- $n_{ii}$
$n_{FN}$ (number of false negatives) = $n_{+j}$ (sum of pixels in the error matrix over column $j$)- $n_{ii}$
$n_{TN}$ (number of true negatives) = $n$ (total number of pixels used to test the classification accuracy)- $n_{TP}$- $n_{FP}$- $n_{FN}$ (Labatut and Cherifi, 2011)

All measures lie among between 0 and 1.

### 6.3.4.2 F-Measure

The F-measure relates to the harmonic mean of PPV and TPR (Witten and Frank, 2005). F-measure can be evaluated using the relation as follows:

$$F_i = 2\frac{PPV_i \times TPR_i}{PPV_i + TPR_i} \qquad (6.13)$$

The F-measure can be inferred as the extent of overlapping concerning the true and estimated classes.

### 6.3.4.3 Jaccard Coefficient

The Jaccard coefficient is well known as Jaccard's coefficient of community (JCC) defined to compare data sets (Jaccard, 1912). The class-specific symmetric measure can be defined as follows:

$$JCC_i = \frac{n_{TP}}{n_{TP} + n_{FP} + n_{FN}} \qquad (6.14)$$

$JCC_i$ is related to the $F_i$ as given in the relation

$$JCC_i = \frac{F_i}{2 - F_i} \qquad (6.15)$$

### 6.3.4.4 Classification Success Index

The individual classification success index (ICSI) is defined for classification purposes. It is a class-specific symmetric measure defined as follows:

$$ICSI_i = 1 - (1 - PPV_i + 1 - TPR_i) = PPV_i + TPR_i - 1 \qquad (6.16)$$

The classification success index (CSI) is an overall measure defined by averaging ICSI over all classes (Koukoulas and Blackburn, 2001).

### 6.3.5 STATISTICAL SIGNIFICANCE OF CLASSIFICATION ACCURACY BY Z-TEST

The Z-test is a statistical test usually performed to check whether two classification algorithms provide similar classification accuracy. The difference in the crop/noncrop classification accuracy between two algorithms is statistically significant ($p \leq 0.05$) if the Z-value is greater than 1.96.

**TABLE 6.2**
**Number of Correctly and Incorrectly Classified Pixels for the Two Algorithms**

| Allocation | Classification 2 | | |
|---|---|---|---|
| Classification 1 | Correct | Incorrect | Sum |
| Correct | a | b | |
| Incorrect | c | d | |
| Sum | | | |

The value of $Z > |1.96|$ indicates the statistically significance of the difference in classification accuracy at the 95% confidence level (Congalton et al., 1983). The statistical significance between two algorithms may be evaluated using these equations:

$$Z = \frac{b-c}{N\sigma} \tag{6.17}$$

$$\sigma = \sqrt{\frac{(b+c)-(b-c)^2/N}{N(N-1)}} \tag{6.18}$$

where:
  $N$ is the total number of training pixels
  $b$ represents the number of sample pixels that were correctly classified by the first algorithm but misclassified by the second algorithm
  Similarly, $c$ represents the sample pixels that were misclassified by first algorithm and found to be correctly classified by the second algorithm (Kanji, 2006)

The number of correctly/wrongly classified sample pixels for two algorithms is presented by Table 6.2.

## 6.4 RESULTS AND DISCUSSION

To evaluate the accuracy, actual land cover types for all the fields were checked on the ground and compared with the pixels or polygons from a classified map developed from a satellite data set. The overall accuracy (OA) and kappa coefficients (κ) of classified images are estimated from error matrices (Congalton and Green, 1999; Jensen, 2005). The OA is found by dividing the sum of correctly classified sample pixels by the total number of reference sample pixels (Lillesand and Kiefer, 1999). The κ is a measurement of how well the associated reference data and the classified map agree with each other. Strong agreement, moderate agreement, and poor agreement occur if κ is greater than 0.80, between 0.40 and 0.80, and less than 0.40, respectively (Jensen, 2005).

Before the discrimination between diverse crop and noncrop, the training samples between the band combinations B2–B3, B3–B4, and B4–B2 were analyzed to check the separation among the classes to be classified. The high spectral reflectance in the red and NIR bands between classes may be one of the reasons for an overall good separation of the classes using the band combination B3–B4. Representation of training pixels in 2D space between bands 3 and 4 is shown in Figure 6.4. The two different methods such as TD and JM were analyzed and equated to check the separability between the crop and noncrop classes. Almost all the classes were found to be well separated using the TD method; however, no good separation among all the classes was found using JM method. Although many classes gave the same or nearly similar results using the JM and TD methods, but for the some classes, JM method gave less separability among the classes. The JM method gave almost the same

# Performance Analysis of Different Predictive Algorithms for the Land Features Modeling

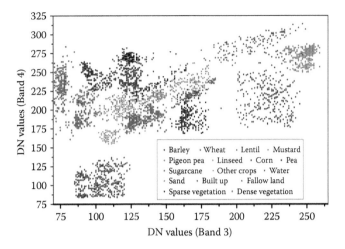

**FIGURE 6.4** Representation of training pixels in 2D space between bands 3 and 4.

**TABLE 6.3**
**Separability Analysis between Diverse Crop and Noncrop Classes Using the TD Method**

|    | 1 | 2 | 3 | 4 | 5 | 6 | 7 | 8 | 9 | 10 | 11 | 12 | 13 | 14 | 15 | 16 |
|----|---|---|---|---|---|---|---|---|---|----|----|----|----|----|----|----|
| 1  |      | 1.99 | 1.97 | 1.98 | 2.00 | 1.98 | 2.00 | 1.90 | 2.00 | 1.82 | 2.00 | 2.00 | 1.98 | 2.00 | 1.97 | 2.00 |
| 2  | 1.99 |      | 1.96 | 1.95 | 1.98 | 1.97 | 2.00 | 1.91 | 2.00 | 1.73 | 2.00 | 2.00 | 1.98 | 2.00 | 1.99 | 2.00 |
| 3  | 1.97 | 1.96 |      | 1.99 | 2.00 | 1.98 | 2.00 | 1.96 | 2.00 | 1.98 | 2.00 | 2.00 | 2.00 | 1.98 | 1.99 | 2.00 |
| 4  | 1.98 | 1.95 | 1.99 |      | 1.97 | 1.71 | 1.97 | 1.95 | 1.99 | 1.98 | 2.00 | 2.00 | 2.00 | 2.00 | 1.97 | 1.99 |
| 5  | 2.00 | 1.98 | 2.00 | 1.97 |      | 1.96 | 1.98 | 1.94 | 1.68 | 1.77 | 2.00 | 2.00 | 2.00 | 1.99 | 1.98 | 1.47 |
| 6  | 1.98 | 1.97 | 1.98 | 1.71 | 1.96 |      | 1.95 | 1.96 | 1.99 | 1.64 | 2.00 | 2.00 | 2.00 | 1.88 | 1.92 | 2.00 |
| 7  | 2.00 | 2.00 | 2.00 | 1.97 | 1.98 | 1.95 |      | 1.30 | 1.93 | 1.99 | 2.00 | 2.00 | 2.00 | 2.00 | 1.84 | 1.96 |
| 8  | 1.90 | 1.91 | 1.96 | 1.95 | 1.94 | 1.96 | 1.30 |      | 1.90 | 1.98 | 2.00 | 2.00 | 2.00 | 2.00 | 1.31 | 1.99 |
| 9  | 2.00 | 2.00 | 2.00 | 1.99 | 1.68 | 1.99 | 1.93 | 1.90 |      | 1.93 | 2.00 | 2.00 | 2.00 | 2.00 | 2.00 | 1.76 |
| 10 | 1.82 | 1.73 | 1.98 | 1.98 | 1.77 | 1.64 | 1.99 | 1.98 | 1.93 |      | 2.00 | 2.00 | 1.99 | 1.99 | 1.92 | 1.97 |
| 11 | 2.00 | 2.00 | 2.00 | 2.00 | 2.00 | 2.00 | 2.00 | 2.00 | 2.00 | 2.00 |      | 2.00 | 2.00 | 2.00 | 2.00 | 2.00 |
| 12 | 2.00 | 2.00 | 2.00 | 2.00 | 2.00 | 2.00 | 2.00 | 2.00 | 2.00 | 2.00 | 2.00 |      | 1.99 | 2.00 | 2.00 | 2.00 |
| 13 | 1.98 | 1.98 | 2.00 | 2.00 | 2.00 | 2.00 | 2.00 | 2.00 | 2.00 | 1.99 | 2.00 | 1.99 |      | 2.00 | 2.00 | 2.00 |
| 14 | 2.00 | 2.00 | 1.98 | 2.00 | 1.99 | 1.88 | 2.00 | 2.00 | 2.00 | 1.99 | 2.00 | 2.00 | 2.00 |      | 1.99 | 2.00 |
| 15 | 1.97 | 1.99 | 1.99 | 1.97 | 1.98 | 1.92 | 1.84 | 1.31 | 2.00 | 1.92 | 2.00 | 2.00 | 2.00 | 1.99 |      | 2.00 |
| 16 | 2.00 | 2.00 | 2.00 | 1.99 | 1.47 | 2.00 | 1.96 | 1.99 | 1.76 | 1.97 | 2.00 | 2.00 | 2.00 | 2.00 | 2.00 |      |

separability as did the TD method for most noncrop classes because of its unique spectral response. The separability analysis using TD and JM methods are presented in Tables 6.3 and 6.4, respectively.

A comparison of kernels based on SVM reveals that the performance of the radial basis function kernel was found more effective as compared to a linear, polynomial of degree 2 and sigmoid kernels. The classification accuracy results were also good using RF and ANN algorithms but were found to have less accuracy in comparison to SVMs except for the sigmoid kernel. Sixteen classes of classification maps generated from the kernel-based SVMs algorithm are shown in Figure 6.5. Visual comparison of kernel-based SVMs, ANN, and RF indicates that the classification maps provided good separations among the classes. However, it was not visually clear how the crops were separated

**TABLE 6.4**
**Separability Analysis between Diverse Crop and Noncrop Classes Using the JM Distance Method**

| | 1 | 2 | 3 | 4 | 5 | 6 | 7 | 8 | 9 | 10 | 11 | 12 | 13 | 14 | 15 | 16 |
|---|---|---|---|---|---|---|---|---|---|---|---|---|---|---|---|---|
| 1 | | 1.98 | 1.70 | 1.88 | 1.99 | 1.95 | 2.00 | 1.89 | 2.00 | 1.71 | 2.00 | 2.00 | 1.97 | 2.00 | 1.91 | 2.00 |
| 2 | 1.98 | | 1.92 | 1.87 | 1.97 | 1.95 | 2.00 | 1.85 | 1.99 | 1.67 | 2.00 | 2.00 | 1.97 | 2.00 | 1.95 | 2.00 |
| 3 | 1.70 | 1.92 | | 1.93 | 1.99 | 1.87 | 2.00 | 1.92 | 2.00 | 1.97 | 2.00 | 2.00 | 1.99 | 1.97 | 1.82 | 2.00 |
| 4 | 1.88 | 1.87 | 1.93 | | 1.95 | 1.34 | 1.95 | 1.89 | 1.98 | 1.92 | 2.00 | 2.00 | 2.00 | 1.99 | 1.90 | 1.97 |
| 5 | 1.99 | 1.97 | 1.99 | 1.95 | | 1.82 | 1.96 | 1.57 | 1.25 | 1.69 | 2.00 | 2.00 | 2.00 | 1.98 | 1.97 | 1.32 |
| 6 | 1.95 | 1.95 | 1.87 | 1.34 | 1.82 | | 1.93 | 1.84 | 1.98 | 1.61 | 2.00 | 2.00 | 1.99 | 1.86 | 1.71 | 2.00 |
| 7 | 2.00 | 2.00 | 2.00 | 1.95 | 1.96 | 1.93 | | 1.25 | 1.89 | 1.98 | 2.00 | 2.00 | 2.00 | 1.99 | 1.78 | 1.82 |
| 8 | 1.89 | 1.85 | 1.92 | 1.89 | 1.57 | 1.84 | 1.25 | | 1.77 | 1.93 | 2.00 | 2.00 | 2.00 | 1.99 | 1.27 | 1.98 |
| 9 | 2.00 | 1.99 | 2.00 | 1.98 | 1.25 | 1.98 | 1.89 | 1.77 | | 1.90 | 2.00 | 2.00 | 1.99 | 2.00 | 1.99 | 1.56 |
| 10 | 1.71 | 1.67 | 1.97 | 1.92 | 1.69 | 1.61 | 1.98 | 1.93 | 1.90 | | 2.00 | 2.00 | 1.96 | 1.98 | 1.87 | 1.95 |
| 11 | 2.00 | 2.00 | 2.00 | 2.00 | 2.00 | 2.00 | 2.00 | 2.00 | 2.00 | 2.00 | | 2.00 | 1.99 | 2.00 | 2.00 | 2.00 |
| 12 | 2.00 | 2.00 | 2.00 | 2.00 | 2.00 | 2.00 | 2.00 | 2.00 | 2.00 | 2.00 | 2.00 | | 1.79 | 2.00 | 2.00 | 2.00 |
| 13 | 1.97 | 1.97 | 1.99 | 2.00 | 2.00 | 1.99 | 2.00 | 2.00 | 1.99 | 1.96 | 1.99 | 1.79 | | 2.00 | 1.99 | 2.00 |
| 14 | 2.00 | 2.00 | 1.97 | 1.99 | 1.98 | 1.86 | 1.99 | 1.99 | 2.00 | 1.98 | 2.00 | 2.00 | 2.00 | | 1.98 | 2.00 |
| 15 | 1.91 | 1.95 | 1.82 | 1.90 | 1.97 | 1.71 | 1.78 | 1.27 | 1.99 | 1.87 | 2.00 | 2.00 | 1.99 | 1.98 | | 1.99 |
| 16 | 2.00 | 2.00 | 2.00 | 1.97 | 1.32 | 2.00 | 1.82 | 1.98 | 1.56 | 1.95 | 2.00 | 2.00 | 2.00 | 2.00 | 1.99 | |

1—Barley, 2—Wheat, 3—Lentil, 4—Mustard, 5—Pigeon pea, 6—Linseed, 7—Corn, 8—Pea, 9—Sugarcane, 10—Other crops, 11—Water, 12—Sand, 13—Built up, 14—Fallow land, 15—Sparse vegetation, 16—Dense vegetation.

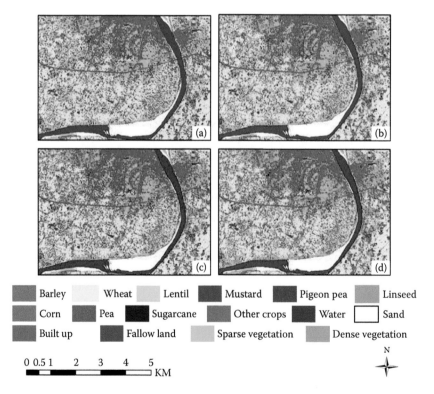

**FIGURE 6.5** (a) SVMs with linear kernel, (b) SVMs with polynomial of degree 2 kernel, (c) SVMs with radial basis kernel, and (d) SVMs with sigmoid function kernel-based classification.

according to different growing stages and the management conditions of the crop fields. Most of the fields had one dominant crop; however, some fields contained a small presence of other classes.

### 6.4.1 Support Vector Machines-Based Summary of Classification Accuracy

A summary of accuracy assessment results for the kernel-based SVM maps is presented in Tables 6.5 through 6.8. The highest OA achieved by the radial basis function kernel classified map was 87.89%, which indicates that the probability of image pixel existence correctly identified

**TABLE 6.5**
**Selected Measures (SVMs with Linear Kernel)**

|  | $TPR_i$ | $TNR_i$ | $PPV_i$ | $NPV_i$ | $F_i$ | $JCC_i$ | $ICSI_i$ |
|---|---|---|---|---|---|---|---|
| Barley | 0.8270 | 0.9898 | 0.8014 | 0.9914 | 0.8140 | 0.6863 | 0.6284 |
| Wheat | 0.9446 | 0.9976 | 0.9692 | 0.9956 | 0.9567 | 0.9170 | 0.9138 |
| Lentil | 0.8097 | 0.9910 | 0.8159 | 0.9906 | 0.8128 | 0.6846 | 0.6256 |
| Mustard | 0.8100 | 0.9895 | 0.7838 | 0.9910 | 0.7967 | 0.6621 | 0.5938 |
| Pigeon pea | 0.7054 | 0.9918 | 0.7902 | 0.9872 | 0.7454 | 0.5941 | 0.4956 |
| Linseed | 0.5223 | 0.9899 | 0.6939 | 0.9792 | 0.5960 | 0.4245 | 0.2162 |
| Corn | 0.8648 | 0.9897 | 0.8446 | 0.9913 | 0.8546 | 0.7461 | 0.7094 |
| Pea | 0.7365 | 0.9914 | 0.8118 | 0.9867 | 0.7723 | 0.6291 | 0.5483 |
| Sugarcane | 0.8226 | 0.9891 | 0.7588 | 0.9926 | 0.7894 | 0.6521 | 0.5814 |
| Other crops | 0.9195 | 0.9873 | 0.8800 | 0.9917 | 0.8993 | 0.8171 | 0.7995 |
| Water | 1.0000 | 1.0000 | 1.0000 | 1.0000 | 1.0000 | 1.0000 | 1.0000 |
| Sand | 0.9909 | 0.9919 | 0.9166 | 0.9992 | 0.9523 | 0.9089 | 0.9075 |
| Built up | 0.8696 | 0.9992 | 0.9849 | 0.9922 | 0.9237 | 0.8582 | 0.8545 |
| Fallow land | 0.9930 | 0.9996 | 0.9930 | 0.9996 | 0.9930 | 0.9861 | 0.9860 |
| Sparse vegetation | 0.9027 | 0.9756 | 0.8038 | 0.9891 | 0.8504 | 0.7397 | 0.7065 |
| Dense vegetation | 0.8820 | 0.9910 | 0.9074 | 0.9881 | 0.8945 | 0.8092 | 0.7894 |

**TABLE 6.6**
**Selected Measures (SVMs with Polynomial Kernel)**

|  | $TPR_i$ | $TNR_i$ | $PPV_i$ | $NPV_i$ | $F_i$ | $JCC_i$ | $ICSI_i$ |
|---|---|---|---|---|---|---|---|
| Barley | 0.7949 | 0.9906 | 0.8083 | 0.9898 | 0.8015 | 0.6688 | 0.6032 |
| Wheat | 0.9446 | 0.9976 | 0.9692 | 0.9956 | 0.9567 | 0.9171 | 0.9138 |
| Lentil | 0.8255 | 0.9894 | 0.7941 | 0.9914 | 0.8095 | 0.6800 | 0.6196 |
| Mustard | 0.8100 | 0.9891 | 0.7780 | 0.9910 | 0.7937 | 0.6579 | 0.5880 |
| Pigeon pea | 0.7143 | 0.9911 | 0.7770 | 0.9876 | 0.7443 | 0.5928 | 0.4913 |
| Linseed | 0.5219 | 0.9899 | 0.6939 | 0.9792 | 0.5957 | 0.4242 | 0.2158 |
| Corn | 0.8463 | 0.9901 | 0.8469 | 0.9901 | 0.8466 | 0.7340 | 0.6932 |
| Pea | 0.7439 | 0.9902 | 0.7935 | 0.9871 | 0.7679 | 0.6232 | 0.5374 |
| Sugarcane | 0.8129 | 0.9864 | 0.7128 | 0.9921 | 0.7596 | 0.6123 | 0.5257 |
| Other crops | 0.9115 | 0.9873 | 0.8795 | 0.9909 | 0.8952 | 0.8103 | 0.7910 |
| Water | 1.0000 | 1.0000 | 1.0000 | 1.0000 | 1.0000 | 1.0000 | 1.0000 |
| Sand | 0.9909 | 0.9919 | 0.9165 | 0.9992 | 0.9523 | 0.9088 | 0.9074 |
| Built up | 0.8694 | 0.9992 | 0.9850 | 0.9922 | 0.9236 | 0.8580 | 0.8544 |
| Fallow land | 0.9930 | 0.9996 | 0.9930 | 0.9996 | 0.9930 | 0.9861 | 0.9860 |
| Sparse vegetation | 0.9066 | 0.9756 | 0.8050 | 0.9895 | 0.8528 | 0.7434 | 0.7116 |
| Dense vegetation | 0.8776 | 0.9943 | 0.9391 | 0.9878 | 0.9073 | 0.8303 | 0.8167 |

## TABLE 6.7
### Selected Measures (SVMs with Radial Basis Kernel)

|  | $TPR_i$ | $TNR_i$ | $PPV_i$ | $NPV_i$ | $F_i$ | $JCC_i$ | $ICSI_i$ |
|---|---|---|---|---|---|---|---|
| Barley | 0.7559 | 0.9910 | 0.8067 | 0.9879 | 0.7805 | 0.6400 | 0.5626 |
| Wheat | 0.9548 | 0.9976 | 0.9694 | 0.9964 | 0.9620 | 0.9268 | 0.9242 |
| Lentil | 0.8175 | 0.9879 | 0.7687 | 0.9910 | 0.7923 | 0.6561 | 0.5861 |
| Mustard | 0.9091 | 0.9902 | 0.8148 | 0.9957 | 0.8594 | 0.7534 | 0.7239 |
| Pigeon pea | 0.6875 | 0.9930 | 0.8105 | 0.9865 | 0.7440 | 0.5923 | 0.4980 |
| Linseed | 0.5841 | 0.9891 | 0.7021 | 0.9818 | 0.6377 | 0.4681 | 0.2862 |
| Corn | 0.8405 | 0.9948 | 0.9133 | 0.9897 | 0.8754 | 0.7784 | 0.7538 |
| Pea | 0.8527 | 0.9906 | 0.8209 | 0.9925 | 0.8365 | 0.7190 | 0.6736 |
| Sugarcane | 0.8505 | 0.9864 | 0.7222 | 0.9937 | 0.7811 | 0.6408 | 0.5727 |
| Other crops | 0.9113 | 0.9926 | 0.9262 | 0.9910 | 0.9187 | 0.8496 | 0.8375 |
| Water | 1.0000 | 1.0000 | 1.0000 | 1.0000 | 1.0000 | 1.0000 | 1.0000 |
| Sand | 0.9955 | 0.9923 | 0.9205 | 0.9996 | 0.9565 | 0.9167 | 0.9160 |
| Built up | 0.8758 | 0.9996 | 0.9926 | 0.9925 | 0.9306 | 0.8701 | 0.8684 |
| Fallow land | 0.9930 | 0.9996 | 0.9930 | 0.9996 | 0.9930 | 0.9861 | 0.9860 |
| Sparse vegetation | 0.8806 | 0.9785 | 0.8194 | 0.9866 | 0.8489 | 0.7375 | 0.7000 |
| Dense vegetation | 0.8857 | 0.9930 | 0.9274 | 0.9886 | 0.9061 | 0.8282 | 0.8131 |

## TABLE 6.8
### Selected Measures (SVMs with Sigmoid Kernel)

|  | $TPR_i$ | $TNR_i$ | $PPV_i$ | $NPV_i$ | $F_i$ | $JCC_i$ | $ICSI_i$ |
|---|---|---|---|---|---|---|---|
| Barley | 0.7478 | 0.9879 | 0.7539 | 0.9875 | 0.7508 | 0.6011 | 0.5017 |
| Wheat | 0.9400 | 0.9976 | 0.9690 | 0.9952 | 0.9543 | 0.9126 | 0.9090 |
| Lentil | 0.7699 | 0.9867 | 0.7407 | 0.9886 | 0.7550 | 0.6064 | 0.5106 |
| Mustard | 0.7934 | 0.9887 | 0.7678 | 0.9902 | 0.7804 | 0.6399 | 0.5612 |
| Pigeon pea | 0.6874 | 0.9895 | 0.7406 | 0.9864 | 0.7130 | 0.5540 | 0.4280 |
| Linseed | 0.5219 | 0.9867 | 0.6345 | 0.9792 | 0.5727 | 0.4013 | 0.1564 |
| Corn | 0.8466 | 0.9893 | 0.8363 | 0.9901 | 0.8414 | 0.7262 | 0.6829 |
| Pea | 0.7363 | 0.9902 | 0.7920 | 0.9867 | 0.7631 | 0.6170 | 0.5283 |
| Sugarcane | 0.7567 | 0.9884 | 0.7299 | 0.9899 | 0.7431 | 0.5911 | 0.4866 |
| Other crops | 0.8749 | 0.9848 | 0.8543 | 0.9872 | 0.8645 | 0.7613 | 0.7292 |
| Water | 1.000 | 1.0000 | 1.0000 | 1.0000 | 1.0000 | 1.0000 | 1.0000 |
| Sand | 0.9909 | 0.9919 | 0.9159 | 0.9992 | 0.9519 | 0.9083 | 0.9068 |
| Built up | 0.8693 | 0.9992 | 0.9849 | 0.9922 | 0.9235 | 0.8579 | 0.8542 |
| Fallow land | 0.9930 | 0.9996 | 0.9930 | 0.9996 | 0.9930 | 0.9861 | 0.9860 |
| Sparse vegetation | 0.8920 | 0.9760 | 0.8050 | 0.9878 | 0.8463 | 0.7335 | 0.6970 |
| Dense vegetation | 0.8817 | 0.9906 | 0.9040 | 0.9881 | 0.8927 | 0.8062 | 0.7857 |

is approximately 88% in the map. In the case of SVMs with linear kernel, $TPR_i$ ranged from 0.5223 for linseed to 0.9446 for wheat in the crop classes and from 0.8696 for built up to 1.0000 for water in the noncrop classes. The $PPV_i$ ranged from 0.6939 for linseed to 0.9692 for wheat in crop classes. It is also ranged from 0.8038 for sparse vegetation to 1.0000 for water in noncrop classes. The 0.9930 of both $TPR_i$ and $PPV_i$ for fallow land indicates that the fallow land was correctly identified on the ground and was accurately classified on the map. In case of SVMs with polynomial of degree 2, the OA and κ were found to be 87.51% and 0.8658, respectively, which were somewhat lower than SVMs with linear kernel, but the results were almost similar.

The 0.8505 $TPR_i$ and 0.7222 $PPV_i$ for sugarcane using SVMs having a radial basis kernel function indicate that 85.05% of the sugarcane areas on the ground were correctly identified as sugarcane, but only 72.22% of the areas named sugarcane on the classified map were found to actually be sugarcane. SVM with sigmoid kernel had a similar but somewhat less accuracy in comparison to all other SVMs kernel functions. The best estimated κ for the SVMs with radial kernel-based classification algorithms was 0.8698, which indicates that the classification achieved an accuracy that is 87% better than would be expected from a random assignment of pixels to classes. The values of all selected measures were found to be excellent for wheat, sand, other crops, water, and fallow land classes, very good for other classes of crop and noncrop, and good results were found for the linseed crop. The overall low $TPR_i$, $PPV_i$, $F_i$, $JCC_i$, and $ICSI_i$ for the linseed crop were mainly due to spectral similarities between lentil, pea, sparse vegetation, and other crops in some of the fields.

### 6.4.2 Artificial Neural Network-Based Summary of Classification Accuracy

The ANN algorithm provided accuracy results similar to the kernel-based SVMs and RF whereas these results were lowest except for SVMs with sigmoid kernel. The classification map produced by the ANN is shown in Figure 6.6. The OA and estimated κ using the ANN algorithm were found to be 85.98% and 0.8496, respectively. For the wheat crop, all algorithms provided $TPR_i$ of nearly 94%, indicating that all of the collected validation samples were also initiate to belong in the same class. The wheat crop was identified easily, whereas linseed was the most difficult to clearly identify among the crop classes. The unique reflectance value of the wheat crop and less mixing of other classes with the wheat crop is the reason for its easy identification. The accuracy for linseed crop was found to be low due to its confusion or the fact that the reflectance values from some of these pixels are similar to the other classes. The OA of wheat was found to be fairly high compared to other crop classes; however, the 1.0000 value of selected measures for water indicated a perfect prediction because of wheat's unique reflectance and no mixing with the other classes in almost all algorithms implemented here. The $TPR_i$ and $NPV_i$ values were found to be high for the linseed

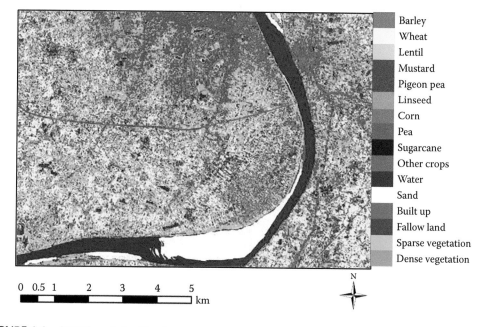

**FIGURE 6.6** ANN-based classification.

**TABLE 6.9**
**Selected Measures (ANN Classification)**

|  | $TPR_i$ | $TNR_i$ | $PPV_i$ | $NPV_i$ | $F_i$ | $JCC_i$ | $ICSI_i$ |
|---|---|---|---|---|---|---|---|
| Barley | 0.7640 | 0.9879 | 0.7580 | 0.9883 | 0.7610 | 0.6142 | 0.5220 |
| Wheat | 0.9347 | 0.9980 | 0.9740 | 0.9948 | 0.9540 | 0.9119 | 0.9087 |
| Lentil | 0.7780 | 0.9875 | 0.7540 | 0.9890 | 0.7658 | 0.6205 | 0.5320 |
| Mustard | 0.7932 | 0.9871 | 0.7439 | 0.9902 | 0.7678 | 0.6231 | 0.5371 |
| Pigeon pea | 0.7501 | 0.9895 | 0.7670 | 0.9891 | 0.7585 | 0.6109 | 0.5171 |
| Linseed | 0.5130 | 0.9887 | 0.6666 | 0.9788 | 0.5798 | 0.4083 | 0.1796 |
| Corn | 0.8465 | 0.9885 | 0.8264 | 0.9900 | 0.8363 | 0.7187 | 0.6729 |
| Pea | 0.7364 | 0.9902 | 0.7920 | 0.9867 | 0.7632 | 0.6171 | 0.5284 |
| Sugarcane | 0.7568 | 0.9876 | 0.7170 | 0.9899 | 0.7364 | 0.5827 | 0.4738 |
| Other crops | 0.8789 | 0.9848 | 0.8550 | 0.9876 | 0.8668 | 0.7649 | 0.7339 |
| Water | 1.0000 | 1.0000 | 1.0000 | 1.0000 | 1.0000 | 1.0000 | 1.0000 |
| Sand | 0.9909 | 0.9919 | 0.9165 | 0.9992 | 0.9522 | 0.9089 | 0.9074 |
| Built up | 0.8693 | 0.9992 | 0.9849 | 0.9922 | 0.9235 | 0.8590 | 0.8542 |
| Fallow land | 0.9930 | 0.9996 | 0.9930 | 0.9996 | 0.9930 | 0.9861 | 0.9860 |
| Sparse vegetation | 0.8989 | 0.9760 | 0.8057 | 0.9887 | 0.8498 | 0.7388 | 0.7046 |
| Dense vegetation | 0.8779 | 0.9947 | 0.9429 | 0.9878 | 0.9092 | 0.8336 | 0.8208 |

crop whereas the $JCC_i$ and $ICSI_i$ measures had relatively low values. The accurate values of all the measures were found for the wheat, water, sand, and fallow land. The accuracy assessment results is summarized in Table 6.9.

### 6.4.3 Random Forest-Based Summary of Classification Accuracy

The map based on the RF classification algorithm is presented in Figure 6.7. The OA and κ were found to be 86.81% and 0.8582, respectively, and were near SVMs and ANN algorithms. The $TPR_i$

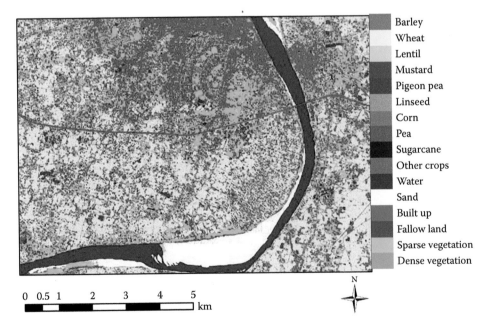

**FIGURE 6.7** RF-based classification.

**TABLE 6.10**
**Selected Measures (RF Classification)**

|  | $TPR_i$ | $TNR_i$ | $PPV_i$ | $NPV_i$ | $F_i$ | $JCC_i$ | $ICSI_i$ |
|---|---|---|---|---|---|---|---|
| Barley | 0.8190 | 0.9879 | 0.7700 | 0.9910 | 0.7937 | 0.6580 | 0.5890 |
| Wheat | 0.9400 | 0.9980 | 0.9740 | 0.9960 | 0.9567 | 0.9170 | 0.9240 |
| Lentil | 0.7780 | 0.9902 | 0.7970 | 0.9891 | 0.7874 | 0.6493 | 0.5750 |
| Mustard | 0.8019 | 0.9891 | 0.7758 | 0.9906 | 0.7886 | 0.6510 | 0.5777 |
| Pigeon pea | 0.6785 | 0.9918 | 0.7835 | 0.9861 | 0.7272 | 0.5714 | 0.4620 |
| Linseed | 0.4960 | 0.9903 | 0.6914 | 0.9781 | 0.5776 | 0.4061 | 0.1874 |
| Corn | 0.8400 | 0.9889 | 0.8305 | 0.9897 | 0.8352 | 0.7171 | 0.6705 |
| Pea | 0.7363 | 0.9898 | 0.7849 | 0.9867 | 0.7598 | 0.6127 | 0.5212 |
| Sugarcane | 0.8037 | 0.9876 | 0.7291 | 0.9918 | 0.7646 | 0.6189 | 0.5328 |
| Other crops | 0.8989 | 0.9864 | 0.8708 | 0.9897 | 0.8846 | 0.7931 | 0.7697 |
| Water | 1.0000 | 1.0000 | 1.0000 | 1.0000 | 1.0000 | 1.0000 | 1.0000 |
| Sand | 0.9864 | 0.9919 | 0.9157 | 0.9988 | 0.9497 | 0.9043 | 0.9021 |
| Built up | 0.8693 | 0.9988 | 0.9780 | 0.9922 | 0.9205 | 0.8526 | 0.8473 |
| Fallow land | 0.9930 | 0.9996 | 0.9930 | 0.9996 | 0.9930 | 0.9861 | 0.9860 |
| Sparse vegetation | 0.9139 | 0.9747 | 0.8010 | 0.9903 | 0.8537 | 0.7448 | 0.7149 |
| Dense vegetation | 0.8776 | 0.9910 | 0.9069 | 0.9877 | 0.8920 | 0.8051 | 0.7845 |

ranged from 0.4960 for linseed to 0.9400 for wheat crop in the crop classes and 0.8693 for built up to 1.0000 for water in the noncrop classes. The $PPV_i$ for crop classes ranged from 0.6914 for linseed to 0.9740 for wheat but from 0.8010 for sparse vegetation to 1.0000 for water in the noncrop classes. The highest OA for wheat crops in the crop classes is because the wheat crop is the major crop in that season. It may be one of the reasons for the wheat crop's high accuracy. This crop has less similarity to sugarcane, pigeon pea, corn, and dense vegetation categories. High accuracy results also found by RF algorithm indicate that there was less mixing among the classes. The corn crop was generally not grown in Varanasi when the image was acquired. However, it was grown for the research purpose in the BHU agriculture farm house and some other places in Varanasi. The variability in corn crop fields was due to the different times of sowing corn in different fields. This could have caused the mixing of the corn crop with sparse and dense crop classes. The overall high $TPR_i$ and $PPV_i$ for fallow land indicate less mixing of built-up and sand classes. The $TPR_i$ (0.4960) and $PPV_i$ (0.6914) values for linseed were found to have the lowest classification algorithm using RF. The summary of accuracy assessment results using the RF algorithm is given in Table 6.10.

### 6.4.4 Postprocessing Summary of Classification Accuracy

To overcome the problem of mixing, the classified maps were postprocessing to eradicate the minor inclusion of other classes within the dominant classes. The different filters such as majority, sieve, and clump and the combination of these filters were used to eradicate minor inclusions and for accuracy enhancement. The comparison of the unfiltered and filtered OA and κ are given in Table 6.11. The majority filter increased the OA and κ values of all types of classification algorithms. Whereas the OA and κ values decreased after using a sieve filter, results for SVMs with a radial basis function were found to be unchanged after the postprocessing. Although the clump filter gave lower OA and κ values than the majority filter, it provided better results than the sieve filter for all classification algorithms. The equal outcomes were found using the majority filter and the combination of the majority and a sieve filter. However, the small decreases in the OA and κ were found using the combination of the majority and a sieve filter for the SVMs with sigmoid kernel.

## TABLE 6.11
## Classification Algorithms OA, κ Results before and after Postprocessing Steps

| Classification Algorithms | | No Post Processing | Majority Filter | Sieve Filter | Clump Filter | Majority + Sieve | Majority + Clump | Majority + Sieve + Clump |
|---|---|---|---|---|---|---|---|---|
| SVMs with linear kernel | OA | 87.63% | 91.46% | 87.59% | 88.26% | 91.46% | 92.10% | 92.14% |
| classification | κ | 0.8670 | 0.9083 | 0.8667 | 0.8740 | 0.9083 | 0.9151 | 0.9155 |
| SVMs with polynomial | OA | 87.51% | 91.43% | 87.48% | 88.15% | 91.43% | 92.51% | 92.58% |
| 2 kernel classification | κ | 0.8658 | 0.9079 | 0.8655 | 0.8727 | 0.9079 | 0.9195 | 0.9203 |
| SVMs with radial basis | OA | 87.89% | 91.76% | 87.89% | 88.67% | 91.76% | 92.47% | 92.51% |
| kernel classification | κ | 0.8698 | 0.9115 | 0.8698 | 0.8784 | 0.9115 | 0.9191 | 0.9195 |
| SVMs with sigmoid kernel | OA | 85.76% | 89.75% | 85.69% | 87.48% | 89.75% | 90.64% | 90.83% |
| classification | κ | 0.8470 | 0.8899 | 0.8463 | 0.8655 | 0.8899 | 0.8995 | 0.9015 |
| ANN classification | OA | 85.98% | 90.36% | 85.92% | 86.95% | 90.36% | 91.28% | 91.42% |
| | κ | 0.8496 | 0.8952 | 0.8489 | 0.8588 | 0.8952 | 0.9048 | 0.9094 |
| RF classification | OA | 86.81% | 90.76% | 86.76% | 87.89% | 90.76% | 91.62% | 91.73% |
| | κ | 0.8582 | 0.9006 | 0.8577 | 0.8695 | 0.9006 | 0.9094 | 0.9104 |

The small increments in the OA and κ were found using the combination of the majority, sieve, and clump filters in comparison to the combination of the majority and a clump filter. The large mixing of the other crop classes in the dominant crop may be the cause of a small decrease in the OA and κ using SVMs with a sigmoid function kernel. Obviously, image filtering improved the values not only for the OA and κ but also for the $TPR_i$ and $PPV_i$ for almost all classes. Nevertheless, unfiltered classification maps provide a clear distinction between the classes.

### 6.4.5 Analyses of Statistical Significance in the Classification Accuracy of Two Algorithms

The almost all the classification accuracy combinations gave Z values of more than 1.96 except for SVMs with a radial basis kernel versus SVMs with a linear kernel ($Z = 1.42$, $p = 0.1556$) and SVMs with a polynomial of degree 2 ($Z = 1.66$, $p = 0.0969$), which are insignificant results. Since Z values are less than the value of 1.96, we do not reject the null hypothesis of no difference. All other combinations were found to be as significantly different with different Z and p values. The significant increases in accuracy were found for mustard, linseed, pigeon pea, pea, and sugarcane crops. The statistical significance between the algorithms using a Z-test is presented in Table 6.12. The combinations using SVMs with radial basis kernel versus SVMs with linear kernel and SVMs with polynomial of degree 2 were significantly more accurate than the other combinations. The Z values were found to be less than 1.96 at the 95% confidence level for these two combinations.

## TABLE 6.12
## Statistical Significance in the Accuracy between Two Different Algorithms by Z-Test

| Classification 1 | Classification 2 | Z-Test | p Value |
|---|---|---|---|
| SVMs with radial basis kernel | SVMs with linear kernel | 1.42 | =0.1556 |
| SVMs with radial basis kernel | SVMs with polynomial of degree 2 | 1.66 | =0.0969 |
| SVMs with radial basis kernel | SVMs with sigmoid kernel | 4.08 | <0.0001 |
| SVMs with radial basis kernel | ANN classification | 4.02 | =0.0001 |
| SVMs with radial basis kernel | RF classification | 3.86 | =0.0001 |

## 6.5 CONCLUSION

The study shows that the kernel-based SVMs, ANN, and RF provided nearly similar results. The unfiltered classified maps provided fairly good crop classification. However, the results were much improved after filtering the classified maps, which eradicated the inclusion of other classes within the dominant class. The results were significant except for SVMs having a radial basis function versus SVMs having a linear kernel and a polynomial kernel of degree 2 by Z-test. The LISS-IV sensor was found to be useful for the discrimination of diverse crops such as corn, lentil, linseed, barley, mustard, pigeon pea, wheat, sugarcane, pea, and others and for noncrops using different supervised classification algorithms. The high spatial resolution of the LISS-IV sensor makes delineating boundaries among various crop fields easier. It also enabled the improvement of crop discrimination by avoiding the overlapping of land covers. Future investigations should focus on the use of more recent algorithms, perhaps by developing a new sensitive algorithm to address these problematic classes. The approaches and algorithms presented for crop discrimination and mapping can be useful for different crops grown in other regions.

## ACKNOWLEDGMENT

The authors thank Professor Rajeev Sangal, Director, Indian Institute of Technology (BHU), Varanasi, Uttar Pradesh, India, for financial support for acquiring the ENVI version 5.1software.

## REFERENCES

Akbari M, Mamanpoush AR, Gieske A, Miranzadeh M, Torabi M, Salem HR (2006). Crop and land cover classification in Iran using Landsat 7 imagery. *International Journal of Remote Sensing* 27: 4117–4135. doi:10.1080/01431160600784192.
Arora MK, Tiwari KC, Mohanty B (2000). Effect of neural network variables on image classification. *Asian Pacific Remote Sensing GIS Journal* 13: 1–11.
Atkinson PM, Tatnall ARL (1997). Introduction neural networks in remote sensing. *International Journal of Remote Sensing* 18: 699–709. doi:10.1080/014311697218700.
Breiman L (2001). Random forests. *Machine Learning* 45: 5–32. doi:10.1023/A:1010933404324.
Brown M, Lewis HG, Gunn SR (2000). Linear spectral mixture models and support vector machines for remote sensing. *IEEE Transactions on Geoscience and Remote Sensing* 38: 2346–2360.
Buechel SW, Philipson WR, Philpot WD (1989). The effects of a complex environment on crop separability with Landsat TM. *Remote Sensing of Environment* 27: 261–272.
Congalton RG, Green K (1999). *Assessing the Accuracy of Remotely Sensed Data: Principles and Practices.* Boca Raton, FL: Lewis Publishers.
Congalton RG, Oderwald RG, Mead RA (1983). Assessing Landsat classification accuracy using discrete multivariate statistical techniques. *Photogrammetric Engineering and Remote Sensing* 49: 1671–1678.
Conrad C, Fritsch S, Zeidler J, Rucker G, Dech S (2010). Per field irrigated crop classification in arid central Asia using SPOT and ASTER data. *Remote Sensing* 2: 1035–1056. doi:10.3390/rs2041035.
Dastorani MT, Moghadamnia A, Piri J, Rico-Ramirez M (2010). Application of ANN and ANFIS models for reconstructing missing flow data. *Environmental Monitoring and Assessment* 166: 421–434.
Foody GM, Arora MK (1997). An evaluation of some factors affecting the accuracy of classification by an artificial neural network. *International Journal of Remote Sensing* 18:799–810. doi:10.1080/014311697218764.
Foody GM, Mathur A (2006). The use of small training sets containing mixed pixels for accurate hard image classification: Training on mixed spectral responses for classification by a SVM. *Remote Sensing of Environment* 103: 179–189. doi:10.1016/j.rse.2006.04.001.
Gupta DK, Kumar P, Mishra VN, Prasad R, Dikshit PKS, Dwivedi SB, Ohri A, Singh RS, Srivastava V (2015). Bistatic measurements for the estimation of rice crop variables using artificial neural network. *Advances in Space Research* 55: 1613–1623.
Gupta DK, Prasad R, Kumar P, Mishra VN (2016). Estimation of crop variables using bistatic scatterometer data and artificial neural network trained by empirical models. *Computers and Electronics in Agriculture* 123: 64–73.
Han D, Chan L, Zhu N (2007). Flood forecasting using support vector machines. *Journal of Hydroinformatics* 9: 267–276.

Huang C, Davis LS, Townshend JRG (2002). An assessment of support vector machines for land cover classification. *International Journal of Remote Sensing* 23: 725–749.

Islam T, Rico-Ramirez MA, Han D, Srivastava PK (2012). Artificial intelligence techniques for clutter identification with polarimetric radar signatures. *Atmospheric Research* 109–110: 95–113.

ITT Industries Inc. (2006). The environment for visualizing images (ENVI), Version4.3. Boulder, CO: ITT Industries.

Jaccard P (1912). The distribution of the flora in the alpine zone. *New Phytologist* 11: 37–50.

Jensen JR (2005). *Introductory Digital Image Processing: A Remote Sensing Perspective*, 3rd ed. Upper Saddle River, NJ: Pearson Education.

Kanji GK (2006). *100 Statistical Tests*, 3rd ed. London: Sage Publications.

Kavzoglu T (2009). Increasing the accuracy of neural network classification using refined training data. *Environmental Modelling and Software* 24: 850–858.

Kavzoglu T, Mather PM (2003).The use of backpropagating artificial neural networks in land cover classification. *International Journal of Remote Sensing* 24: 4907–4938. doi:10.1080/0143116031000114851.

Koukoulas S, Blackburn GA (2001). Introducing new indices for accuracy evaluation of classified images representing semi-natural woodland environments. *Photogrammetric Engineering and Remote Sensing* 67: 499–510.

Kumar P, Gupta DK, Mishra VN, Prasad R (2015). Comparison of support vector machine, artificial neural network and spectral angle mapper algorithms for crop classification using LISS IV data. *International Journal of Remote Sensing* 36: 1604–1617. doi:10.1080/2150704X.2015.1019015.

Labatut V, Cherifi H (2011). Accuracy measures for the comparison of classifiers. In: Al-Dahoud Ali (Ed.), *The 5th International Conference on Information Technology*, May 2011, Amman, Jordan. Al-Zaytoonah University of Jordan, pp. 1,5.

Lillesand TM, Kiefer RW (1999). Remote Sensing and Image Interpretation. New York: Wiley.

Mas JF, Flores JJ (2008). The application of artificial neural networks to the analysis of remotely sensed data. *International Journal of Remote Sensing* 29: 617–663. doi:10.1080/01431160701352154.

Mishra VN, Rai PK (2016). A remote sensing aided multilayer perceptron-Markov chain analysis for land use and land cover change prediction in Patna district (Bihar), India. *Arabian Journal of Geosciences* 9(4): 249.

Mountrakis G, Im J, Ogole C (2011). Support vector machines in remote sensing: A review. *ISPRS Journal of Photogrammetry and Remote Sensing* 66: 247–259. doi:10.1016/j.isprsjprs.2010.11.001.

Nidamanuri RR, Zbell B (2012). Existence of characteristic spectral signatures for agricultural crops–Potential for automated crop mapping by hyperspectral imaging. *Geocarto International* 27: 103–118. doi:10.1080/10106049.2011.623792.

Pal M (2005). Random forest classifier for remote sensing classification. *International Journal of Remote Sensing* 26: 217–222. doi:10.1080/01431160412331269698.

Pal M (2009). Kernel methods in remote sensing: A review. *ISH Journal of Hydraulic Engineering* 15: 194–215.

Paola JD, Schowengerdt RA (1995). A review and analysis of backpropagation neural networks for classification of remotely-sensed multi-spectral imagery. *International Journal of Remote Sensing* 16: 3033–3058.

Petropoulos GP, Kontoes C, Keramitsoglou I (2011). Burnt area delineation from a uni-temporal perspective based on Landsat TM imagery classification using support vector machines. *International Journal of Applied Earth Observation and Geoinformation* 13: 70–80.

Pouteau R, Collin A (2013). Spatial location and ecological content of support vectors in an SVM classification of tropical vegetation. *Remote Sensing Letters* 4: 686–695.

Richards JA (1999). *Remote Sensing Digital Image Analysis*. Berlin, Germany: Springer-Verlag.

Rodriguez-Galiano VF, Ghimire B, Rogan J, Chica-Olmo M, Rigol-Sanchez JP (2012). An assessment of the effectiveness of a random forest classifier for land-cover classification. *ISPRS Journal of Photogrammetry and Remote Sensing* 67: 93–104.

Ryerson RA, Curran PJ, Stephens PR (1997). Agriculture. In: Philipson W.R. (Ed.), Manual of photographic interpretation., 2nd ed. American Society for Photogrammetry and Remote Sensing, Bethesda, MD, pp. 365–397.

Schowengerdt RA (2006). *Remote Sensing: Models and Methods for Image Processing*, 2nd ed. Burlington, NJ: Elsevier, pp. 414–416.

Shao Y, Lunetta R (2012). Comparison of support vector machine, neural network, and CART algorithms for the land-cover classification using limited training data points. *ISPRS Journal of Photogrammetry and Remote Sensing* 70: 78–87.

Song C, Woodcock CE, Seto KC, Lenney MP, Macomber SA (2001). Classification and change detection using Landsat TM data: When and how to correct atmospheric effects? *Remote Sensing of Environment* 75: 230–244.

Sonobe R, Tani H, Wang X, Kobayashi N, Shimamura H (2014). Random forest classification of crop type using multi-temporal TerraSAR-X dual-polarimetric data. *Remote Sensing Letters* 5: 157–164. doi:10.1080/2150704X.2014.889863.

Srivastava PK, Han D, Rico-Ramirez MA, Bray M, Islam T (2012). Selection of classification techniques for land use/land cover change investigation. *Advances in Space Research* 50: 1250–1265.

Swain PH, Davis SM (1978). *Remote Sensing: The Quantitative Approach*. New York: McGraw Hill.

Turker M, Arikan M (2005). Sequential masking classification of multi-temporal Landsat7 ETM+ images for field-based crop mapping in Karacabey, Turkey. *International Journal of Remote Sensing* 26: 3813–3830. doi:10.1080/01431160500166391.

Turker M, Ozdarici A (2011). Field based crop classification using SPOT4, SPOT5, IKONOS and QuickBird imagery for agricultural areas: A comparison study. *International Journal of Remote Sensing* 32: 9735–9768. doi:10.1080/01431161.2011.576710.

Vapnik VN (1998). *Statistical Learning Theory*. New York: Wiley.

Wheeler SG, Misra PN (1980). Crop classification with Landsat multispectral scannerdata II. *Pattern Recognition* 12: 219–228.

WitDe AJW, Clevers JGPW (2004). Efficiency and accuracy of per field classification for operational crop mapping. *International Journal of Remote Sensing* 25: 4091–4112. doi:10.1080/0143116031000161958.

Witten IH, Frank E (2005). *Data Mining: Practical Machine Learning Tools and Techniques*, 2nd ed. Burlington, MA: Morgan Kaufmann.

Yang C, Everitt JH, Fletcher RS, Murden D (2007). Using high resolution Quick Bird imagery for crop identification and area estimation. *Geocarto International* 22: 219–233. doi:10.1080/10106040701204412.

Yang C, Everitt JH, Murden D (2011). Evaluating high resolution SPOT 5 satellite imagery for crop identification. *Computers and Electronics in Agriculture* 75: 347–354. doi:10.1016/j.compag.2010.12.012.

# 7 Urban Growth and Management in Lucknow City, the Capital of Uttar Pradesh

*Akanksha Balha and Chander Kumar Singh*

## CONTENTS

7.1 Introduction ................................................................................................................. 109
7.2 Urbanization as a Phenomenon .................................................................................. 110
7.3 Scenario of Urbanization in Developing and Developed Nations ............................. 110
7.4 Linkage between Urban Growth, Ecology, and Growth Management ..................... 111
7.5 Application of Remote Sensing, Geographic Information System, and Spatial Metrics in Urban Growth Studies .......................................................................................... 112
7.6 Case Study of Lucknow ............................................................................................. 115
    7.6.1 Study Area ....................................................................................................... 115
    7.6.2 Data and Software Used ................................................................................... 115
    7.6.3 Methodology .................................................................................................... 115
        7.6.3.1 Preparation of Land Use Land Cover of Lucknow District ............... 115
        7.6.3.2 Analysis of Urban Sprawl .................................................................. 116
    7.6.4 Analysis and Results ........................................................................................ 118
    7.6.5 Conclusion ....................................................................................................... 119
References ............................................................................................................................ 120

**ABSTRACT** Urbanization or urban expansion is a long-term phenomenon driven by many socioeconomic factors like population growth, migration, infrastructural development, and so on. Urban expansion, which is believed to be caused by economic growth as one of the factors, results in various disparities, for example, in terms of fragmented growth and lack of facilities at outskirt expanded regions. Urban growth management plays a pivotal role in synchronizing the increasing urbanization with the overall development of the area. Urban growth management is an essential part of urban planning because besides causing unplanned urban expansion, rapid and haphazard urbanization leads to a disruption of ecosystem functioning also. It causes deforestation, reduced infiltration, increased soil erosion, and many other environmental problems. The advancement of remote sensing and GIS in recent decades has made it easier for researchers to carry out urban growth studies. Having a huge significance on demography and environment, urban growth and its management has been studied widely in developing as well as developed nations. A case study of Lucknow district situated in Uttar Pradesh state of India is analyzed in this chapter. The change in urban growth in the study area over the period of 2000–2017 has been characteristically examined.

## 7.1 INTRODUCTION

Urbanization, also referred as *urban expansion* (Sudhira et al., 2004) or *urban sprawl* (Sudhira et al., 2004) is a continuously growing phenomenon (Spence et al., 2009). It is caused mainly due to population growth, large-scale immigration, rise in economy, and infrastructural plans. It is characterized by dispersed growth of built-up areas at the outskirts of cities due to lack of planning (Bhat et al., 2015). Such areas lack basic amenities (Hasse and Lathrop, 2003; Ji et al., 2006; Ramachandra et al. (2012a)) and infrastructural facilities, for example, sanitation, drinking water, health, and electricity. Therefore, to provide proper services and facilities to these areas, management of urban growth and planning becomes necessary.

Urban sprawl results into land fragmentation, causing the alteration of the ecosystem's structure and functioning (Bhat et al., 2015). Hence, in addition to catering to the needs of the people of the sprawl areas, urban planning and management also need to care for the ecology that is degraded due to unplanned and haphazard urban development. This chapter mentions the way urban areas expand, how urban growth affects the ecology of an area, and the capabilities of remote sensing, geographic information system (GIS) and spatial metrics in analyzing and measuring urban dynamics.

## 7.2 URBANIZATION AS A PHENOMENON

In urban studies, urban area and urban (growth) center have different meanings (Ranpise et al., 2016). *Urban land* is defined by the population density and the occupation of the people and their lifestyles (Bryant, 2003; Tsai, 2005; Weeks, 2010) whereas *urban center* is characterized by the business or economic activities (Dobbs and Remes, 2012; Kotharkar et al., 2014; Cortright, 2015). Urban centers lie mainly near post offices or market zones (Ranpise et al., 2016).

The rapidly and continuously expanding characteristic of urbanization leads to suburbanization (Champion, 2001), a phenomenon that is common in developing areas (Ranpise et al., 2016) and is known to result into shifting the urban growth center (Zhou, 1997; Hermelin, 2007) over a period of time. Initially, the process begins with the rise in population on the fringes of urban areas, which further takes the form of decentralization and then moves backward, terming it as *counterurbanization* (Zhou, 1997; Grigorescu et al., 2012). Various urban growth studies (Brockerhoff, 1999; Guest and Brown, 2005; Tran et al., 2008) have revealed that sub-urbanization resulting in the shifting of urban centers is a common phenomenon in developing nations, for example, in Brazil, South Africa, Southeast Asia, and India. However, there is a huge dearth of any scientific assessment or documentation on shifting of urban centers (Ranpise et al., 2016). The shift of urban center can be understood by studying the change in population density and economic activities at the fringes of urban land (Zhou, 1997) temporally.

## 7.3 SCENARIO OF URBANIZATION IN DEVELOPING AND DEVELOPED NATIONS

Every country across the globe has witnessed urbanization (Sudhira et al., 2004) due to the increase in population, upgrade in economy, higher immigration rate, and rise in infrastructural facilities (Ranpise et al., 2016). However, urban growth studies suggest different scenarios of urbanization as a process in developed and developing countries. In developing nations, rapid urban expansion leads to increase in the urban population, which poses various challenges for urban planners and policy-makers (Yuan et al., 2005; Pacione, 2007). The challenges include lack of services available to the growing urban population, increasing rural-urban migration, proliferation of informal settlements, and epidemics and environmental degradation (Rakodi, 1995; Brown, 2001). On the other hand, in developed countries, it is observed that urbanization is more likely to promote national economic growth (Collier, 2016). Evidently, developing nations should consider and adapt this idea from their developed nations' counterparts to work on sustainable and smart urban development that can lead to economic growth and socioeconomic development. The study of present and future urban growth patterns with the help of accurate and timely spatial information of urban growth can play a major role to achieve smart and sustainable urban development (Herold et al., 2002). Conclusively, geospatial information regarding urbanization will be helpful at the planning and policy-making level to promote smart and equitable urban, environmental, and economic development (United Nations, 2015).

Kamusoko (2017) has pointed out that although many urban growth studies have been performed in developed nations (Batty et al., 1999; Barnes et al., 2001; Epstein et al., 2002), not much significance has been given to quantify urban landscapes in developing countries regardless of their rapid urbanization. One of the reasons for the lack of such studies in developing countries is the high cost involved in acquiring the data, mainly from land surveys and aerial photography, the source that aids in maintaining accurate and timely geospatial information (Conitz, 2000). Nonetheless, advancement in the remote sensing technology in the form of high- and medium-resolution satellite data (Ikonos, Quickbird, WorldView, Landsat satellite series, etc.) with good spatial coverage has boosted the mapping of urban land use/land cover (LULC) (Ward et al., 2000; Guindon et al., 2004; Yuan et al., 2005; Lu and Weng, 2007). A few of the urban growth studies conducted in developing countries particularly in China and India include Yeh and Li, 2001; Cheng and Masser, 2003; Jothimani, 1997; Lata et al., 2001; Sudhira et al., 2003.

## 7.4 LINKAGE BETWEEN URBAN GROWTH, ECOLOGY, AND GROWTH MANAGEMENT

To cope with the situation by providing natural resources such as safe and clean water and basic amenities such as sanitation and electricity, urban planners and experts require in-depth study and analysis of urban sprawl and its growth rate and pattern. Not doing so leads to the lack of basic amenities in newly urbanized areas and leads to their degradation (Sudhira et al., 2004).

Urban development has always been accompanied with changes in the surrounding landscape, which ultimately alters the structure and functioning of the ecosystem of the region (Mitsova et al., 2011). Consequently, it also impacts the ecosystem services and disrupts the benefits arising from them (Costanza et al., 1997), for example, reduced soil erosion, groundwater recharge, infiltration of storm water run off, and aesthetic and recreational values (Osborne and Kovacic, 1993; Dosskey, 2001; Randolph, 2004). Alarm regarding this has caused planners to put more efforts into protecting natural areas in urban environment (Mitsova et al., 2011).

Although the situation was realized quite late only after the destruction of many green areas, decision-makers who initiated the planning and management led to smart growth with an underlying idea of green infrastructure. It aims to reduce the negative effects of urban sprawl by protecting and preserving open spaces, agricultural lands, and ecosensitive areas (Smart Growth Network, 1996). With this approach, green infrastructure will be able to cope with the urban development pressure (Mitsova et al., 2011). The approach recognizes the urgent need for regulations and policies at every level ranging from regional to global to protect open, degraded, and vulnerable spaces.

However, any growth management approach needs to be comprehended carefully before implementation because contrasting observations with respect to growth management and urban expansion have been noted. Some studies suggest that growth management reduces the outward extension of urban areas, whereas others argue that growth management has no impact on urban expansion (Bhagat, 2011). Few studies, however, reveal a direct relationship between growth management and outward expansions of urban areas. Deriving a relationship between growth management and urban expansion is crucial because it impacts land use policies; however, difference in opinion on such crucial element exists because of the lack of data availability, difference in statistical methods, or models of urban spatial expansion.

That is why the authorities who have at their disposal various planning approaches like wetland restoration, special area zoning, transferable development rights, and outreach programs (USEPA, 2005) to construct, restore, and maintain green infrastructure must refer to scientific research and tools to identify critical areas to be conserved prior to making regulations and policies (Mitsova et al., 2011). Also, for comprehensive smart urban planning, various aspects like present and future environmental regulations, different land use scenarios, and green infrastructure need to be taken into account. In addition to this, it has been suggested that to deal with the cumulative effects caused by development, conservation at a site level alone does not help much, and therefore community-level approaches need to be adapted along with local conservation regulations and practices (Arendt, 2004).

## 7.5 APPLICATION OF REMOTE SENSING, GEOGRAPHIC INFORMATION SYSTEM, AND SPATIAL METRICS IN URBAN GROWTH STUDIES

As mentioned in the previous sections, before making policies and regulations, urban land-use policy-makers need to incorporate a scientific approach in their decision-making procedures to have an in-depth understanding of the urban dynamics and to identify the critical areas from urban development to be conserved. The scientific approach can be attained with the utilization of remote sensing and GIS technology.

Remote sensing technology aids in the acquisition of high- and medium-resolution satellite images that help in mapping and analyzing urban growth in developed countries (Lo and Choi, 2004; Yuan et al., 2005; Bagan and Yamagata, 2012). Various classification approaches like segmentation and object-based classifications (Guindon et al., 2004), neural networks (Seto and Liu, 2003), hybrid methods that incorporate soft and hard classifications (Lo and Choi, 2004), expert systems (Stefanov et al., 2001), support vector machines (Ghosh et al., 2014), and random forests (Cao et al., 2009) have been used to prepare urban LULC.

Although it has been observed that sometimes medium- or low-resolution satellite data have coarse spatial resolution, which is a hindrance to performing detailed analysis of urban landscapes, the data can surely be useful in mapping basic land-use cover types in urban areas (Kamusoko, 2017). Because urban dynamics is a phenomenon that varies spatially as well as temporally, it should be evaluated using both the aspects. Satellite data of different dates help in assessing temporal urban dynamics (Bharath et al., 2012) in wide spatial coverage and a cost-effective manner (Yang and Liu, 2005). Such data help in modeling and analyzing the growing pattern and extent of the sprawl over a given period of time (Sudhira et al., 2004). With the inclusion of parameters considered to be driving forces, the data even help in predicting future urban sprawl for studying future spatial patterns and trends for land use as changing urban sprawl brings changes in land use land cover. Often, growing urban expansion is analyzed using socioeconomic indicators that could not be explained in a spatial context (Ji et al., 2006).

Numerous studies have quantified urban sprawl with the help of spatial data (Sudhira et al., 2004; Bhatta, 2009; Taubenbock et al., 2009; Bhatta et al., 2010; Ramachandra et al., 2012b), methods based on land-use classification (Herold et al., 2002; Sudhira et al., 2004; Yang and Liu, 2005; Terzi and Bolen, 2009; Bharath et al., 2012; Ramachandra et al., 2013). Landscape or spatial metrics, measures for ecological quantifications, have recently started to be used by many researchers Sudhira et al., 2004; Bhatta, 2009; Ramachandra et al., 2013 to understand urban landscape and the urban growth phenomenon. Ramachandra et al. (2012a) used the density gradient metric approach (constituting sprawl density, population densities, etc.) to understand urban dynamics (Bhat et al., 2015). A brief description of the spatial metrics used in the case study mentioned shortly is given in Table 7.1.

Thus, remote sensing and GIS technology along with spatial metrics have been able to help researchers and others understand the structure, function, and change in the urban region at the landscape level (ICIMOD, 1999; Civco et al., 2002).

In the next section, the chapter presents a case study of a metropolitan city in India, describing the urban sprawl and various urban structures and forms considered in the study area.

## TABLE 7.1
## Brief Description of the Spatial Metrics Used in the Study

| Spatial Metric | Metrics Group Category | Formula | Description | Units | Range |
|---|---|---|---|---|---|
| Edge Density (ED) | Area and edge metrics | $ED = \dfrac{\sum_{k=1}^{m} e_{ik}}{A}(10{,}000)$<br><br>where:<br>$e_{ik}$ is the total length (m) of edge in landscape involving patch type (class) $i$; includes landscape boundary and background segments involving patch type $i$.<br>$A$ is the total landscape area (m²). | ED equals the sum of the lengths (m) of all edge segments involving the corresponding patch type divided by the total landscape area (m²), multiplied by 10,000 (to convert to hectares). | Meters per hectare | ED ≥ 0, without limit |
| Largest Patch Index (LPI) | Area and edge metrics | $LP = \dfrac{\max(a_{ij})\Big|_{j=1}^{a}}{A}(100)$<br><br>where:<br>$a_{ij}$ is the area (m²) of patch $ij$,<br>$A$ is the total landscape area (m²). | LPI equals the area (m²) of the largest patch of the corresponding patch type divided by total landscape area (m²), multiplied by 100 to convert to a percentage. It is a simple measure of dominance. | Percentage | 0 < LPI ≦ 100 |
| Perimeter-Area Fractal Dimension (PAFRAC) | Shape metrics | $PAFRAC = \dfrac{2}{\left[\dfrac{n_i\left[\sum_{j=1}^{n} \ln p_{ij} \cdot \ln a_{ij}\right] - \left(\sum_{j=1}^{n} \ln p_{ij}\right)\left(\sum_{j=1}^{n} \ln a_{ij}\right)}{\left(n_i \sum_{j=1}^{n} \ln p_{ij}^2\right) - \left(\sum_{j=1}^{n} \ln p_{ij}\right)^2}\right]}$<br><br>where:<br>$a_{ij}$ is the area (m²) of patch $ij$;<br>$p_{ij}$ is the perimeter (m) of patch $ij$;<br>$n_i$ is the number of patches in the landscape of patch type (class) $i$. | PAFRAC equals 2 divided by the slope of the regression line obtained by regressing the logarithm of patch area (m²) against the logarithm of patch perimeter (m). That is, 2 divided by the coefficient b1 derived from a least squares regression fit to the following equation: ln(area) = b0 + b1ln(perim). | None | 1 ≦ PAFRAC ≦ 2 |

*(Continued)*

## TABLE 7.1 (Continued)
## Brief Description of the Spatial Metrics Used in the Study

| Spatial Metric | Metrics Group Category | Formula | Description | Units | Range |
|---|---|---|---|---|---|
| Number of Patches (NP) | Aggregation metrics | $NP = n_i$<br>where:<br>$n_i$ is the number of patches in the landscape of patch type (class) $i$. | NP equals the number of patches of the corresponding patch type (class). | None | $NP \geq 1$, without limit |
| Percentage of Like Adjacencies (PLADJ) | Aggregation metrics | $$PLADJ = \frac{g_{ii}}{\sum_{k=1}^{m} g_{ik}}$$<br>where:<br>$g_{ii}$ is the number of like adjacencies (joins) between pixels of patch type (class) $i$ based on the double-count method.<br>$g_{ik}$ is the number of adjacencies (joins) between pixels of patch types (classes) $i$ and $k$ based on the double-count method. | PLADJ equals the number of like adjacencies involving the focal class divided by the total number of cell adjacencies involving the focal class; multiplied by 100 (to convert to a percentage). | Percentage | $0 \leq PLADJ \leq 100$ |
| Euclidean Nearest-Neighbor Distance (ENN) | Aggregation metrics | $ENN = h_{ij}$<br>where:<br>$h_{ij}$ is the distance (m) from patch $ij$ to nearest neighboring patch of the same type (class) based on patch edge-to-edge distance computed from cell center to cell center. | It is the distance (m) to the nearest neighboring patch of the same type based on shortest edge-to-edge distance. Note that the edge-to-edge distances are from cell center to cell center. | Meters | $ENN > 0$, without limit. |

## 7.6 CASE STUDY OF LUCKNOW

### 7.6.1 Study Area

The study area for the present study is the Lucknow district (Figure 7.1) situated in Uttar Pradesh state of India. Geographically, the district is located at 26.78 N and 80.89 E at 123 m above mean sea level and covers an area of 2528 km$^2$.

The Lucknow district is situated in the middle of the Indus-Gangetic plain with Gomti River as its chief geographical feature. The river flows through the district dividing it into trans-Gomti and Cis-Gomti regions. In the middle of the district lies Lucknow city, one of the fastest developing cities and capital of the fourth largest and the most populated state of the country, Uttar Pradesh. Lucknow is the eleventh most populous city and twelfth most populous urban agglomeration of the country. Lucknow possesses a humid subtropical climate with cool, dry winters and dry, hot summers. It receives an average rainfall of 896.2 mm.

### 7.6.2 Data and Software Used

In the present study, high-resolution satellite images of LISS IV and LISS III merged for 2017 and 2007, respectively, and Landsat 7 ETM+ (resampled at 5 m) for year 2000 were used. ERDAS IMAGINE 2014 was used for preprocessing the satellite images and for preparing LULC of Lucknow district for different years. FRAGSTATS v 4.2.1 (McGarigal, 2002) was used to quantify and analyze different urban structures in the study area on the basis of different spatial metrics. FRAGSTATS v4.2.1 is a spatial pattern analysis program that computes various statistical metrics (or parameters) for each patch and class (patch type) in the landscape and for the landscape as a whole.

### 7.6.3 Methodology

#### 7.6.3.1 Preparation of Land Use Land Cover of Lucknow District

The satellite images of previously mentioned specifications were preprocessed and used to prepare LULC of Lucknow for 2000, 2007, and 2017. A maximum likelihood algorithm of supervised classification was used to prepare the LULC. Seven LULC classes were identified: water, built up, dense vegetation, sparse vegetation/plantations, cropland, fallow land, and open land (Table 7.2).

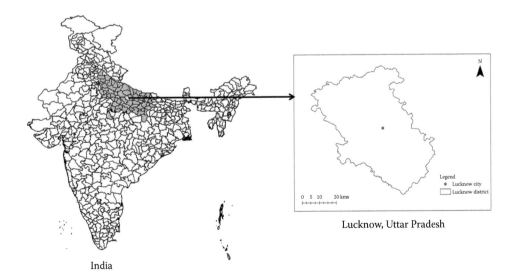

**FIGURE 7.1** Study area.

## TABLE 7.2
### Description of the Land Use/Land Cover Classes Identified in the Study Area, Mentioning the Different Features

| S. No | LULC Class | Description |
|---|---|---|
| 1. | Water | River, canals, ponds, and other natural or man-made water bodies |
| 2. | Built up | Residential, commercial, and industrial area, public facilities zone, transportation area, and other impervious land |
| 3. | Dense vegetation | Forest area and densely placed tree areas |
| 4. | Sparse vegetation | Urban tree canopy, green spaces (e.g., parks, lawns, golf courses; includes plantations) |
| 5. | Cropland | Agricultural land having crops and other growing items |
| 6. | Fallow land | Harvested agricultural land |
| 7. | Open land | Open nongreen area, barren land; includes unusable land |

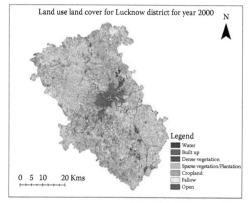
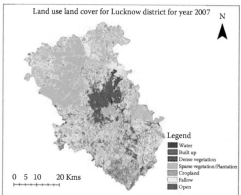
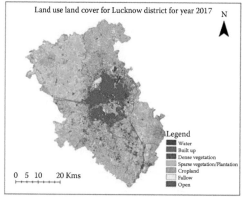

**FIGURE 7.2** Land use land cover (LULC) maps for Lucknow district for years 2000, 2007, and 2017.

LULC maps of the Lucknow district for different years can be viewed in Figure 7.2. Accuracy assessment of all the LULC images have been done with accuracy of more than 85% in all the images achieved.

### 7.6.3.2 Analysis of Urban Sprawl

The maps of the built-up area for years 2000, 2007 and 2017 were extracted using LULC maps of the respective years. These were used to analyze urban expansion in the study area (Figure 7.3). The area covered (in km$^2$) by built-up and other LULC classes in the study area is presented in Table 7.3.

# Urban Growth and Management in Lucknow City, the Capital of Uttar Pradesh

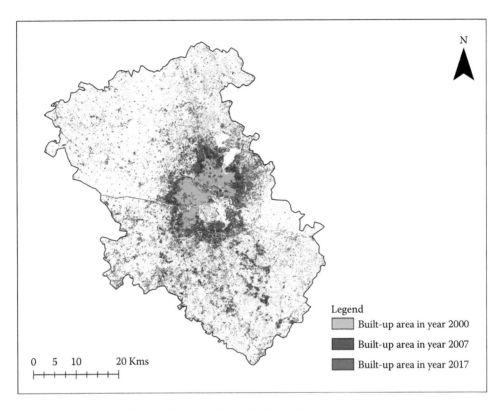

**FIGURE 7.3** Increasing built-up (urban) area from 2000 to 2017.

**TABLE 7.3**
**Area (in km²) Covered by Different Land Use/Land Cover Classes in Lucknow District over the Period of 2000–2017**

| S. No | LULC Class | 2000 | 2007 | 2017 |
| --- | --- | --- | --- | --- |
| 1. | Water | 11.86 | 15.12 | 15.13 |
| 2. | Built up | 304.94 | 326.07 | 552.19 |
| 3. | Dense vegetation | 114.93 | 107.53 | 97.64 |
| 4. | Sparse vegetation/ Plantation | 452.43 | 610.43 | 423.83 |
| 5. | Cropland | 1123.10 | 445.56 | 986.50 |
| 6. | Fallow land | 261.35 | 916.5 | 354.71 |
| 7. | Open land | 259.39 | 106.79 | 98.00 |

In the present study, FRAGSTATS v4.2.1 was used at the class level to compute different spatial metrics to quantify the structure of the built-up class in the study area. It has helped in analyzing the spatial heterogeneity of the class. The temporal variations of the spatial metrics computed for the built-up class are shown in the Figure 7.4. The variations in metrics measure urbanization in the study area temporally.

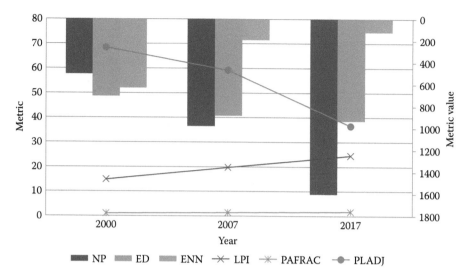

**FIGURE 7.4** Variation of spatial metrics derived using FRAGSTATS for built-up class over the period of time. (*Note: Spatial metrics shown by the line chart type refer to primary axis, and metrics shown by the bar chart type refer to the secondary axis.*)

### 7.6.4 Analysis and Results

The change detection matrix (Table 7.4) provides data for the inter-land-use class area conversion (km²) between the years 2000 and 2017. It shows the to-and-fro changes among the land-use class area between the two years. In the matrix, any given value shows how much area was converted from one land-use class (column) to another (row). The values in the diagonal row show the area that remains unchanged in a class. The last row and column of the matrix show the total class area in the study area in the earlier (2000) and later (2017) years, respectively. The total class area values of the matrix for both the years can be confirmed with data given in Table 7.3. Hence, in addition to only the class area, the change detection matrix also signifies the contribution of different classes in the conversion of one particular class to another. In the present study, Table 7.4 reveals that from 2000 to

**TABLE 7.4**
**Change Detection in Area (km²) between Land Use/Land Cover Classes from 2000 to 2017**

| LULC Classes | Water | Built Up | Dense Vegetation | Sparse Vegetation | Cropland | Fallow Land | Open Land | |
|---|---|---|---|---|---|---|---|---|
| Water | 11.86 | 0 | 0 | 0.07 | 1.11 | 1.67 | 0.42 | 15.13 |
| Built up | 0 | 304.94 | 0 | 62.15 | 24.02 | 75.61 | 85.47 | 552.19 |
| Dense vegetation | 0 | 0 | 97.64 | 0 | 0 | 0 | 0 | 97.64 |
| Sparse vegetation | 0 | 0 | 11.02 | 227.07 | 115.70 | 9.65 | 60.39 | 423.83 |
| Cropland | 0 | 0 | 2.24 | 94.17 | 862.18 | 27.91 | 0 | 986.5 |
| Fallow land | 0 | 0 | 4.03 | 57.90 | 111.42 | 111.28 | 70.08 | 354.71 |
| Open land | 0 | 0 | 0 | 11.07 | 8.67 | 35.23 | 43.03 | 98.00 |
| | 11.86 | 304.94 | 114.93 | 452.43 | 1123.10 | 261.35 | 259.39 | 2528 |

2017, open land and fallow land were the significant contributors to urban expansion. Areal statistics in Table 7.3 clearly reveal the extent of the drastically expanding urban in the study area from 2000 to 2017. Figure 7.3 displays the built-up expansion and its dispersed growth over the years. Dense vegetation in the study area declined due to its conversion to sparse vegetation, fallow land, and agricultural land. Over the period of years, sparse vegetation was converted to agricultural land (cropland and fallow land) and open land.

In addition to the LULC class area and interclass conversion, we looked into the quantification of the structure of urban class also in this study. Hence, spatial metrics mentioned in Table 7.1 were calculated at class level to measure the built-up area in the Lucknow district using different aspects of landscape, for example, area and edge, shape, aggregation, and so on. Figure 7.4 clearly reflects the increase in NP, symbolizing the number of urban patches. It reveals that the number of patches of urban area increased over the period of time, that is, the rate of urbanization is increasing with time. Similarly, LPI, indicating the largest patch index of urban area, shows the highest value in 2017, indicating that the dominance of urban area has continuously increased. Increasing edge density (ED) also points to the increasing size of the urban patches and therefore to relatively more dominance of urban areas in the landscape over the years. The increasing values of perimeter-area fractal dimension (PAFRAC) imply that the pattern of urban areas has become more regular with the time period. This may imply that earlier fragmented small urban patches expanded over the years and reached comparatively regular shapes. The decreasing percentage of like adjacencies (PLADJ) values show that in earlier years, the urban areas were more compact and interior, whereas over the years, the urbanization led to fragmented dispersed growth. The same can be seen in Figure 7.3, where the green patches representing the year 2000 are denser and close to each other in comparison with blue patches in 2010 and red patches in 2017. Red patches seem to be most fragmented among all the three. Euclidean nearest-neighbor distance (ENN), indicating the minimum distance between patches, showed a decreasing trend which means that the distance between the urban patches decreased over the years. It also implies upward trend in urbanization leading to the dominance of urban areas and thus, the distance between its patches tend to decline.

Hence, the spatial metrics help us to conclusively measure the spatial heterogeneity of urban growth in the study area that over the years; the number as well as the size of the built up patches have increased, leading to the decreased distance between the patches. With increasing patch size, their shapes have become more regular. Also, the growth has been more fragmented and dispersed with time.

### 7.6.5 Conclusion

In addition to presenting a brief overview on urban growth and its management, the chapter has also focused on an analysis of an urban landscape as a case study. The analysis is intended to help in understanding the application of various tools such as remote sensing, GIS, and spatial metrics in urban growth studies. The present work has analyzed urban growth in the Lucknow district over a span of almost two decades. Increasing urban expansion was the one characteristic feature for the temporal study; the use of a change detection matrix in analysis has provided more insight into the changing pattern of LULC in general and urban or built-up area in particular. FRAGSTATS played an interesting role by analyzing the urban growth patterns using spatial metrics or parameters. Also, as depicted by Figure 7.4, the interpretation of the results derived from FRAGSTATS seems agree with the urban spatial extent derived from LULC maps. The analysis is intended to be helpful to urban planners and decision makers because spatial metrics provide the spatial details (area, size, aggregation, etc.) of the urban patches so that suitable development plans and rules can be made.

Therefore, we recommend the fusion of spatial metrics and remote sensing-GIS to explore more dimensions in urban studies to develop an in-depth understanding of urban landscape.

## REFERENCES

Arendt, R. 2004. Linked landscapes: Creating greenway corridors through conservation subdivision design strategies in the northeastern and central United States. *Landscape Urban Plan*, 68: 241–269.

Bagan, H., and Yamagata, Y. 2012. Landsat analysis of urban growth: How Tokyo became the world's largest megacity during the last 40 years. *Remote Sensing of Environment*, 127: 210–222.

Barnes, K.B., Morgan III, J.M., Roberge, M.C., and Lowe, S. 2001. Sprawl development: Its patterns, consequences, and measurement. Towson University, Towson, MD. Retrieved from http://www.chesapeake.towson.edu/landscape/urbansprawl/download/Sprawl white paper.pdf.

Batty, M., Xie, Y., and Sun, Z. 1999. The dynamics of urban sprawl. Working Paper Series, Paper 15, Centre for Advanced Spatial Analysis, University College, London. Retrieved from http://www.casa.ac.uk/working papers/.

Bhagat, R. B. 2011. Emerging pattern of urbanisation in India. *Economic and Political Weekly*, 10–12.

Bharath, H.A., Bharath, S., Sreekantha, S., Sannadurgappa, D., and Ramachandra, T. V. 2012. Spatial patterns of urbanization in Mysore: Emerging Tier II City in Karnataka. *Online Proceedings of National Remote Sensing Users Meet*, NRSC, Hyderabad, India, February 16–17, 2012.

Bhat, V., Aithal, B.H., and Ramachandra, T.V. 2015. Spatial patterns of urban growth with globalisation in India's silicon valley. *Organized By Department of Civil Engineering, Indian Institute of Technology (Banaras Hindu University)*, Varanasi-221005 Uttar Pradesh, India, 98.

Bhatta, B. 2009. Analysis of urban growth pattern using remote sensing and GIS: A case study of Kolkata, India. *International Journal of Remote Sensing*, 30(18): 4733–4746.

Bhatta, B., Saraswati, S., and Bandyopadhyay, D. 2010. Urban sprawl measurement from remote sensing data. *Applied Geography*, 30: 731–740.

Brockerhoff, M. 1999. Urban growth in developing countries: A review of projections and predictions. *Population and Development Review*, 25: 757–778.

Brown, A. 2001. Cities for the urban poor in Zimbabwe: Urban space as a resource for sustainable development. *Development of Practice*, 11: 263–281.

Bryant, C. 2003. The impact of urbanization on rural land us. *Encyclopedia of Life Support System*. Retrieved from http://www.eolss.net.

Cao, X., Chen, J., Imura, H., and Higashi, O. 2009. A SVM-based method to extract urban areas from DMSP-OLS and SPOT VGT data. *Remote Sensing of Environment*, 113: 2205–2209.

Champion, T. 2001. Urbanization, suburbanization, counter-urbanization and re-urbanization. In R. Paddison (Ed.), *Handbook of Urban Studies*, pp. 143–160. doi:10.4135/9781848608375.n9.

Cheng, J., and Masser, I., 2003. Urban growth pattern modelling: A case study of Wuhan City, PR China. *Landscape Urban Plan*, 62: 199–217.

Civco, D.L., Hurd, J.D., Wilson, E.H., Arnold, C.L., and Prisloe, M., 2002. Quantifying and describing urbanizing landscapes in the Northeast United States. *Photogrammetric Engineering and Remote Sensing*, 68 (10): 1083–1090.

Collier, P. 2016. African urbanization: An analytic policy guide. Accessed February 26, 2016, from http://www.theigc.org/wp-content/uploads/2016/01/African-UrbanizationJan2016_Collier_ Formatted-1.pdf.

Conitz, M.W. 2000. GIS applications in Africa: Introduction. *Photogrammetric Engineering and Remote Sensing*, 66: 672–673.

Cortright, J. 2015. City report. Retrieved from http://cityobservatory.org.

Costanza, R., d'Arge, R., de Groot, R., Farber, S., Grasso, M., Hannon, B., Limburg, K., et al., 1997. The value of the world's ecosystem services and natural capital. *Nature* 387: 253–260.

Dobbs, R., and Remes, J. 2012. Trends. The shifting urban economic landscape: What does it mean for cities? Retrieved from http://siteresources.worldbank.org.

Dosskey, M. 2001. Toward quantifying water pollution abatement in response to installing buffers on crop land. *Journal of Environmental Management*, 28(5): 577–598.

Epstein, J., Payne, K., and Kramer, E. 2002. Techniques for mapping suburban sprawl. *Photogrammetric Engineering and Remote Sensing*, 63(9): 913–918.

Ghosh, A., Sharma, R., and Joshi P. K. 2014. Random forest classification of urban landscape using Landsat archive and ancillary data: Combining seasonal maps with decision level fusion. *Applied Geography*, 48: 31–41.

Grigorescu, I., Mitrica, B., Mocanu, I., and Ticana, N. 2012. Urban sprawl and residential development in the Romanian metropolitan areas. *Romanian Journal of Geography*, 56: 43–59.

Guest, A.M., and Brown, S.K. 2005. Population distribution and suburbanization. In D. L. Poston and M. Micklin (Eds.), *Handbook of Sociology and Social Research* (Chapter II, 59–86). Berlin, Germany: Springer. doi:10.1007/0-387-23106-4_3.

Guindon, B., Zhang, Y., and Dillabaugh, C. 2004. Landsat urban mapping based on a combined spectral-spatial methodology. *Remote Sensing of Environment*, 92: 218–232.

Hasse, J.E., and Lathrop, R.G. 2003. Land resource impact indicators of urban sprawl. *Applied Geography*, 23: 159–175.

Hermelin, B. 2007. The urbanization and suburbanization of the service economy: Producer services and specialization in Stockholm. *Geografiska Annaler. Series B, Human Geography*, 89: 59–74.

Herold, M., Scepan, J., and Clarke, K.C. 2002. The use of remote sensing and landscape metrics to describe structures and changes in urban land uses. *Environment and Planning*, 34: 1443–1458.

ICIMOD, 1999. Integration of GIS, remote sensing and ecological methods for biodiversity inventory and assessment. In *Issues in Mountain Development*. Retrieved from http://www.icimodgis.net/web/publications/Issues Mountain 6 1999.pdf.

Ji, W., Ma, J., Twibell, R.W., and Underhill, K. 2006. Characterizing urban sprawl using multi-stage remote sensing images and landscape metrics. *Computers, Environment and Urban Systems*, 30(6): 861–879.

Jothimani, P. 1997. Operational urban sprawl monitoring using satellite remote sensing: Excerpts from the studies of Ahmedabad, Vadodara and Surat, India. *18th Asian Conference on Remote Sensing*, October 20–24, 1997, Malaysia.

Kamusoko, C. 2017. Importance of remote sensing and land change modeling for urbanization studies. In *Urban Development in Asia and Africa*, pp. 3–10. Springer Singapore.

Kotharkar, R., Bahadure, P., and Sarda, N. 2014. Measuring compact urban form: A case of Nagpur City, India. *Sustainability*, 6: 4246–4272. doi:10.3390/su6074246.

Lata, K.M., Rao, C.S., Prasad, V.K., Badrinath, K.V.S., and Raghavaswamy, V. 2001. Measuring urban sprawl: A case study of Hyderabad. *GIS Development*, 5(12): 26–29.

Lo, C.P., and Choi, J. 2004. A hybrid approach to urban land use/cover mapping using Landsat 7 enhanced thematic mapper plus (ETM+) images. *International Journal of Remote Sensing*, 25: 2687–2700.

Lu, D., and Weng, Q. 2007. A survey of image classification methods and techniques for improving classification performance. *International Journal of Remote Sensing*, 28: 823–870.

McGarigal, K. 2002. FRAGSTATS: Spatial pattern analysis program for categorical maps. Accessed February 10, 2008, from http://www.umass.edu/landeco/research/fragstats/fragstats.html.

Mitsova, D., Shuster, W., and Wang, X. 2011. A cellular automata model of land cover change to integrate urban growth with open space conservation. *Landscape and Urban Planning*, 99(2): 141–153.

Osborne, L., and Kovacic, D.A., 1993. Riparian vegetated buffer strips in water quality restoration and stream management. *Freshwater Biology*, 29: 243–258.

Pacione, M. 2007. Sustainable urban development in the UK: Rhetoric or reality? *Geography*, 92: 246–263.

Rakodi, C. 1995. *Harare—Inheriting a Settler-Colonial City: Change or Continuity?* Chichester, UK: Wiley.

Ramachandra, T.V., Bharath H.A., and Durgappa, D.S. 2012a. Insights to urban dynamics through landscape spatial pattern analysis. *International Journal of Applied Earth Observation and Geoinformation*, 18: 329–343.

Ramachandra, T.V., Bharath H.A., Vinay, S., Joshi, N.V., Kumar, U., and Rao, K.V. 2013. Modelling urban revolution in Greater Bangalore, India. *Proceeding of 30th Annual In-House Symposium on Space Science and Technology ISRO-IISc Space Technology Cell*, Indian Institute of Science, Bangalore, November 7–8, 2013.

Ramachandra. T.V., Bharath, S., and Bharath H.A. 2012b. Peri-urban to urban landscape patterns elucidation through spatial metrics. *International Journal of Engineering Research and Development*, 2(12): 58–81.

Randolph, J. 2004. *Environmental Land Use Planning and Management*. Washington, DC: Island Press.

Ranpise, R.S., Kadam, A.K., Gaikwad, S.W., and Meshram, D.C. 2016. Appraising spatio-temporal shifting of urban growth center of Pimpri-Chinchwad Industrialized City, India Using Shannon Entropy Method. *Studies*, 4: 343–355.

Seto, K.C., and Liu, W. 2003. Comparing ARTMAP neural network with the maximum-likelihood classifier for detecting urban change. *Photogrammetric Engineering & Remote Sensing*, 69: 981–990.

Smart Growth Network. 1996. Principles of smart growth. Accessed August 26, 2010, from http://www.smartgrowth.org/about/principles/.

Spence, M., Annez, P.C., and Buckley, R.M. (Eds.) 2009. Urbanization and growth. Retrieved from http://siteresources.worldbank.org.

Stefanov, W.L., Ramsey, M.S., and Christensen, P.R. 2001. Monitoring urban land cover change: An expert system approach to land cover classification of semiarid to arid centers. *Remote Sensing of Environment*, 77: 173–185.

Sudhira, H.S., Ramachandra, T.V., and Jagadish, K.S. 2003. Urban sprawl pattern recognition and modelling using GIS. Paper presented at *Map India, 2003*, New Delhi, India. January 28–31, 2003.

Sudhira, H.S., Ramachandra, T.V., and Jagadish, K.S. 2004. Urban sprawl: Metrics, dynamics and modelling using GIS. *International Journal of Applied Earth Observation and Geoinformation*, 5(1): 29–39.

Taubenbock, H., Wegmann, M., Roth, A., Mehl, H., and Dech, S. 2009. Urbanization in India spatiotemporal analysis using remote sensing data. *Computers, Environment and Urban Systems*, 33: 179–188.

Terzi, F., and Bolen, F. 2009. Urban sprawl measurement of Istanbul. *European Planning Studies*, 17(10): 1559–1570.

Tran, T.N.Q., Fanny, Q., de Claude, M., Vinh, N.Q., Nam, L.V., and Truong, T.H. (Eds.). 2008. Trends of urbanization and suburbanization in Southeast Asia. *Regional Conference Trends of Urbanization and Suburbanization in Southeast Asia (CEFURDS, LPED)*, Ho Chi Minh City, December 9–11, 2008. Ho Chi Minh City: Ho Chi Minh City General Publishing House, 328.

Tsai, Y.H. 2005. Quantifying urban form: Compactness versus "sprawl." *Urban Studies*, 42: 141–161. doi:10.1080/0042098042000309748.

United Nations. 2015. World urbanization prospects: The 2014 revision. Highlights (ST/ESA/SER.A/352). Accessed on September 28, 2015, from http://esa.un.org/unpd/wup/Highlights/WUP2014-Highlights.pdf.

U.S. Environmental Protection Agency (USEPA). 2005. Riparian buffer width, vegetative cover, and nitrogen removal effectiveness: A review of current science and regulations. EPA/600/R-05/118. Office of Research and Development, National Risk Management Research Laboratory, Ada, OK.

Ward, D., Phinn, S.R., and Murray, A.T. 2000. Monitoring growth in rapidly urbanizing areas using remotely sensed data. *The Professional Geographer*, 52(3): 371–386.

Weeks, J.R. 2010. Defining Urban Areas. In T. Rashed, and C. Jurgens (Eds.), *Remote Sensing of Urban and Suburban Areas, Remote Sensing and Digital Image Processing*, 10: 33–45. Amsterdam, the Netherlands: Springer. doi:10.1007/978-1-4020-4385-7_3.

Yang, X., and Liu, Z. 2005. Use of satellite-derived landscape imperviousness index to characterize urban spatial growth. *Computers, Environment and Urban Systems*, 29: 524–540.

Yeh, A.G.O., and Li, X., 2001. Measurement and monitoring of urban sprawl in a rapidly growing region using entropy. *Photogrammetric Engineering and Remote Sensing*, 67(1): 83.

Yuan, F., Saway, K.E., Loeffelholz, B.C., and Bauer, M.E. 2005. Land cover classification and change analysis of the Twin Cities (Minnesota) metropolitan area by multitemporal Landsat remote sensing. *Remote Sensing of Environment*, 98: 317–328.

Zhou, Y.X. 1997. On the suburbanization of Beijing. *Chinese Geographical Science*, 7: 208–219.

# 8 Change in Volume of Glaciers and Glacierets in Two Catchments of Western Himalayas, India since 1993–2015

*Mohd Soheb, AL. Ramanathan,
Manish Pandey, and Sarvagya Vatsal*

## CONTENTS

8.1 Introduction ............................................................................................................. 123
8.2 Study Region............................................................................................................ 124
8.3 Data and Methods.................................................................................................... 125
    8.3.1 Catchment and Glacier Areas...................................................................... 125
    8.3.2 Thickness and Volume Estimation .............................................................. 125
8.4 Results and Discussion ............................................................................................ 127
    8.4.1 Total Glaciated Area in Stok and Matoo Village Catchments .................... 127
    8.4.2 Ice Reserved in Stok and Matoo Village Catchments ................................ 127
8.5 Conclusion ............................................................................................................... 128
Acknowledgments............................................................................................................. 129
References......................................................................................................................... 129

**ABSTRACT** Estimation of glacier volume is important for assessing water reserve stored in the form of ice. The livelihood dependency on the cryospheric system in Ladakh region makes it even more important to understand the change in ice volume. Glacier and glacieret volume estimation was performed in two catchments (18 glaciers/glacierets) to understand the volume change for a period of over two decades. It was found that the glaciers/glacierets of these catchments had decreased with time. However, glacierets in these regions showed a higher retreat, almost twice the amount, as had glaciers. The Matoo village catchment witnessed comparatively higher retreat over two decades. By the end of August 2015, total ice reserve in these catchments was found to be 0.107 and 0.116 Gt in Stok and Matoo village catchment, respectively.

## 8.1 INTRODUCTION

Estimation of the volume of ice in glaciers is very important in terms of regional water supply and availability of water resources in the future. Several approaches to estimate volume of glaciers have been proposed, for example, volume–area (V–A) relations (e.g., Chen and Ohmura, 1990; Bahr, 1997) and slope-dependent estimations of ice thickness (Haeberli and Hölzle, 1995). More methods have also been developed recently. Remote sensing technology has become a valuable tool for understanding and documenting glaciers' response to changing climate (Kuhn, 2007; Solomon et al., 2007). The free available data from the Landsat series has a significant role to play in studying glaciers on a large scale using remote sensing techniques (Kulkarni, 1994; Braun et al., 2007). Since it is impossible to study glaciers on a large scale using field-based methods for several reasons such as logistics and political and climatic conditions (Racoviteanu et al., 2008); therefore, a number of indirect techniques have been developed for such studies.

Ladakh is one of the most dependent regions on meltwater from snow and glaciers. Ironically, very limited have been carried out in this region on such systems, making it even more urgent to monitor the dynamics of the glaciers in this region. The present study was carried out using the two adjacent village catchments to estimate the change in area and volume of glaciers/glacierets from 1993 to 2015 with the help of remote sensing and geographic information system (GIS) techniques using certain established empirical relationships and scaling factors.

## 8.2 STUDY REGION

Stok and Matoo village catchments are located on the northeastern slope of the Zanskar range in the Leh district of Jammu and Kashmir, India (Figure 8.1). The Stok village catchment consists of 5 glaciers and 2 glacierets whereas the Matoo village catchment has a total number of 4 glaciers and 11 glacierets (Table 8.1). Of 65 and 104 km² of the total area, the glaciated region is only 4.25 and

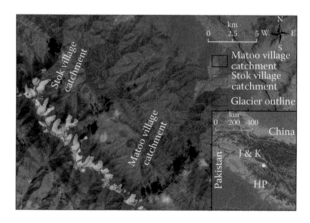

**FIGURE 8.1** Study area for the present study. Stok and Matoo catchments are presented in pink and blue boundary, and yellow boundary presents glacier/glacieret outlines.

**TABLE 8.1**
**Brief Details of the Study Area**

|  | Stok | Matoo |
|---|---|---|
| State/Country | Jammu and Kashmir/India | |
| Region/District | Ladakh/Leh | |
| Basin | Chhabe Nama | |
| Range | Zanskar range | |
| Drainage system | Indus | |
| Climate | Cold-arid | |
| Catchment area | 65 km² | 104 km² |
| Total glaciated area | 4.25 km² (6.5%) | 5.03 km² (4.8%) |
| Number of glaciers | 5 | 4 |
| Number of glacierets | 2 | 11 |

5.03 km² (equivalent to 6.5% and 4.8%) in comparison to the Stok and Matoo village catchments, respectively. The majority of the glaciers/glacierets is oriented toward the northeast. The region is westerly dominated and receives most of the precipitation in the winter months as snow. Annual precipitation and average temperature is 102 mm and 5.2°C (IMD, n.d.), which makes the region one of the coldest and most arid places in India. These two basins feed a total of 579 households, that is, 300 (1471 individuals) and 279 (1165 individuals) in Stok and Matoo village catchments, respectively (LAHDC, Leh, n.d.). Meltwater from the snow cover and glaciers/glacierets of these catchments are the only source of water for irrigation and domestic use.

## 8.3 DATA AND METHODS

For the present study, satellite data from Landsat series (5, and 8) and ASTER GDEM were used. Satellite scenes were obtained from NASA's earthexplorer portal (NASA, 2015). Scenes before 1993 had either higher cloud cover or higher snow cover. Therefore, scenes from 1993 to 2015 were selected to perform the estimations for required parameters (Table 8.2). After requisition, the data were preprocessed and brought to the same geospatial standard, that is, spheroid and datum-WGS 1984, projection system-Universal Transverse Mercator (UTM), to make comparisons possible. While extracting glacier parameters, the subpixel level of accuracy was maintained for all scenes. For ascertaining the precise snout position of glaciers/glacierets, submeter accurate Google Earth™ images with embedded Digital Elevation Model (DEM) were also consulted.

### 8.3.1 CATCHMENT AND GLACIER AREAS

The catchments were automatically delineated with the help of the Hydrology tool, a module in Spatial Analyst Tool in ArcGIS version 10.1 and ASTER GDEM data. The glacier outlines were manually delineated for the scenes. The area of the two catchments and glaciers (for the two time periods) were calculated. Digitization of the shapes (catchments and glaciers) and calculation of the area were performed in ArcGIS 10.1. Furthermore, the glaciers were divided in two categories: (1) glaciers for an area larger than 0.5 km² and (2) glacierets for an the area less than 0.5 km² and sustained for at least two consecutive hydrological years. It has been found that almost all glacierets in the area existed before the beginning of the study period, that is, 1993 and hence those are ice bodies.

### 8.3.2 THICKNESS AND VOLUME ESTIMATION

Various indirect methods were established for thickness and volume estimation, and four of those approaches were used for the present study. These methods are very commonly used for such estimations. A mean of all four methods was taken for further estimations/calculations. The following methods were used to estimate the volume change in glaciers/glacierets of the two village catchments under study.

**TABLE 8.2**
**Details of the Satellite Data Used for the Present Study**

| Date | Scene ID | Sensor | Mission | Path/Row | Pixel (m) | Pixel after Pan Sharpening (m) |
| --- | --- | --- | --- | --- | --- | --- |
| September 2, 1993 | LT51470361993245ISP00 | TM | Landsat 5 | 147/36 | 30 | — |
| August 30, 2015 | LC81470362015242LGN00 | OLI | Landsat 8 | 147/36 | 30 | 15 |
| October 17, 2011 | ASTGDEMV2_0N34E077 | ASTER | TERRA | — | 30 | — |
| October 17, 2011 | ASTGDEMV2_0N33E077 | ASTER | TERRA | — | 30 | — |

*Method 1*:
This method was proposed by Brückl (1970) and used in Swiss glaciers by Müller et al. (1976). Here, $h$ is the mean thickness in meters, $A$ is the area of the glacier/glacieret in km², and $V$ is the volume of the glacier:

$$h = 5.2 + 15.4 \times A^{0.5} \tag{8.1}$$

$$V = h \times A \tag{8.2}$$

*Method 2*:
The empirical relationship between area and thickness was first developed by Chaohai and Sharma (1988) for glaciers in the Himalayan region that have been used by many, including Kulkarni et al. (2007) and Bisht et al. (2015), to estimate glaciers' volume in the Himalayan region.

$$h = -11.32 + 53.21 \times A^{0.3} \tag{8.3}$$

$$V = h \times A \tag{8.4}$$

where:
$h$ is the depth/thickness (in meters)
$A$ is the area
$V$ is the volume of the glacier/glacieret

*Methods 3 and 4*:
These two methods are after Frey et al. (2014). The two methods used here follow the same equations (eq. 8.5 and 8.6) used by Frey et al. (2014). However, the difference in the methods lies in the use of scaling parameters i.e. $c$ and $\gamma$. In method 3, the scaling parameters used have been determined by (Chen and Ohmura, 1990) using the measurements from 63 glaciers whereas that for the method 4 have been obtained from (Bahr, 1997) which is based on theoretical study. The scaling parameters are given in Table 8.3.

$$h = cA^{\gamma-1} \tag{8.5}$$

$$V = cA^{\gamma} \tag{8.6}$$

where:
$h$ is the thickness (in meters)
$c$ and $\gamma$ are scaling parameters
$A$ is the area (in m²)
$V$ is the volume of the glacier/glacieret (Table 8.3)

**TABLE 8.3**
**Scaling Parameters Used in Equations 8.5 and 8.6**

| Method | c | γ | Source |
|---|---|---|---|
| 3 | 0.2055 | 1.360 | Chen and Ohmura (1990) |
| 4 | 0.191 | 1.375 | Bahr (1997) |

## 8.4 RESULTS AND DISCUSSION

### 8.4.1 TOTAL GLACIATED AREA IN STOK AND MATOO VILLAGE CATCHMENTS

The calculation of glacier area for 1993 and 2015 includes 18 glaciers/glacierets. Only seven are present in the Stok village catchment with the remainder in the Matoo village catchment. The total surface area of the glaciated region in 1993 and 2015 is given in Figure 8.2 and Table 8.4. The surface area is based on manually digitization of Landsat image. The catchmentwide change in surface area of glaciers is observed to be −0.27 km² and −0.29 km², equivalent to 6.7% and 8.7% of the total glacier area, for the Stok and Matoo village catchments, respectively, in over 22 years. On the other hand, glacierets in these catchments were observed to have a slightly higher change in surface area. The glacierets vacated an area of −0.07 km² and −0.32 km², which is around 12.5% and 14.8% of the total glacieret surface area for Stok and Matoo, respectively. The rate of change in area of the Stok catchment was found to be ~12,374 and ~3105 m²a⁻¹ in glaciers/glacierets. In the case of the Matoo village catchment, the rate of change in area was found to be ~13,600 and ~14,900 m²a⁻¹. Overall, the glaciated region in catchments of the Stok and Matoo villages was vacated around 0.34 and 0.62 km², equivalent to 7.4% and 11%, respectively, since 1993 to 2015.

### 8.4.2 ICE RESERVED IN STOK AND MATOO VILLAGE CATCHMENTS

An attempt has been made to estimate the total ice reserve stored in glaciers/glacierets of the catchments of Stok and Matoo villages for the years 1993 to 2015 (Figure 8.3). In the Stok village catchment, the total ice reserve in the glacier was found to be 0.109 and 0.099 gigatons (Gt) whereas total ice reserve in the glacierets was found to be 0.010 and 0.008 Gt in 1993 and 2015, respectively

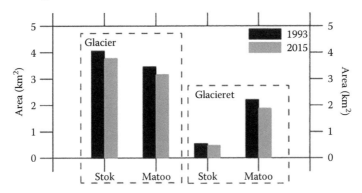

**FIGURE 8.2** Catchmentwide change in surface area of glaciers/glacierets of Stok and Matoo village catchments.

**TABLE 8.4**
**Area Change and Surface Area of Glaciers/Glacierets in 1993 and 2015**

|  | Area 1993 km² | Area 2015 km² | Area Change (km²) | Area Change % | Rate of Change m² a⁻¹ |
|---|---|---|---|---|---|
| Glaciers in Stok | 4.05 | 3.78 | 0.27 | 6.71 | 12,374.5 |
| Glaciers in Matoo | 3.45 | 3.15 | 0.29 | 8.66 | 13,620.9 |
| Glacierets in Stok | 0.54 | 0.47 | 0.07 | 12.53 | 3105.7 |
| Glacierets in Matoo | 2.20 | 1.88 | 0.32 | 14.84 | 14,907.0 |

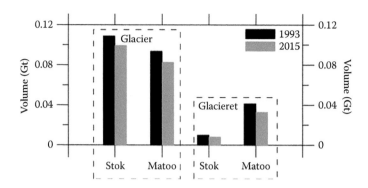

**FIGURE 8.3** Total ice reserve of Glacier and Glacieret of Stok and Matoo village catchments in 1993 and 2015.

**TABLE 8.5**
**Volume Change of Glaciers and Glacierets between 1993 and 2015 in Stok and Matoo Catchment**

|  | Change in Volume | | Rate of Change Tons a$^{-1}$ |
| --- | --- | --- | --- |
|  | (Gt) | % |  |
| Glaciers in Stok | 0.009 | 8.87 | 439,387.22 |
| Glaciers in Matoo | 0.011 | 11.85 | 504,484.37 |
| Glacierets in Stok | 0.001 | 19.67 | 88,525.64 |
| Glacierets in Matoo | 0.008 | 20.71 | 389,720.11 |

(Figure 8.3). The total reserve in the Matoo village catchment was found to be 0.093 and 0.083 Gt for 1993 and 2015, respectively, while in the glacierets, the total ice reserve was found to be 0.041 and 0.033 Gt for 1993 and 2015, respectively. In over two decades, the Stok village catchment had a loss of ~0.011 Gt of ice with an average of ~527000 tons a$^{-1}$ while the Matoo village catchment lost ~0.019 Gt, which is equivalent to around 894,000 tons a$^{-1}$. Glaciers/glacierets in these catchments showed different results with glacierets having a comparatively higher retreat, almost twice in terms of percentage values, than the Stok glaciers/glacierets, which retreated around 8.87% and 19.67%, respectively, while in the Matoo catchment, the retreat was found to be 11.85% and 20.71%, respectively. By the end of the 2015 hydrological year, more volume of ice in glaciers was found to be in the Stok village catchment (0.099 Gt) than in the Matoo village catchment (0.083 Gt) whereas the volume of ice in glacierets of Matoo (0.033) was larger amount than that in the Stok, which was 0.008 km$^3$ (Table 8.5). Overall, in 2015, the village catchment of Matoo had a larger reserve of ice (0.115 Gt) than that of Stok (0.107 km$^3$).

## 8.5 CONCLUSION

Area and volume were calculated for the 18 glaciers/glacierets in Stok and Matoo village catchments based on their manual onscreen digitization for the years 1993 and 2015. It was observed that the glaciers/glacierets in these two catchments are retreating gradually with time. The glacierets show a comparatively higher retreat than do the glaciers in these regions. In the present study, it was observed that the glaciated region in Stok and Matoo village catchments has vacated an area of 0.34 and 0.62 km$^2$, equivalent to ~7% and 11%, respectively, of the total glaciated area of these catchments in over two decades. In terms of volume, the glacierets in these catchments witnessed a retreat of almost twice as much as that in glaciers. Overall, Matoo village catchment showed a comparatively higher retreat, that is, 11.85% and 20.71% in glaciers/glacierets.

## ACKNOWLEDGMENTS

The authors express their gratitude to Jawaharlal Nehru University and its School of Environmental Sciences for providing the lab facility. They also thank Dr. Pratima Pandey for her suggestions and comments.

## REFERENCES

Bahr, D.B. 1997. Width and length scaling of glaciers. *Journal of Glaciology* 43 (145): 557–562. doi:10.3198/1997JoG43-145-557-562.

Bisht, P., S.N. Ali, A.D. Shukla, S. Negi, Y.P. Sundriyal, M.G. Yadava, and N. Juyal. 2015. Chronology of late quaternary glaciation and landform evolution in the upper Dhauliganga valley (Trans Himalaya), Uttarakhand, India. *Quaternary Science Reviews* 129: 147–162. doi:10.1016/j.quascirev.2015.10.017.

Braun, M., T.V. Schuler, and R. Hock. 2007. Comparison of remote sensing derived glacier facies maps with distributed mass balance modelling at Engabreen, Northern Norway. *IAHS Publications–Series of Proceedings and Reports* 318: 126–134.

Brückl, E. 1970. Eine methode zur volumbestimmung von Gletschern Auf Grund Der Plastizitätstheorie. *Archiv Für Meteorologie, Geophysik Und Bioklimatologie Serie A* 19 (1): 317–328. doi:10.1007/BF02274644.

Chaochai, L., and C.K. Sharma. 1988. Report on first expedition to glaciers and glacier lakes in the Pumqu (Arun) and Poiqu (Bhote-Sun Kosi) River Basins, Xizang (Tibet), China: Sino-Nepalese investigation of glacier lake outburst floods in the Himalayas. Science Press. Retrieved from https://books.google.co.in/books?id=pAGnAAAACAAJ.

Chen, J., and A. Ohmura. 1990. Estimation of Alpine glacier water resources and their change since the 1870s. *Hydrology in Mountainous Regions V* 193: 127–136.

Frey, H., H. Machguth, M. Huss, C. Huggel, S. Bajracharya, T. Bolch, A. Kulkarni, A. Linsbauer, N. Salzmann, and M. Stoffel. 2014. Estimating the volume of glaciers in the Himalayan-Karakoram region using different methods. *The Cryosphere* 8 (6): 2313–2333. doi:10.5194/tc-8-2313-2014.

Haeberli, W., and M. Hölzle. 1995. Application of inventory data for estimating characteristics of and regional climate-change effects on mountain glaciers: A pilot study with the European Alps. *Annals of Glaciology* 21 (1): 206–212.

Indian Meteorological Department, n.d. Indian Meteorological Department, Department of Science and Technology, Government of India.

Kuhn, M. 2007. Using glacier changes as indicators of climate change. *Zeitschrift Fur Gletscherkunde and Glaziolgeologie* 41: 7–28.

Kulkarni A.V. 1994. A conceptual model to assess effect of climatic variations on distribution of Himalayan Glaciers. Global Change Studies-Scientific Results from ISRO Geosphere Biosphere Programme, ISRO-GBP-SR-42-94.

Kulkarni, Anil V., I.M. Bahuguna, B.P. Rathore, S.K. Singh, S.S. Randhawa, R.K. Sood, and S. Dhar. 2007. Glacial retreat in Himalaya using Indian remote sensing satellite data. *Current Science* 92 (1): 69–74. doi:10.1117/12.694004.

LAHDC, Leh. n.d. *Statistical Handbook, 2014–2015*. Ladakh Autonomous Hill Development Council, Leh, Government of J&K, India.

Müller, F., Caflisch, T., Müller, G. 1976. Firn und Eis der Schweizer Alpen: Gletscher. Geographisches Institut ETH Zürich, Publ. Nr. 57. 2: 174p.

NASA. 2015. Landsat data, NASA EOSDIS Land Processes DAAC, USGS Earth Resources Observation and Science (EROS) Center. Sioux Falls, South Dakota. Retrieved from https://lpdaac.usgs.gov.

Racoviteanu, A.E., M.W. Williams, and R.G. Barry. 2008. Optical remote sensing of glacier characteristics: A review with focus on the Himalaya. *Sensors* 8 (5): 3355–3383.

Solomon, S., D. Quin, M. Manning, Z. Chen, M. Marquis, K.B. Averyt, M. Tignor, and H.L. Miller. 2007. Climate Change 2007: The Physical Science Basis, Contribution of Working Group I to the Fourth Assessment Report of the Intergovernmental Panel on Climate Change. Cambridge University Press, Cambridge, United Kingdom and New York, NY.

# 9 Analysis of Drainage Morphometry and Tectonic Activity in the Dehgolan Basin Kurdistan, Iran, Using Remote Sensing and Geographic Information System

*Payam Sajadi, Amit Singh, Saumitra Mukherjee, Harshita Asthana, P Pingping. Luo, and Kamran Chapi*

## CONTENTS

| | | |
|---|---|---|
| 9.1 | Introduction | 132 |
| 9.2 | Study Area | 133 |
| | 9.2.1 General | 133 |
| | 9.2.2 Geological Settings | 133 |
| | 9.2.3 Tectonic Setting | 133 |
| 9.3 | Materials and Methods | 135 |
| | 9.3.1 Stream Order | 135 |
| | 9.3.2 Stream Number | 136 |
| | 9.3.3 Stream Length | 136 |
| | 9.3.4 Drainage Density | 136 |
| | 9.3.5 Hypsometric Curve | 136 |
| | 9.3.6 Hypsometric Integral | 136 |
| | 9.3.7 Stream-Length Gradient Index (Hack's Index) | 136 |
| | 9.3.8 Drainage Basin Asymmetry | 137 |
| | 9.3.9 Elongation Ratio | 137 |
| | 9.3.10 River Sinuosity | 137 |
| | 9.3.11 Basin Slope | 137 |
| 9.4 | Results | 137 |
| | 9.4.1 Morphometric Indices | 137 |
| |     9.4.1.1 Stream Order | 137 |
| |     9.4.1.2 Stream Number | 137 |
| |     9.4.1.3 Stream Length | 137 |
| |     9.4.1.4 Drainage Density | 138 |
| | 9.4.2 Morphotectonic Indices | 139 |
| |     9.4.2.1 Hypsometric Curve and Hypsometric Integral | 139 |
| |     9.4.2.2 Stream Length-Gradient Index (Hack's Index) | 142 |

| | 9.4.2.3 | Drainage Basin Asymmetry | 142 |
|---|---|---|---|
| | 9.4.2.4 | Elongation Ratio | 142 |
| | 9.4.2.5 | River Sinuosity | 145 |
| | 9.4.2.6 | Basin Slope | 146 |
| 9.5 | Discussion | | 146 |
| 9.6 | Conclusion | | 148 |
| Acknowledgments | | | 148 |
| References | | | 148 |

**ABSTRACT** Integration of morphometric and morphotectonic analysis with remote sensing and geographic information system (GIS) techniques is useful in quantifying the possible tectonic activity within a specific area. Such an integrated analysis was conducted in this study for the Qorveh–Dehgolan basin using 11 parameters. While morphometric parameters included stream order ($U$), stream number ($N_u$), stream length ($L_u$), and drainage density ($D_d$), morphotectonic indices include stream length-gradient index ($Sl$), drainage basin asymmetry ($Af$), hypsometric curve ($HC$), hypsometric integral ($HI$), elongation ratio ($R_e$), stream sinuosity ($K$), and basin slope ($S_b$). The study area is part of the Sanandaj–Sirjan zone belonging to the Zagros orogenic belt. Because the study area is mainly a low relief basin, relevant indices like sinuosity ($K$), basin slope ($S_b$), and elongation ratio ($R_e$) were primarily used to account appropriately for such a physiographic condition. The whole area was divided into eight microwatersheds to have comparative results. Analysis of all 11 parameters revealed that basins that are located at a lower altitude are actually old where depositional and geomorphic processes control the area, whereas tectonic activity plays an important role in watershed hydrology of mountainous parts of the study area.

## 9.1 INTRODUCTION

*Morphometry* refers to the measurement and analysis of the earth surface characteristics, shape, and landform dimensions (Horton, 1945; Strahler, 1952; Pareta and Pareta, 2011). Study of these landforms that are the result of interaction between structural activities and the surface processes falls under the domain of tectonic geomorphology (Keller and Pinter, 2002). Quantitative analysis of the drainage network was introduced by Horton (1945) as an empirical relationship of landform configuration and its controlling agents such as hydrology, climate, vegetation, and soil characteristics. In general, the drainage pattern and the physiographic characteristic of the catchment controlled by structural activity include active tectonics (Holbrook and Schumm, 1999). Morphometric and morphotectonic analyses are useful for investigation of tectonically active faults in a particular area as a sensitive tool to lithology, climate, and geology. (Strahler, 1964; Hack, 1973; Cox, 1994; Keller and Pinter, 2002; Burbank and Anderson, 2001). Accordingly, the application of a few indices became popular among geomorphologists in morphotectonic studies viz hypsometric curve ($HC$), hypsometric integral ($HI$), elongation ratio $(R_e)$, valley floor width–valley height ratio ($Vf$), basin shape ($B_s$), and mountain-front sinuosity ($J$). Strahler (1952) introduced hypsometric analysis as a key parameter to understand the stage of basin development and rate of upliftment in a small-scale catchment. Hack (1973) introduced the stream-length index ($Sl$) as an important morphometric parameter to identify structural controls on stream power and erosion processes. In addition, integration of morphometric studies with remote sensing and GIS technique has become very popular in order to speed up processing and to analyze parameters to quantify tectonic activity (Ghosh and Mistri, 2012). The current study tries to integrate morphometric-morphotectonic analysis with remote sensing and GIS techniques to study the possible role of tectonic activity in the Qorveh–Dehgolan basin. The main watershed was digitized from Landsat 8 image and then divided into eight macrowatersheds to compare them based on tectonic activity. Finally, 11 morphometric and morphotectonic parameters with respect to basin characteristics were selected and calculated to find the most affected watershed by structural control.

## 9.2 STUDY AREA

### 9.2.1 GENERAL

The Dehgolan basin is located in Kurdistan province (Northwest Iran) at the east of Sanandaj city between 47° 08′ 00″ and 48° 12′ 00″ E longitude and 35° 00′ 00″ to 36° 00′ 00″ N latitude. From the east connected to the Qorveh plain with NW–SE trend with the geographic coordinate of 47° 10′ 00″ °47 to 47° 42′ 00″(E) and 35° 05′ 00″ to 35° 25′ 00N (Figure 9.1). Shanooreh low ridge and Ghorveh city are situated in the northeastern part of the study area. From the south, the Dehgolan basin is bounded by Darband Kabud and the Darband Mountains, Zajian, Abdolrahman, and Bandabad; the northern part starts at the Sheida Mountain and highland ridges and goes to the lowlands of Akhi kamal, Sarab-e-Srokhe, Jaghe, Dolatabad, Slarabad, Tilku, Amirabad, Bandel, and villages. Finally, from the west, the Dehgolan basin is limited to Khatoon Shishe ray, Khosrokesh, Kujeh, Bawariz, Youef Siah highlands, and Sanandaj city. The whole basin area is about 2250 km$^2$. The area with the minimum elevation is located at the northern boundary with 1700 m ASL whereas the maximum elevation is Kholaneh Mountain with 2800 m ASL in the southeast.

### 9.2.2 GEOLOGICAL SETTINGS

The study area is geologically located at the northern part of the Sanandaj-Sirjan zone (SSZ) belonging to the Zagros orogenic belt (ZOB). ZOB (Figure 9.1) is the result of subsequent Neo-Tethyan closure from the continental collision of the Arabian plate and the central Iran microcontinent (Berberian and King, 1981; Alavi, 1994; Ghasemi and Talbot, 2006). It is a part of Alpine-Himalaya extending 2000 km from the NW–SE direction of the Taurus Mountains in the southeast of Turkey to the southern part of Iran at Bandar–Abas (Alavi, 1994; Arfania and Shahriari, 2009). Alavi (1994) divided the ZOB into three distinctive parallel tectonic zones with a NE–SW trend after Stöcklin (1968): (1) the Urumieh-Dokhtar magmatic arc at the northeast, (2) Sanandaj- Sirjan zone at the center, and (3) the Zagros fold-thrust belt (ZSFB) at the southwest (Berberian and King, 1981; Alavi, 1994).

### 9.2.3 TECTONIC SETTING

The Sanandaj–Sirjan zone is one of the most complicated structures in Iran, and its absolute tectonic model is still controversial. Although Mohajjel and Sahandi (1999) considered Sanandaj-Sirjan to be a large-scale parallel-closed fold, Alavi (1994) believed that this zone is a large composite duplex structure with a northeast dipping imbricate system with shear sense indicators showing NE–SW transportation that displaced several units of metamorphosed and nonmetamorphosed Phanerozoic rocks in kilometric scales. An associated negative Bouguer gravity anomaly in the southwestern part of this zone indicates an increase in crustal thickness of about 10–15 km due to southwestward displacement and accumulation of rock units (Alavi, 1994). Tectonic activity has not been equally distributed in this zone because Pre-Cimmerian metamorphism exposes the southeastern part, and the northern part is dominated by volcanic activity that had gone under Laramide activity from Middle Cimmerian (Aghanabati, 2004). By this, Eftekharnejad (1981) divided SSZ into two main parts: (1) the south Sanandaj Sirjan zone with metamorphosed and deformed rock from Middle Triassic—Late Triassic and (2) the north Sanandaj-Sirjan zone consisting of volcanic and felsic plutons (greenschist facies) such as the Alvand, Borojerd, Arak, and Malayer plutons. The northwest of the zone is separated due to the occurrence of a high-angle dipping fault that influenced the older rock there (Alavi, 1994).

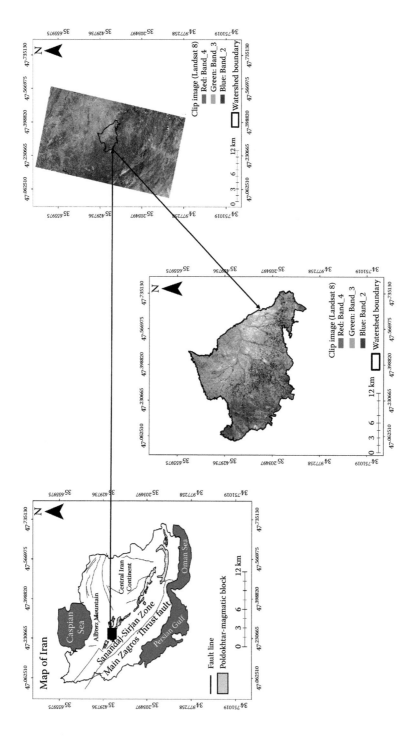

**FIGURE 9.1** Location map of the study area.

## 9.3 MATERIALS AND METHODS

A complete analysis of various data sets is needed when attempting to quantify tectonic activity within such a vast area. Hence, a multidisciplinary approach was followed to integrate morphometric, geological, and lithological data with the GIS-based techniques. A drainage network was digitized from a satellite image (Landsat 8, June 2013) using basic image-processing operations, visual image interpretation, and ArcGIS 10.1 software. The main watershed was divided into eight macro-watersheds based on their physiographic characteristics to find the region of the basin most affected by tectonic activity. The current study consists of two analyses: morphometric and morphotectonic. All 11 calculated parameters and their equations are given in Table 9.1.

The morphometric and morphotectonic analyses were integrated to highlight the watersheds most susceptible to ongoing tectonic activity. The morphometric indices analyzed were stream order $(U)$, stream number $(N_u)$, stream length $(L_u)$, and drainage density $(D_d)$.

### 9.3.1 Stream Order

Identifying stream order $(U)$ is the first step in every morphometric study. According to Strahler (1964), headstream tributaries are ranked as first order, two first-order streams join each other to form a second-order stream, and second-order streams join each other to form third-order streams, and so forth. U designation was done following Strahler's (1964) hierarchical system using ArcGIS (Table 9.1, Equation 9.1).

### TABLE 9.1
### Morphometric and Morphotectonic Parameters and Their Equation

| Number | Parameters | Equation | References | Calculated Value |
|---|---|---|---|---|
| | | **Morphometric Parameters** | | |
| 9.1 | Stream order $(U)$ | Hierarchic order | Strahler, 1964; Rai et al., 2014 | 4th |
| 9.2 | Stream number $(N_u)$ | Total number of stream | Horton, 1945; Rai et al., 2014 | 149 |
| 9.3 | Stream length $(L_u)$ | Total length of stream $(k_m)$ | | 676 |
| 9.4 | Drainage Densit $(D_d)$ | $D_d = \dfrac{\sum L_u}{A}$ | Horton, 1945; Nag, 1998; Pareta and Pareta, 2011 | 0.45 |
| | | **Morphotectonic Parameters** | | |
| 9.5 | Hypsometric curve $(HC)$ | Strahler's (1964) method | Strahler, 1952; Pedrera et al., 2009; Pike and Wilson, 1971 | 0–0.5 |
| 9.6 | Hypsometric integral $(HI)$ | $HI = (h_{mean} - h_{min})/(h_{max} - h_{min})$ | Strahler, 1952; Pedrera et al., 2009; Pike and Wilson, 1971 | 0–0.5 |
| 9.7 | Stream length-gradient $(Sl)$ Index | A) $Sl = (\Delta H/\Delta l) * L$ <br> B) $Sl_{Section} = \dfrac{(H_1 - H_2)}{\log_e L2 - \log_e L1}$ | Hack, 1973; Troiani and Della Seta, 2008 | <300 <br> 300–500 <br> >500 |
| 9.8 | Basin asymmetry $(Af)$ | $Af = 100(A_r/A_t)$ | Keller and Pinter, 2002; El Hamdouni et al., 2008 | 20–80 |
| 9.9 | Elongation ratio $(R_e)$ | $R_e = \dfrac{2}{L_b} * \sqrt{(A/\pi)}$ | Schumm, 1956; Sinha-Roy, 2001 | 0–0.75 |
| 9.10 | Stream sinuosity $(K)$ | $K = \dfrac{L}{L_m}$ or $\dfrac{\text{Reach Length}}{\text{Valley Length}}$ | Mueller, 1968; Rosgen, 1996 | 1–1.5 |
| 9.11 | Basin slope $(S_b)$ | $S_b = \dfrac{H}{L_b}$ | Schumm, 1956; Shankar and Dharanirajan, 2014 | 0.01–0.03 $(m)$ |

### 9.3.2 STREAM NUMBER

A total number of streams in given order is known as $N_u$. The total of streams in all orders was calculated using ArcGIS (Table 9.1, Equation 9.2).

### 9.3.3 STREAM LENGTH

A total of the length of streams in all orders is known as $L_u$. The $L_u$ for all orders was computed using ArcGIS (Table 9.1, Equation 9.3).

### 9.3.4 DRAINAGE DENSITY

Horton (1945) defined drainage density $(D_d)$ as the ratio of the total length of streams in all orders per basin area as kilometer per square kilometer, or km/km². $D_d$ was calculated based on Horton's equation and mapped using the line density method with help of ArcGIS (Table 9.1, Equation 9.4).

The morphotectonic indices included stream length-gradient index $(Sl)$, drainage basin asymmetry $(Af)$, $HC$, $HI$, elongation ratio $(R_e)$, stream sinuosity $(K)$, and basin slope $(S_b)$.

### 9.3.5 HYPSOMETRIC CURVE

Strahler (1952) defined hypsometric analysis as the study of ground surface area distribution or a horizontal cross-sectional area of a landmass with respect to altitude change. To make this analysis, plotting elevation on an ordinate and area on the abscissa was suggested. The curve in this study was obtained with the aid of hypsometric toolbox extension in ArcGIS 10.1 software using ASTER GDEM (30 meters resolution) (Table 9.1, Equation 9.5).

### 9.3.6 HYPSOMETRIC INTEGRAL

The $HI$ is defined as the elevation variations for a given area below the $HC$ and can be calculated by the integration of the $HC$ function for a range of 0 to 1 (Strahler, 1952; Pike and Wilson, 1971). It is equivalent to the ratio of area under the $HC$ to the entire area. The $HC$ was imported into Mathematica software to find curve functions; finally, the function was integrated to calculate the $HI$ (Table 9.1, Equation 9.6).

### 9.3.7 STREAM-LENGTH GRADIENT INDEX (HACK'S INDEX)

For evaluation of the possible relationship among rock resistance, stream power, and tectonic activity, the stream-length gradient was introduced by Hack (1973). The stream-length gradient was calculated over the main trunk for each section using following equation:

$$Sl = (\Delta H / \Delta l) * L \qquad (9.7a)$$

where $Sl$ is the stream-length index, $\Delta H$ is the elevation change in a given section, and $L$ is the linear length of the most distant point of a given section into the headwater of the channel. The ratio $\Delta H/\Delta l$ is the stream section slope that Hack (1973) defined as the natural tangent to the channel slope at a particular point. He also noted that while calculating $Sl$ value in longer reaches, the real value at each reach point is the secant of the channel instead of the tangent (slope) where the difference increases as the channel length increases and leads to an error in calculating $Sl$. Therefore, in order to minimize this error, he also proposed the following equation:

$$Sl_{\text{Section}} = \frac{(H_1 - H_2)}{\log_e L2 - \log_e L1} \quad (9.7b)$$

Where, $H_1$ and $H_2$ are elevations at higher and lower points, respectively, and $L2$ and $L1$ are linear distance from head water to each reach end, respectively. In this study, the $Sl$ index was calculated using the equation 9.7b to increase the accuracy of results.

### 9.3.8 Drainage Basin Asymmetry

Drainage basin asymmetry ($Af$) is calculated using the asymmetry factor $A_f$ using the equation proposed by Keller and Pinter (2002) (Table 9.1, Equation 9.8).

### 9.3.9 Elongation Ratio

Schumm (1956) defined elongation ratio ($R_e$) as the ratio between the diameters of the circle of the same area of the basin to the maximum basin length (Table 9.1, Equation 9.9). $R_e$ was calculated in ArcGIS and mapped for a better understanding of each basin.

### 9.3.10 River Sinuosity

Mueller (1968) defined river sinuosity ($K$) as the ratio of length reach measured along the channel to reach length along the valley (Table 9.1, Equation 9.10). $K$ was calculated using the equation proposed by Mueller with the help of ArcGIS software.

### 9.3.11 Basin Slope

The slope map of Dehglan basin was derived from ASTER DEM using a spatial analyst tool available in ArcGIS software (Table 9.1, Equation 9.11).

## 9.4 RESULTS

### 9.4.1 Morphometric Indices

#### 9.4.1.1 Stream Order
Stream order ($U$) was based on Strahler's method and is shown in Figure 9.2. The result revealed that the drainage network order is 4, that is, the highest order of stream in the Dehgolan basin is the fourth order.

#### 9.4.1.2 Stream Number
According to Horton's law of stream number ($N_u$), the number of streams geometrically decreases as stream order increases. The total number of first-order streams in the study area was found to be highest with 103 counts, whereas the number of the streams of the fourth order was found to be only 4. The number of streams of each order was individually calculated for each watershed and is presented in Table 9.2. The logarithm of stream number versus the stream order for individual watersheds was also plotted and is illustrated in Figure 9.3.

#### 9.4.1.3 Stream Length
Stream length ($L_u$) is a most important factor that indicates the hydrologic characteristics of the basin. Horton's law of stream length shows that the total length of the stream decreases as the stream order increases. The calculation and tabulation of the stream length of various orders are presented in Table 9.3. They revealed that the drainage network in the Dehgolan basin follows

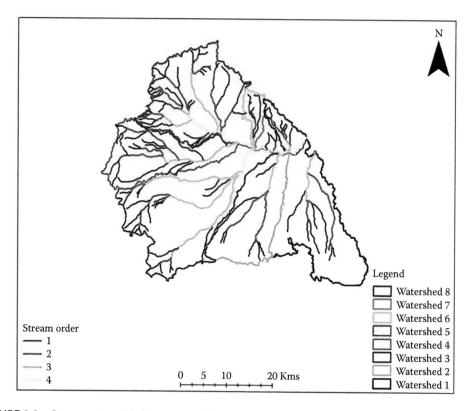

**FIGURE 9.2** Stream order of drainage network in the Qorveh–Dehgolan basin.

**TABLE 9.2**
**Stream Number in Given Order**

| | Stream Number ($N_u$) | | | |
|---|---|---|---|---|
| Watershed No | Order 1 | Order 2 | Order 3 | Order 4 |
| 1 | 8 | 4 | 1 | — |
| 2 | 7 | 2 | 1 | — |
| 3 | 22 | 6 | 3 | 1 |
| 4 | 17 | 3 | 1 | — |
| 5 | 34 | 10 | 5 | 1 |
| 6 | 5 | 1 | 1 | — |
| 7 | 3 | 1 | — | 1 |
| 8 | 7 | 2 | 1 | 1 |

the Horton's law of stream length with first-order streams having the highest length of 360.51 km, whereas the fourth-order streams have only 23 km length. However, the calculated stream length for individual watersheds showed an anomaly within watersheds 3, 4, and 8. Log of stream length versus stream orders for eight watersheds was also plotted and illustrated in Figure 9.4.

### 9.4.1.4 Drainage Density

Horton (1945) introduced drainage density ($D_d$) as an index of channel space closeness. It was found that drainage density is sensitive to climate, lithology, vegetation, and geology of the area (Strahler, 1952; Melton, 1957). Low drainage density is an indicator of low relief areas with highly permeable

# Analysis of Drainage Morphometry and Tectonic Activity

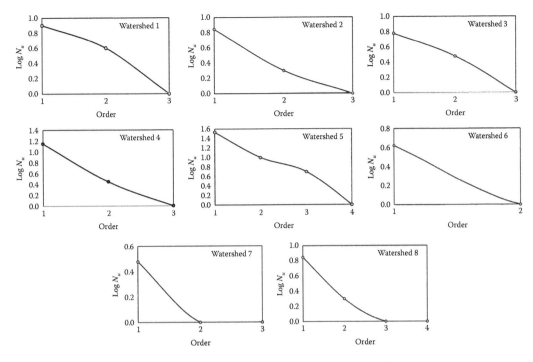

**FIGURE 9.3** Log stream number and stream order relationship for individual watersheds.

**TABLE 9.3**
**Stream Length in Given Order**

| | Stream Length ($L_u$) Km | | | |
|---|---|---|---|---|
| Watershed Number | Order 1 | Order 2 | Order 3 | Order 4 |
| 1 | 44.22 | 19.55 | 8.21 | — |
| 2 | 36.97 | 24.66 | 9.09 | — |
| 3 | 75.07 | 22.91 | 67.76 | 8.12 |
| 4 | 84.31 | 42.13 | 45.26 | — |
| 5 | 82.48 | 40.17 | 18.25 | 8.48 |
| 6 | 15.75 | 7.77 | 2.31 | — |
| 7 | 8.03 | 5.69 | 3.69 | 1.89 |
| 8 | 13.64 | 7.38 | 9.84 | 6.89 |

subsurface soil, whereas high drainage density indicates a high relief area with less vegetation and impermeable soil. The total drainage density in the Dehgolan basin was found to be 0.45, which shows that the basin is mainly marked by a flat topography as also observed in the drainage density map (Figure 9.5).

### 9.4.2 Morphotectonic Indices

#### 9.4.2.1 Hypsometric Curve and Hypsometric Integral

Hypsometric analysis is very useful tool to address the tectonically active basin from inactive ones (Strahler, 1952). Analysis of *HC* and *HI* can be used not only to infer erosional stage of the basin but also to understand the associated tectonic, climatic, and lithologic influences on the system (Moglen et al., 1998; Willgoose and Hancock, 1998). Strahler (1952) categorized the *HI* based on their shape into three categories: Class I: $HI > 0.5$ with convex shape indicating the younger

**140**          Geospatial Applications for Natural Resources Management

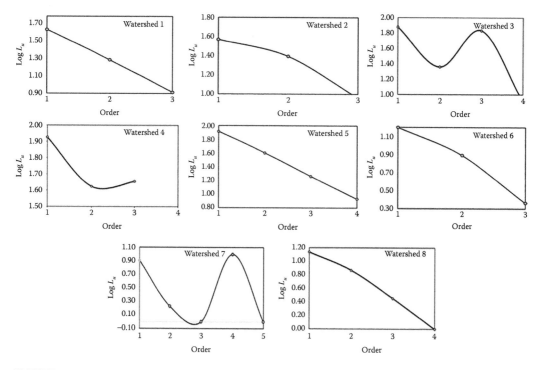

**FIGURE 9.4** Log stream length and stream order relationship.

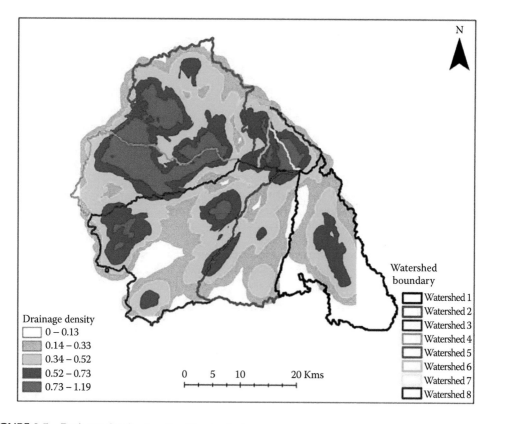

**FIGURE 9.5** Drainage density map for all watersheds.

# Analysis of Drainage Morphometry and Tectonic Activity

stage of basin development, Class II: $0.4 < HI < 0.5$ with concave–convex (S-shaped) indicating the equilibrium or intermediate stage in which denudation and deposition processes are balanced, Class III: $HI < 0.4$ with concave shape indicating old stage or peneplain. The *HC* for the study area are shown in Figure 9.6. Also, the *HI* for individual watersheds and their associated classes are tabulated in Table 9.4. The results showed that watersheds 1 and 2 are in the peneplain stage while 75% of basins including watersheds 3, 4, 5, 6, 7, and 8 fall in Class II with an S-shaped curve referring to the intermediate stage of basin development.

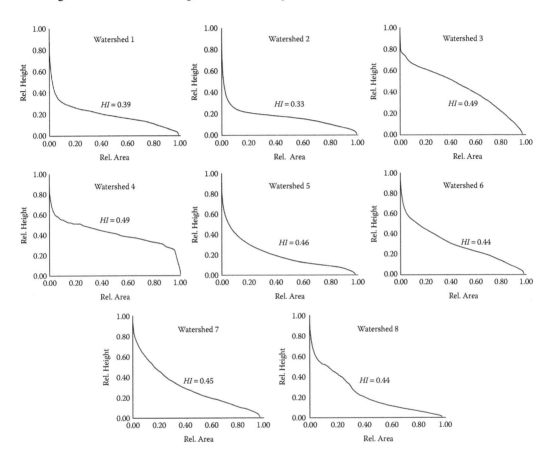

**FIGURE 9.6** Hypsometric curve (*HC*) and hypsometric integral (*HI*) for eight watersheds.

**TABLE 9.4**
***HI* for Eight Watersheds and Their Classes**

| Watershed Number | *HI* Value | *HI* Class |
|---|---|---|
| 1 | 0.39 | 3 |
| 2 | 0.33 | 3 |
| 3 | 0.49 | 2 |
| 4 | 0.49 | 2 |
| 5 | 0.46 | 2 |
| 6 | 0.44 | 2 |
| 7 | 0.45 | 2 |
| 8 | 0.44 | 2 |

### 9.4.2.2 Stream Length-Gradient Index (Hack's Index)

The *Sl* index is a powerful index sensitive to channel slope differences that may also be used to indicate the stream power. The *Sl* index perturbation generally can be related to two processes: either structural control (faulting, tectonic activity) or lithology. If a river passes from hard resistant rocks into soft rocks or vice versa, the channel slope will change abruptly, leading to corresponding changes in *Sl* value. This anomaly can be interpreted as a lithological variation or could be indicative of a possible tectonic activity. In other words, a river passing through a homogeneous lithology and the *Sl* value changing it abruptly reveal the role of structural control over drainage basin (Hack, 1973; Keller and Pinter, 2002; Zhang et al., 2011; Dar et al. 2013). The *Sl* value is expected to increase if a stream flows over an active uplifting and its value may decrease if there are features produced by the action of strike faulting (Keller and Pinter, 2002). *Sl* values were calculated and classified into three groups (Table 9.5) following the classification provided by El Hamdouni et al. (2008): Class I ($Sl > 500$), Class II ($300 < Sl < 500$), and Class III ($Sl < 300$). The *Sl* values were overlaid on 3D-DEM (Figure 9.7) to find possible manifestations of tectonic activity within each watershed. It was observed that watersheds 1, 2, 5, 6, and 7 fall into Class III with no anomaly related to tectonic activity. Extension of fault line on the eastern and southeastern boundary of watersheds 3, 4, and 8 resulted in a sudden increase in *Sl* value, suggesting a possible tectonic-related activity within these watersheds.

### 9.4.2.3 Drainage Basin Asymmetry

Basin asymmetry ($Af$) for eight watersheds was computed and tabulated (Table 9.6). $Af$ values closer or equal to 50 indicate an equilibrium condition in the basin with no tilting. $Af > 50$ indicates a basin tilted left toward its main trunk and $Af < 50$ indicates a basin tilted right toward its main trunk. $Af$ values were computed and segregated into three classes (El Hamdouni et al., 2008; Dehbozorgi et al., 2010): Class I ($Af \geq 65$ or $Af < 35$), Class II ($35 \leq Af < 43$ or $57 \leq Af < 65$), and Class III ($43 \leq Af < 57$). $Af$-50 was also computed as the difference between the calculated asymmetry factor and the absolute value to identify the highest and lowest rates of tilting within a basin (El Hamdouni et al., 2008). $Af$ values were found to vary from 23.32 to 56.92. Results also showed that watersheds 1 and 2 belong to Class III where no tilting happened; watersheds 3, 4, 5, 6, and 7 belong to Class II and watershed 8 belongs to Class I. It was also noted that watersheds 3 and 8 had the highest and lowest values of basin asymmetry, respectively.

### 9.4.2.4 Elongation Ratio

The elongation ratio ($R_e$) value was computed and classified using Schumm (1956) classification (Table 9.7). In general, under certain climatic and geological conditions, $R_e$ varies from 0.6 to 1.0. Values closer to 1 indicate a circular basin with low relief and gentle slope. Lower values of $R_e$ imply

**TABLE 9.5**
***Sl* Value for Eight Watersheds and Their Classes**

| Watershed Number | *Sl* Value | *Sl* Class |
|---|---|---|
| 1 | <300 | 3 |
| 2 | <300 | 3 |
| 3 | 300–500 | 2 |
| 4 | 300–500 | 2 |
| 5 | <300 | 3 |
| 6 | <300 | 3 |
| 7 | <300 | 3 |
| 8 | 300–500 | 2 |

# Analysis of Drainage Morphometry and Tectonic Activity

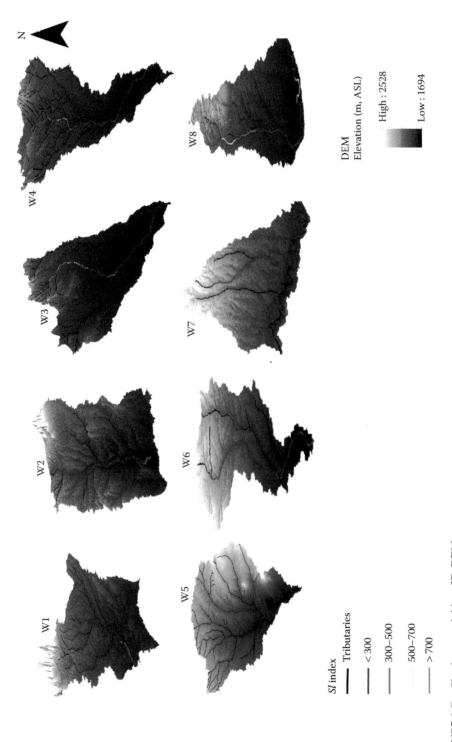

**FIGURE 9.7** *SL* values overlaid on 3D-DEM.

## TABLE 9.6
### Basin Asymmetry (Af) Value for Eight Watersheds and Their Classes

| Watershed Number | Ar (km²) | At (km²) | Af | Af-50 | Class |
|---|---|---|---|---|---|
| 1 | 178.06 | 307.43 | 56.92 | 7.92 | 3 |
| 2 | 113.77 | 227.29 | 33.34 | −16.66 | 1 |
| 3 | 254.18 | 34.57 | 62.75 | 12.75 | 2 |
| 4 | 85.17 | 425.42 | 39.17 | −10.83 | 2 |
| 5 | 191.18 | 343.34 | 55.68 | 5.68 | 3 |
| 6 | 13.58 | 34.57 | 39.30 | −10.70 | 2 |
| 7 | 11.79 | 23.48 | 46.62 | −3.38 | 3 |
| 8 | 9.90 | 42.46 | 23.32 | −26.68 | 1 |

## TABLE 9.7
### Schumm (1956) Classification of Elongation Ratio

| Watershed Number | Schumm Value | Classification | The Stage of Basin |
|---|---|---|---|
| 1 | <0.5 | More elongated | Class 1 |
| 2 | 0.5–0.7 | Elongated | Class 2 |
| 3 | 0.7–0.8 | Less elongated | |
| 4 | 0.8–0.9 | Oval | Class 3 |
| 5 | 0.9–0.10 | Circular | |

## TABLE 9.8
### Elongation Ratio ($R_e$) of Eight Watersheds and Their Classes

| Watershed Number | Bl (km) | A (km²) | Pi (π) | $R_e$ | Class |
|---|---|---|---|---|---|
| 1 | 28.00 | 307.43 | 3.14 | 0.71 | 3 |
| 2 | 25.52 | 227.29 | 3.14 | 0.71 | 3 |
| 3 | 39.68 | 34.57 | 3.14 | 0.59 | 2 |
| 4 | 33.23 | 425.42 | 3.14 | 0.47 | 1 |
| 5 | 34.18 | 343.34 | 3.14 | 0.63 | 2 |
| 6 | 13.33 | 34.57 | 3.14 | 0.52 | 2 |
| 7 | 11.78 | 23.48 | 3.14 | 0.53 | 2 |
| 8 | 18.75 | 42.46 | 3.14 | 0.39 | 1 |

a highly elongated shape with a steep slope and suggest the superiority of structural control over lithological control (Lykoudi and Angelaki, 2004) whereas a higher ratio indicates an inactive area with a gentle slope (El Hamdouni et al., 2008; Singh and Singh, 1997). Tabulated $R_e$ values observed for various watersheds in the study area are given in Table 9.8, and the elongation map of each of the watershed has been individually mapped and is illustrated in Figure 9.8. It was observed that watersheds 1 and 2 belong to Class III, suggesting a circular shape of basin, whereas watersheds 3, 5, 6, and 7 belong to Class II indicating an elongated shape of basin. Watersheds 4 and 8 fall in Class I, suggesting a highly elongated shape of these basins.

# Analysis of Drainage Morphometry and Tectonic Activity

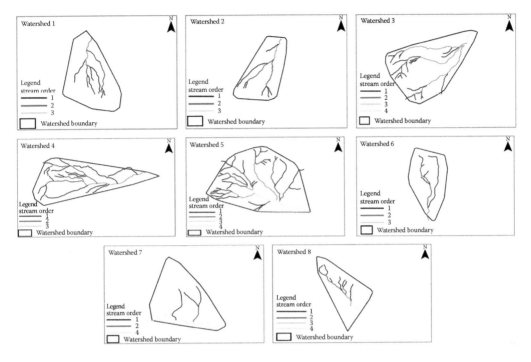

**FIGURE 9.8** Elongation map for eight watersheds.

## 9.4.2.5 River Sinuosity

The pattern of a river channel can reveal the nature of instability that exists within it. *River instability* refers to the river's not being in balance with its environmental conditions as it tries to reach a new equilibrium through certain modifications (Panizza and Panizza, 1996). In general, a river channel can appear in two patterns: straight or meandering, but Morisawa (1985) explained the intermediate case as a sinuous river. The sinuosity index, which ranges from 1.0–1.5, can be used to quantify this scenario. Straight rivers have an index value of 1, and the value ranges for sinuous rivers are between 1.0 and 1.5, and a sinuosity value of > 1.5 refers to a meandering stream (Kunze, 2004). Sinuosity value for all watersheds was calculated and tabulated (see Table 9.9). The index suggests that watersheds 1, 2, 5, 6, and 7 are sinuous with a gentle slope while watersheds 3, 4, and 8 face a sudden rise in sinuosity value (1.21) in third-order streams where the basin slope rises to 55° (Figure 9.9).

### TABLE 9.9
### River Sinuosity ($K$) Value in Given Order for All Watersheds

| Watershed Number | Stream Sinuosity ($K$) | | | |
|---|---|---|---|---|
| | Order 1 | Order 2 | Order 3 | Order 4 |
| 1 | 1.082 | 1.090 | 1.110 | — |
| 2 | 1.064 | 1.062 | 1.121 | — |
| 3 | 1.066 | 1.035 | 1.210 | 1.381 |
| 4 | 1.102 | 1.044 | 1.334 | — |
| 5 | 1.056 | 1.084 | 1.141 | 1.174 |
| 6 | 1.076 | 1.062 | 1.095 | — |
| 7 | 1.016 | 1.062 | 1.047 | 1.472 |
| 8 | 1.088 | 1.043 | 1.086 | 1.478 |

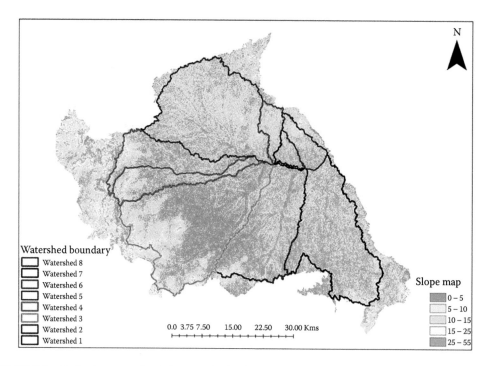

**FIGURE 9.9** Basin slope map for eight watersheds.

#### 9.4.2.6 Basin Slope

Basin slope ($S_b$) is a function of basin relief ($H$), basin elevation ($Z$), and contour length (Wentworth, 1930). *Higher basin slope* refers to a high relief area, whereas *lower basin slope* indicates a lower relief. Basin slope was classified into five classes, and a map was prepared (Figure 9.9). Slope map showed that basin slope in the study area varies from 0° at lower altitude basins (1, 2, 6, and 7) to 55° at higher altitude basins (3, 4, 5, and 8).

### 9.5 DISCUSSION

Results from four morphometric and seven morphotectonic parameters revealed interesting insights into the Qorveh–Dehgolan basin.

Figure 9.10 shows that watersheds 1, 2, 5, 6, and 7 are mainly covered by low-resistance lithology (marl, shale, sandstone, and conglomerate ~70%). The concave-shaped $HC$ associated with low $HI < 0.4$ supports this fact and highlights a peneplain stage of basin development in these watersheds. These watersheds are located near the study area outlet where the slope is almost gentle. Because they are located at the lower region, depositional processes and associated geomorphic features dominate (low-level piedmont fan and valley terrace deposits). $Sl$ values overlaid on 3D-DEM did not reveal any anomaly or deformational processes in these watersheds. These are primarily circular basins with sinuous streams. The morphometric analysis also suggests a low relief area without any conspicuous anomaly. Overall integration of morphometric and morphotectonic analysis suggests that ongoing tectonic processes do not appear to have a major effect on watersheds 1, 2, 5, 6, and 7.

Watershed 3, on the other hand, was found to be associated with a relatively resistant lithology (metavolcanics, phyllites, slate, and metalimestone ~40%). However, an extension of a fault line is observed at the southern and southeastern parts of this watershed (Figure 9.10). The $HC$ and $HI$ also suggest that this basin is relatively young. In the upstream regions of watershed 3, where

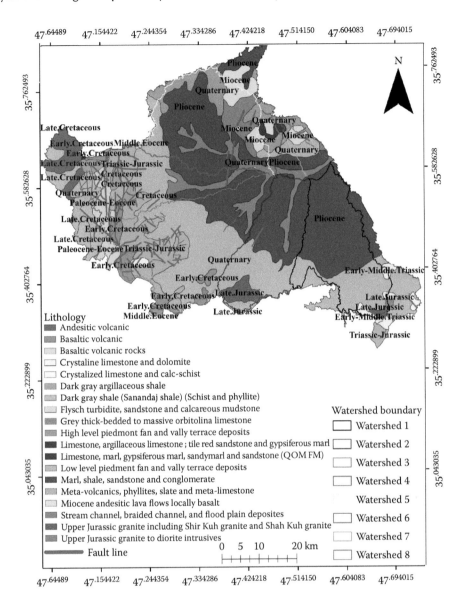

**FIGURE 9.10** Lithology map of study area overlaid on watershed boundary.

second-order streams join to produce third-order streams, the river bends, and slope values close to about 40° are observed. The *Sl* overlaid on 3D-DEM illustrated an unusual basement uplift at this bending point. Interestingly, river sinuosity (*K*) increased in the third-order stream. The morphometric analysis showed that stream length increases from second order to third order despite Horton's law of stream length. This deviation from Horton's law can be explained by the existence of a thrust fault line that forces the river to bend and travel a longer distance in the third order rather than second order (Singh and Singh, 1997; Waikar and Nilawar, 2014; Moges and Bhole, 2015).

The geology of watershed 4 is very similar to that of watershed 3. A fault line extending toward the watershed leads to a bending in the river near a headstream. A sudden increase in *Sl* values was observed, which when overlaid on 3D-DEM showed an unusual deformation over the area. Moreover, there is also an anomaly observed in the stream length of third-order streams, and drainage density was found to increase to 1.19, which is higher than the value of drainage density for the whole basin (0.45). Overall, the results suggest that tectonic activity has prominent control over the drainage in this watershed.

Watershed 8 falls into an extreme range for all indicators of tectonic-related structural activity, that is, elongation ratio and basin asymmetry. Watershed 8 has a highly resistant lithology with a thrust fault extension at the boundary of the basin. Morphometric analysis also revealed an anomaly in stream length at higher stream orders where 3D DEM illustrated possible structural deformation of basement. The integrated result from morphometric and morphotectonic parameters suggests an overall dominance of tectonic activity and associated influences in this basin.

## 9.6 CONCLUSION

The present chapter is the first attempt to provide a synoptic investigation into tectonic activity within the Qorveh–Dehgolan basin using an integrated approach involving morphometric and morphotectonic analysis with the help of remote sensing and GIS. Recent studies of tectonic activity in such mountainous terrain focus on particular indices such as mountain-front sinuosity index ($J$), ratio of $Vf$ or $Bs$, whereas this study used alternate indices that are usually applicable in a low relief area with smaller rivers. The whole basin was divided into eight macrowatersheds to find evidence of tectonic activity (structural deformations) within the basin and the regions most affected. Overall, 11 parameters were computed, including $HC$, $HI$, stream length gradient, basin asymmetry, elongation ratio, sinuosity, stream length, stream order, and drainage density. Results show that among all eight watersheds, only two watersheds (1, 2) are old and the others are almost at the intermediate or young stage of basin development. It was also observed that moving toward the mountainous regions of the study area (where watersheds 3, 4, and 8 are located), lithology was found to be harder, more resistant, and the regions associated with various anomalies (in both morphometric and morphotectonic indices). Existence of both fault lines and homogeneous lithology suggested the possible ongoing tectonic activity in these watersheds. Because the study reported here is a first attempt to integrate morphometric-morphotectonic analysis with remote sensing and GIS to quantify the influence of tectonic activity in the Qorveh–Dehgolan basin, it is a valuable baseline assessment to use in further hydrological analyses. Although lack of published resources for the study area was the main constraint for the authors in correlating their findings with previous results, an attempt was made to analyze a comprehensive list of parameters for encompassing all lithological, hydrological, and tectonic (structural) aspects. Further analysis of subsurface conditions and associated structural deformations with their influence on surface hydrology within the Qorveh–Dehgolan basin is highly recommended to obtain a clear picture of all potential factors affecting the drainage network behavior in this area.

## ACKNOWLEDGMENTS

The authors would like to acknowledge NASA EOSDIS LP DAAC and the U.S. Geological Survey (USGS) as the source of ASTER GDEM and Landsat satellite imagery used in this study.

## REFERENCES

Aghanabati, A. 2004. Geology of Iran. 1st Edition. Geological Survey of Iran (in Persian). Tehran, Iran, p. 586.

Alavi, M. 1994. Tectonics of the Zagros orogenic belt of Iran: New data and interpretations. *Tectonophysics* 229 (3–4): 211–238. doi:10.1016/0040-1951(94)90030-2.

Arfania, R. and S. Shahriari. 2009. Role of southeastern Sanandaj-Sirjan zone in the tectonic evolution of Zagros orogenic belt, Iran. *Island Arc* 18 (4): 555–576. doi:10.1111/j.1440-1738.2009.00680.x.

Berberian, M. and G. C. P. King. 1981. Towards a paleogeography and tectonic evolution of Iran: Reply. *Canadian Journal of Earth Sciences* 18 (11): 1764–1766. doi:10.1139/e81-163.

Burbank, D. and R. Anderson. 2001. *Tectonic Geomorphology.* Malden, MA: Blackwell Science, pp. 273.

Cox, R. T. 1994. Analysis of drainage-basin symmetry as a rapid technique to identify areas of possible Quaternary tilt-block tectonics: An example from the Mississippi Embayment. *Geological Society of America Bulletin* 106 (5):571–581.

Dar, R. A., R. Chandra, and S. A. Romshoo. 2013. Morphotectonic and lithostratigraphic analysis of intermontane karewa basin of Kashmir Himalayas, India. *Journal of Mountain Science* 10 (1): 1–15. doi:10.1007/s11629-013-2494-y.

Dehbozorgi, M., M. Pourkermani, M. Arian, A. A. Matkan, H. Motamedi, and A. Hosseiniasl. 2010. Quantitative analysis of relative tectonic activity in the Sarvestan area, central Zagros, Iran. *Journal of Geomorphology* 121 (3–4): 329–341. doi:10.1016/j.geomorph.2010.05.002.

Eftekharnejad, J. 1981. Tectonic division of Iran with respect to sedimentary basins. *Journal of Iranian Petroleum Society* 82: 19–28.

El Hamdouni, R., C. Irigaray, T. Fernández, J. Chacón, and E.A. Keller. 2008. Assessment of relative active tectonics, southwest border of the Sierra Nevada (southern Spain). *Journal Geomorphology* 96 (1–2): 150–173. doi:10.1016/j.geomorph.2007.08.004.

Ghasemi, A. and C. J. Talbot. 2006. A new tectonic scenario for the Sanandaj-Sirjan zone (Iran). *Journal of Asian Earth Sciences* 26 (6): 683–693. doi:10.1016/j.jseaes.2005.01.003.

Ghosh, S. and B. Mistri. 2012. Hydrogeomorphic significance of sinuosity index in relation to river instability: A case study of Damodar River, West Bengal, India. *International Journal of Advances in Earth Sciences* 1 (2): 49–57.

Hack, J. T. 1973. Stream-profile analysis and stream-gradient index. *Journal of Research of the US Geological Survey* 1 (4): 421–429.

Holbrook, J. and S. A. Schumm. 1999. Geomorphic and sedimentary response of rivers to tectonic deformation: A brief review and critique of a tool for recognizing subtle epeirogenic deformation in modern and ancient settings. *Tectonophysics* 305 (1): 287–306.

Horton, R. E. 1945. Erosional development of streams and their drainage basins; hydrophysical approach to quantitative morphology. *Geological Society of America Bulletin* 56 (3): 275–370.

Keller, E. A. and N. Pinter. 2002. *Active Tectonics: Earthquakes, Uplift, and Landscape.* Upper Saddle River, NJ: Prentice Hall, p. 362.

Kunze, B. 2004. Personal communication. NRCS Area Office, Brookings Area Office, Brookings, SD 57006-1910.

Lykoudi, E. and M. Angelaki. 2004. Contribution of the morphometric parameters of an hydrographic network to the investigation of the neotectonic activity: An application to the upper acheloos. *Proceedings of the 10th International Congress, Thessaloniki, April, Bulletin of the Geological Society of Greece* 36 (2): 1084–1092.

Melton, M. A. 1957. An analysis of the relations among elements of climate, surface properties, and geomorphology; Office of Naval Research Project NR 389–042, Technical Report 11, Department of Geology, Columbia University New York, p.102.

Moges, G. and V. Bhole. 2015. Morphometric characteristics and the relation of stream orders to hydraulic parameters of river goro: An ephemeral river in Dire-dawa, Ethiopia. *Universal Journal of Geoscience* 1: 13–27. doi:10.13189/ujg.2015.030102.

Moglen, G. E., E. A. Eltahir, and R. L. Bras. 1998. On the sensitivity of drainage density to climate change. *Water Resources Research* 34 (4): 855. doi:10.1029/97WR02709.

Mohajjel, M. and M. Sahandi. 1999. Tectonic evolution of Sanandaj-Sirjan Zone (Farsi with English abstract). *Geosciences* 8 (31–32): 28–49.

Morisawa, M. 1985. Geomorphology text book: Rivers, forms and process, Chapter 5. Structural and lithological control, p. 390.

Mueller, J. E. 1968. An introduction to the hydraulic and topographic sinuosity indices 1. *Annals of the Association of American Geographers* 58 (2): 371–385. doi:10.1111/j.1467-8306.1968.tb00650.x.

Nag, S. 1998. Morphometric analysis using remote sensing techniques in the Chaka sub-basin, Purulia district, West Bengal. *Journal of the Indian Society of Remote Sensing* 26 (1): 69–76.

Panizza, M. and M. Paniiza. 1996. *Environmental Geomorphology.* Amsterdam, The Netherlands: Elsevier.

Pareta, K. and U. Pareta. 2011. Quantitative morphometric analysis of a watershed of Yamuna basin, India using ASTER (DEM) data and GIS. *International Journal of Geomatics and Geosciences* 2 (1): 248.

Pedrera, A., J. V. Pérez-Peña, J. Galindo-Zaldívar, J. M. Azañón, and A. Azor. 2009. Testing the sensitivity of geomorphic indices in areas of low-rate active folding (Eastern Betic Cordillera, Spain). *Geomorphology* 105 (3): 218–231.

Pike, R. J. and S. E. Wilson. 1971. Elevation-relief ratio, hypsometric integral, and geomorphic area-altitude analysis. *Geological Society of America Bulletin* 82 (4): 1079–1084.

Rai, P. K., K. Mohan, S. Mishra, A. Ahmad, and V. N. Mishra. 2014. A GIS-based approach in drainage morphometric analysis of Kanhar River Basin, India. *Applied Water Science* 1–16.

Rosgen, D. L. 1996. *Applied River Morphology*. Pagosa Springs, CO: Wildland Hydrology Books, p. 486.

Shankar, S. and K. Dharanirajan. 2014. Drainage morphometry of flood prone rangat watershed, middle Andaman, India—A geospatial approach. *International Journal of Innovative Technology and Exploring Engineering* 3: 15–22.

Singh, S. and M. C. Singh. 1997. Morphometric analysis of kanhar river basin. *National Geographical Journal of India* 43: 31–43.

Sinha-Roy, S. 2001. Neotectonic significance of longitudinal river profiles: An example from the Banas drainage basin, Rajasthan. *Geological Society of India* 58 (2): 143–156.

Stöcklin, J. 1968. Salt deposits of the Middle East. *Geological Society of America Special Papers* 88: 158–181. doi:10.1130/SPE88-p157.

Strahler, A.N. 1964. Quantitative geomorphology of drainage basins and channel networks. In Chow, V.T. (Ed.) *Handbook of Applied Hydrology*. New York: McGraw-Hill, pp. 439–476.

Strahler, A. N. 1952. Hypsometric (area-altitude) analysis of erosional topography. *Geological Society of America Bulletin* 63(11):1117–1142. doi:10.1130/0016-7606(1952)63[1117:HAAOET]2.0.CO;2.

Troiani, F. and M. Della Seta. 2008. The use of the stream length–Gradient index in morphotectonic analysis of small catchments: A case study from Central Italy. *Geomorphology* 102 (1): 159–168.

Waikar, M. L. and A. P. Nilawar. 2014. Morphometric analysis of a drainage basin using geographical information system: A case study. *International Journal of Multidisciplinary and Current Research* 2: 179–184.

Wentworth, C. K. 1930. A simplified method of determining the average slope of land surfaces. *American Journal of Science* 117: 184–194. doi:10.2475/ajs.s5-20.117.184.

Willgoose, G. and G. Hancock. 1998. Revisiting the hypsometric curve as an indicator of form and process in transport-limited catchment. *Earth Surface Processes and Landforms* 23 (7): 611–623.

Zhang, W., Y. S. Hayakawa, and T. Oguchi. 2011. DEM and GIS based morphometric and topographic-profile analyses of Danxia landforms. *Geomorphometry*, 121–124.

# 10 Fog—A Ground Observation-Based Climatology and Forecast over North India

*Sanjay Kumar Srivastava, Rohit Sharma, Kamna Sachdeva, and Anu Rani Sharma*

## CONTENTS

| | | |
|---|---|---|
| 10.1 | Introduction | 152 |
| 10.2 | Study Area | 152 |
| 10.3 | Dataset and Methodology | 154 |
| | 10.3.1 Fog Climatology | 154 |
| | 10.3.2 Time Series and Trend Analysis | 155 |
| | 10.3.3 Geospatial Analysis | 155 |
| | 10.3.4 Autoregressive Integrated Moving Average Modeling | 156 |
| | 10.3.5 Fog Relationship with Aerosol Optical Depth | 156 |
| 10.4 | Results and Discussions | 157 |
| | 10.4.1 Long-Term Climatology of Fog | 157 |
| | 10.4.2 Duration and Variability of Fog | 158 |
| | 10.4.3 Intensity and Persistence of Fog | 160 |
| | 10.4.4 Trend Analysis | 161 |
| | 10.4.5 Decadal Time Series and Trend Analysis over IGP | 164 |
| | 10.4.6 Spatial Variability | 164 |
| | 10.4.7 Autoregressive Integrated Moving Average Modeling | 167 |
| | 10.4.8 Relationship between Fog Occurrence and Aerosol Optical Depth | 168 |
| 10.5 | Conclusion | 169 |
| References | | 171 |

**ABSTRACT** Fog has been believed to create numerous hazards, economic loss, and inconvenience to local inhabitants. Fog hinders visibility, causing financial losses in transportation delays and human loss in accidents. The purpose of the study was to understand the behavior of fog with respect to the climatology of areas prone to it and its relationship with other meteorological parameters and pollutants. Long-term climatology and time series analysis can be very useful in understanding the behavior of fog. In the case study discussed here, trend analysis and the autoregressive integrated moving average (ARIMA) model forecasting were performed for observations from 1970 to 2010 (40 years) to assess the climatology related to and the future trend of fog over the national capital region of the city of Ghaziabad. Furthermore, the study was elaborated and extended to the whole Indo–Gangetic Plain (IGP). The study found that there has been a significant positive trend over the decades in the frequency of fog over IGP. Additionally, the relationship of fog with aerosol optical depth (AOD) retrieved through the MODIS satellite shows a positive correlation in fog frequency. The results would show a new prospective for taking various proactive measures in better understanding the fog climatology.

## 10.1 INTRODUCTION

The obscurity near the surface layer of the atmosphere due to the presence of suspended water droplets, reducing the horizontal visibility to less than 1000 m was termed *fog* (Glickman, 2000). During peak winter months (December and January), fog was one of the most troublesome weather phenomena that significantly influences human activity, causing severe delays in aviation, marine, and surface transportation by reducing visibility up to zero m (Collier, 1970; Badrinath et al., 2009; Srivastava et al., 2017). The need to understand the complexity and physical–chemical characteristics of fog has been increasing as has examining potential methods of fog forecasting since the safety of the population and various economic factors were strongly affected by this weather phenomenon (Michaelides and Gultepe, 2008). Surface fog formation, intensity, duration, and extent were all extremely difficult to predict, but tools such as synoptic forecasting models, numerical prediction models, satellite remote sensing, and analysis of climatological data can significantly enhance the understanding (Srivastava et al., 2016). Over the Indian region, widespread fog forms across the IGP during winter months every year (India Meteorological Department, 1982; Dutta et al., 2004; Lewis et al., 2004; Mishra and Mohapatra, 2004; Singh and Kant, 2006; Jenamani et al., 2007; Singh et al., 2007; Suresh et al., 2007; Ram and Mohapatra, 2008; Badrinath et al., 2009; Mishra et al., 2010; Jenamani and Kalsi, 2012; Mathur et al., 2012; Singh and Dey, 2012; Laskar et al., 2013). The IGP was at a low elevation with generally gradual sloping surfaces of the Indus and Ganges river basins and was highly prone to prolonged, widespread dense fog during winter season. Furthermore, a high concentration of aerosol content in the lower atmosphere over the study area enhances the formation and persistence of fog through additional condensation nuclei (Ramanathan et al., 2005).

Every year, the IGP of India witnesses fog formation during the postmonsoon and winter months (Badrinath et al., 2009). Fog has remarkable spatial and temporal heterogeneity and variability over this region of India. These spatial continuous data, however, were usually not readily available and often difficult and expensive to acquire, especially from mountainous, deep marine, and difficult and remote locations. The geographical information system (GIS) has proved to be the most practical tool (Saraf et al., 2011) in monitoring and modeling various techniques used in environmental studies. Several studies have attempted to examine and understand the mechanism behind this spatial and temporal variability. However, our understanding of the mechanism still remains incomplete regarding numerous complex processes, influence formation, development, and dissipation of fog (Gultepe et al., 2007; Tardif and Rasmussen, 2007). Due to this, our ability to accurately forecast fog over IGP was extremely limited. Timely and accurate forecasts of fog were necessary for the agencies responsible for road, rail, and air transportation, search and rescue operations, industrial and agriculture production, citizens' health and welfare.

Here, an attempt was made initially to examine long-term climatological characteristic of fog using ground-based observations of meteorological parameters over an urban location of Ghaziabad in NCR and later extending the study to examine long-term climatological characteristic across the IGP of Indian subcontinent. The results were significant and highlighted the period of occurrence, dispersal, duration, and intensity of the fog during the winter months over the IGP. Later, an attempt was made to examine the spatial and temporal variability of winter fog over the IGP with the trend in frequency, formation, and forecast of the fog events using the ARIMA statistical model over the study region. Furthermore, the relationship between the aerosol optical depth (AOD) and fog frequency was evaluated.

## 10.2 STUDY AREA

The study reported here focused on the national capital region (NCR) and covered one of the biggest industrial cities of northern India, Ghaziabad, and was later extended to the whole IGP. Ghaziabad city is located at latitude 28°43′N and longitude 77°22′E with an elevation of 214 m above mean

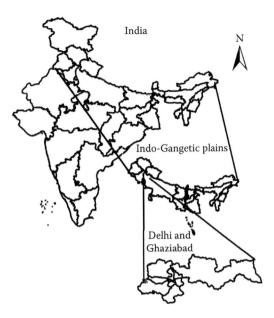

**FIGURE 10.1** Study area—IGP and Ghaziabad.

sea level (AMSL). It is situated 32 km northeast of Delhi in one of the most populous states of the republic of India, that is, Uttar Pradesh. It has two major rivers in close vicinity that act as major sources of moisture. The Yamuna River flows 12 km to the west of Ghaziabad, and the Hindon River flows 3 km to its east (Figure 10.1). The city is surrounded by a large number of industrial tools, including kilns. These smoke-emitting chimneys were major sources of smoke and condensation nuclei that affect visibility in the city under a favorable wind profile. The average annual rainfall over Ghaziabad is 795.5 mm, 78% of which falls during the summer monsoon. The winter season (December–February) was the coldest period, characterized by low temperatures, poor visibility (including radiation fog), dry winds, and a stable atmosphere.

Furthermore, the study was extended across the IGP (Figure 10.1), covering a large area from 20°N to 32°N and 60°E to 100°E because this is the region that experiences widespread fog during the winter season every year. The IGP has a low elevation and generally sloping surfaces of the Indus and Ganges river basins and is critical for agriculture production and food security in the south Asian region. IGP is among the most agriculturally productive regions of the world. Characterized by favorable climate, fertile soil, and abundant water supply, the IGP is considered as the *breadbasket* of South Asia, providing food and livelihood security for the hundreds of millions of inhabitants. However, it was difficult to generalize the climate of the region that has been categorized under four broad seasons: winter (December–February), premonsoon (April–May), monsoon (June–September), and postmonsoon (October–November). The key component that controls the climate of the IGP is the southwest monsoon that brings significant rainfall all across the plain. During the winter season due to its physiography and the extra tropical systems such as western disturbances, IGP becomes highly prone to prolonged widespread dense fog episodes. For the present study, as per data availability, 27 ground meteorological observatories were chosen (Figure 10.2); these locations spread across the IGP and were representative of the entire study area. Furthermore, to reveal potential relationships between fog frequency and the surrounding environment, these sites were classified into three zones: west IGP, which typically constitutes Punjab, Himachal Pradesh, Haryana, Delhi, and adjoining areas; central IGP, mainly constituting Uttar Pradesh, Bihar, Jharkhand; and east IGP, covering West Bengal and northeast India.

**FIGURE 10.2** Ground meteorological observatories in IGP.

## 10.3 DATASET AND METHODOLOGY

### 10.3.1 Fog Climatology

For Ghaziabad, the meteorological data for the period 1971–2010 (40 years) for the months October to February was retrieved from the government aviation meteorological division in Ghaziabad. Uniformity checks and data gaps were filled according to standard procedures. The monthly mean fog frequency, standard deviation, and coefficient of variation were evaluated using these descriptive statistical graphical tools. To have a better understanding of hidden characteristics and periods of significant change, the data were divided into four decades that is, 1971–1980, 1981–1990, 1991–2000, and 2001–2010. The contributions of various meteorological parameters to the change were also analyzed using the actual meteorological data. The frequency, time of onset, time of dispersal, duration, and intensity of the fog were correlated with meteorological data to better understand the formation and dissipation of fog on Ghaziabad.

As an extension to the present study, meteorological data for IGP for the duration (1971–2010) were retrieved for 27 locations from government of India registers for areas across the IGP. The data were obtained specifically for winter months (December–January) being the peak fog months. The data were recorded using meteorological sensors from automatic weather stations and conventional meteorological instruments. Furthermore, these data were also subjected to consistency checks, and data gaps were filled according to standard procedures. Detailed statistical analysis was performed on mean frequency and variability of fog behavior over the region retrieved monthly.

## 10.3.2 TIME SERIES AND TREND ANALYSIS

To determine the long-term climatology and characteristics of fog over Ghaziabad and across IGP, the time series and trend analysis were carried out using inferential statistical techniques. The analysis revealed that there has been a significant change in the frequency and trend of fog over the years in both study areas. Furthermore, temporal variability was examined using Mann–Kendall statistical test (MK test) for December and January. The MK test is the most common test used for trend analysis of hydrometeorological data (Yue et al., 2002; Karmeshu, 2012; Narayanan et al., 2013; Narayanan et al., 2016a, 2016b). The limitation of the MK test is that in the case of positive autocorrelation present in a series, it shows a positive trend even when a trend is not present (Bayazit and Onoz 2007). Hamed and Rao (1998) suggested a modified MK test, which calculates the autocorrelation between the ranks of the data after removing the apparent trend due to positive autocorrelation. A positive (negative) value of MK test statistics indicates an upward (downward) trend. Furthermore, the magnitude of trend was estimated using Theil and Sen's median slope estimator. This method gave a robust estimate of trend (Yue and Hashino 2003) because it was not influenced by outliers. The trends that have statistical significance might not have practical significance and vice versa (Theil, 1950; Yue and Hashino, 2003); hence, the practical significance of trend was assessed by estimating Theil and Sen's median slope. Diurnal variability and average daily persistence of fog over the entire IGP were evaluated and analyzed using advanced descriptive statistical techniques.

## 10.3.3 GEOSPATIAL ANALYSIS

Many different areas of research and application require spatial data for planning, decision-making, and further analysis. Hot spot analysis was one of the methods used to conduct a spatial data clustering. Basically, *hot spot* refers to an area with usually high occurrence of point incident. At times, point observations were transferred into area measurements and then a hot spot was defined as an area showing a high quantity or intensity. Hot spot analysis basically aims to assist identification of locations with unusual high concentration of occurrences. The hot spot analysis tool assesses whether high or low values cluster spatially. It calculates the Getis-Ord GI* (Gi*) statistic, which was a Z-score for each feature in the data set. The resultant Z-score identifies where features with high or low values cluster spatially. This works by looking at each feature within the context of neighboring features. A feature with a high value may be important and significant but may not be a statistically significant hot spot. For a feature to be a statistically significant hot spot, it has to have a high value and be surrounded by other features with high value as well. The local sum for a feature and its neighbor was compared proportionally to the sum of all features. When the local sum was different from the expected local sum and that difference was too large to be because of the result of random chance, statistically significant Z-score results. The Gi* statistic return for each feature in the data set was a Z-score. To be a statistically significant positive Z-score, the larger the Z-score was, the more intense was the clustering of high value. These were hot spots. Similarly, for a statistically significant negative Z-score, the smaller the Z-score was, the more intense was the clustering of low value. These were cold spots. The Gi* local statistic was given by this equation:

$$G_i^* = \frac{\sum_{j=1}^{n} w_{i,j} x_j - \bar{X} \sum_{j=1}^{n} w_{i,j}}{S \sqrt{\frac{\left[n \sum_{j=1}^{n} w_{i,j}^2 - \left(\sum_{j=1}^{n} w_{i,j}\right)^2\right]}{n-1}}}$$

where:

$x_j$ is the attribute value for feature $j$
$w_{ij}$ is the spatial weight between feature $i$ and $j$
$n$ is equal to the total number of features

$$\bar{X} = \frac{\sum_{j=1}^{n} x_j}{n}$$

$$S = \sqrt{\frac{\sum_{j=1}^{n} x_j^2}{n} - (\bar{X})^2}$$

$G_i^*$ statistic is a Z-score, so no further calculation was required. A high Z-score and high p-value for a feature indicates a spatial clustering of high values. A low negative Z-score and small p-value indicates a spatial clustering of low values. The higher the Z-score, the more intense is the clustering. A Z-score near zero indicates no apparent spatial clustering.

### 10.3.4 Autoregressive Integrated Moving Average Modeling

The ARIMA model has been beneficial in the prediction of hydrometeorological parameters (Dickey and Fuller, 1979; Boochabun et al., 2004; Box et al., 2007; Chattopadhyay and Chattopadhyay, 2010; Chattopadhyay et al., 2011; Hyndman and Athanousopoulos, 2015; Narayanan et al. (2016b)). Globally, such climate models have been used extensively to predict patterns of climatic variables (IPCC, 2007, 2014). However, global model forecasts have certain limitations (Sun et al., 2009). The ARIMA model was based on statistical concepts and predicts future values as a product of several past observations and random errors (Yurekli et al., 2007). This method was preferred over other time series analysis methods because it has both auto regression and moving average concepts included in a single method. Furthermore, it includes model identification and diagnostic checks. In the present study, the ARIMA model has been used to predict the future frequency of fog during the peak winter months over IGP.

### 10.3.5 Fog Relationship with Aerosol Optical Depth

Increased fog frequencies in urban areas may be the result of increased air pollution emanating from different primary and secondary sources (Mohan and Payra, 2009). The increased pollution load and specifically aerosol particles act as cloud condensation nuclei initiating the fog formation at lower levels. Increased aerosol concentration further enhances the presence of water aerosol under favorable meteorological conditions and high relative humidity (Hansen et al., 2007; Yasmeen et al., 2012). To assess the relationship of the aerosol particle and the fog frequency, satellite data sets (MODIS) were procured and were correlated with fog frequency for the individual months of January and December for the decade 2000–2010. The MODIS level 3 product containing statistics derived from over 100 scientific parameters from level 2 atmosphere products was used. *MODIS-Terra MOD08_M3V3* ocean- and land-corrected AOD daily means for the months of January and December for the duration 2000 to 2010 were retrieved for the Punjab state of IGP. AOD gives a measure of radiation extinction due to aerosol scattering and absorption. Furthermore, these AOD values were correlated with the meteorological data obtained from the governmental registers of the cities of Jalandhar, Ambala, Amritsar, Bhatinda, and Ludhiana.

## 10.4 RESULTS AND DISCUSSIONS

### 10.4.1 LONG-TERM CLIMATOLOGY OF FOG

Figures 10.3 and 10.4 illustrate the monthly mean and the decadal mean frequency of the fog occurrence over Ghaziabad for the months of October, November, December, January, and February from 1970 to 2010. It was observed that the frequency of fog was at its peak during the month of December with 22.9 days. The mean, standard deviation, and coefficient of variation of fog occurrences for these months under study is presented in Tables 10.1 and 10.2. There has been a certain consistency of fog occurrence in the months of December–January, whereas its occurrence in October, November, and February shows fluctuations. It can be seen (Figure 10.4) that during the month of October, the mean frequency of fog for the decade 1971–1980 was just 3.6 days, whereas it significantly increased to 17.3 days in the decade 2001–2010. Furthermore, during November, a remarkable increase of 136% was observed, and an increase of 119% was observed for December, January, and February. The average monthly visibility (<1 km) from January to December (12 months) (Figure 10.5) was evaluated and with the average poor visibility days range value of 15.58 days, January was identified as the prominent month whereas with lowest value of 1.42 days in April represents the least fog episodes over IGP. The highest average numbers of fog days were specific to the coldest months, January

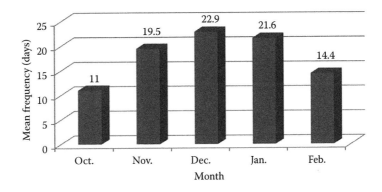

**FIGURE 10.3** Monthly mean frequency of fog over Ghaziabad (October–February).

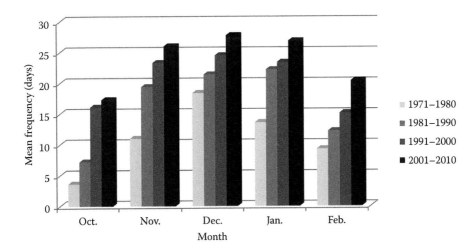

**FIGURE 10.4** Decadal frequency of fog.

## TABLE 10.1
### Climatological Data over Ghaziabad

| Month | Mean | Standard Deviation | Coefficient of Variation | Standard Error |
|---|---|---|---|---|
| October | 11.05 | 7.11 | 64.38 | 1.12 |
| November | 19.93 | 8.41 | 42.20 | 1.33 |
| December | 23.03 | 6.33 | 27.49 | 1.00 |
| January | 21.65 | 6.74 | 31.14 | 1.07 |
| February | 14.25 | 6.01 | 42.19 | 0.95 |

## TABLE 10.2
### Fog Statistics over IGP

| | Mean | Standard Error | Standard Deviation | Variance |
|---|---|---|---|---|
| January | 15.58 | 1.25 | 6.78 | 45.95 |
| February | 7.39 | 0.92 | 4.94 | 24.39 |
| March | 2.69 | 0.43 | 2.29 | 5.25 |
| April | 1.42 | 0.16 | 0.85 | 0.73 |
| May | 1.96 | 0.22 | 1.19 | 1.43 |
| June | 3.25 | 0.95 | 5.14 | 23.38 |
| July | 2.29 | 0.19 | 1.03 | 1.07 |
| August | 2.31 | 0.21 | 1.09 | 1.18 |
| September | 2.41 | 0.28 | 1.49 | 2.24 |
| October | 4.26 | 0.68 | 3.66 | 13.42 |
| November | 7.82 | 1.19 | 6.38 | 40.75 |
| December | 13.27 | 1.47 | 7.92 | 62.67 |

(~16 days) and December (~13 days). This information reveals that the monthly fog occurrence for the last 40 years was not very consistent. The maximum variance in poor visibility frequency was during the winter season and the least variance was observed during premonsoon season.

### 10.4.2 Duration and Variability of Fog

Duration and variability of fog is one of the important factors in terms of safety and also determines the duration of sunshine over a specific area. Figure 10.6a shows the plotted frequency of each duration category as a percentage of all fog events over Ghaziabad for the months of October to February. The most common fog event duration during October with a frequency of 19.8% was 0–1 h. The frequency decreased to 6.6% for fog events with a duration of 7–8 h. The two most common fog event durations during November, both with a frequency of 9.5%, were 5–6 h and 20–24 h. The most common fog event duration during December with a frequency of 14.8% was 10–11 h. During January, the most common fog event duration with a frequency of 18.8% was 20–24 h. Finally, the most common fog event duration during February was 1–2 h with a frequency of 14.4%. Fog events lasting for almost the entire day occurred most often in January with a frequency of 18.8%, followed by 10.9% during December and 9.5% during November. The duration was directly related to the availability of solar radiation (length of day). The analysis of the results reveals that shorter duration fog events were more frequent in October and February and that longer duration fog events were more frequent during January, November, and December.

# Fog—A Ground Observation-Based Climatology and Forecast over North India

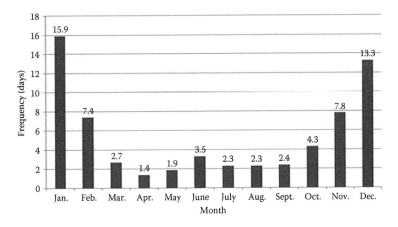

**FIGURE 10.5** Average monthly visibility (less than 1 km in days).

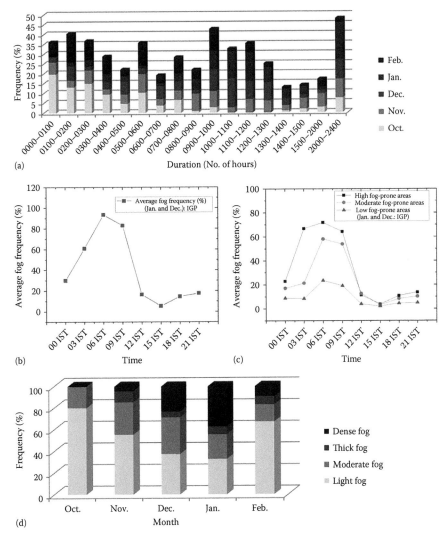

**FIGURE 10.6** (a) Percentage frequency of duration, (b) diurnal variability of fog (January and December), (c) diurnal variability of fog (clusterwise), and (d) percentage frequency of intensity.

To ascertain the most fog-prone period in a day during peak winter months of January and December over IGP, fog diurnal variability were examined; see Figure 10.6c. It can be observed from Figure 10.6b that average fog frequency was more than 80%; that is, on more than 80% of the occasions between 0600 and 0900 h, the visibility over IGP was poor due to fog during the peak winter months of January and December. The percentage dropped rapidly after 0900 h to less than 20% by 1200 h, indicating a rapid improvement in visibility after 0900 h. The average fog frequency percentage reached its lowest value close to 10% by 1500 h. It remained below 20% from 1200 h until around midnight. Figure 10.6c illustrates the diurnal variability of fog frequency in percentages for all the types of fog-prone areas (clusterwise). It was observed that over the high fog cluster areas, the percentage reached above 60 by 0300 h and remained above 60 until 0900 h. The figure reveals that on 60%–70% of occasions, the visibility over high fog-prone areas of IGP remained poor due to fog during January and December. Over the moderate fog cluster areas of IGP, the peak remained between 50% and 60% at 0600–0900 h.

### 10.4.3 Intensity and Persistence of Fog

Fog was classified into four categories depending on the surface visibility: (1) "light" when the reported visibility was less than 1000 m but more than 500 m, (2) "moderate" when the horizontal visibility was less than 500 m but more than 200 m, (3) "thick" when the visibility was less than 200 m but more than 50 m, and (4) "dense" when the visibility was less than or equal to 50 m. Figure 10.6d shows the percentage frequency of fog under different categories for the months between October and February. The figure shows that during October, 80% of fog events was classified as light fog and 20% was moderate. Thick fog and dense fog never occurred during the month of October. During the month of November, the percentage frequency of light fog fell to 55% with the frequency of moderate fog increasing to 30%. Thick and dense fog was seen on a few occasions during this month. During December, light fog frequencies further fell to 37% whereas moderate and thick fog increased in frequency to 34% and 28%, respectively. However, during January there was a significant hike in the percentage frequency of dense fog, which went up to 37%. Furthermore, a reversal in behavior was observed in February when the percentage frequency of light fog increased to 67% with moderate and dense fog reducing significantly to 16% and 9%, respectively. Data analysis reveals that incidents of dense fog were higher during peak winter months December and January) whereas light fog was more prevalent before and after the peak winter months (i.e., October and February).

The average daily persistence of fog over different locations of IGP was important to reduce the negative impact of fog on day-to-day activities of one of the most densely populated regions of the world. Figure 10.7 shows the average persistence of fog in a day during January and December over various locations of IGP. It was observed that over central IGP, the average fog duration in a day for

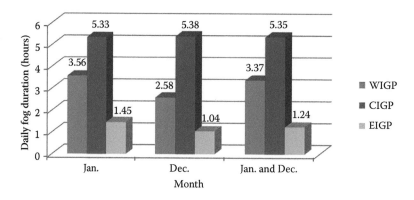

**FIGURE 10.7** Average daily fog persistence (h).

January and December was 5h 33min, 5h 38min respectively, and for both the months together was 5h 35min. During the month of December, the maximum average daily duration over Gorakhpur was approximately 9 h whereas it fell to approximately 7 h in the month of January (Kendall, 1975). IGP, the average fog duration in a day for each month of January and December and both months together was 3 h 56 min, 2 h 58 min, and 3 h 27 min, respectively. During the month of December, the maximum average daily duration over Jalandhar was approximately 5 h 12 min whereas it increased to approximately 5 h 40 min in the month of January. Furthermore, it was observed that the average daily fog duration was maximum over central IGP and was about 55% more than average daily fog duration over west IGP. The average fog duration fell to just less than 1 hour in a day over northeastern India except over Jorhat where the average duration was 3–4 h. The study revealed that December and January were the heavy fog months over IGP. The factors responsible for frequent, persistent, and intense fog include the topography; synoptic features such as passage of western disturbance, pollution, and irrigation network, which includes rivers, canals, and other water bodies; land use and other meteorological parameters like low temperature, high humidity, and so on. The increasing aerosols amount, the surface temperature and higher relative humidity condition were found favoring the formation of fog (Das et al., 2008). One of the physical mechanisms responsible for fog formation involves the cooling of air to its dew point when the surface beneath cools. This mechanism results in radiation fog because radiation is the means by which surface can cool down quickly. It happens at a much greater extent if the sky is clear.

### 10.4.4 Trend Analysis

Trend analysis was carried out in order to estimate the uncertain pattern of fog in the past and to predict its future pattern. Then a nonparametric MK test was performed on the time series of fog observations for the months of October to February during the period 1971–2010. The results are presented in Table 10.3, showing that the trend was positive with values of 0.50 for October, 0.47 for November, 0.30 for December, 0.39 for January, and 0.37 for February. This indicates that the increasing trend of fog occurrence reached a maximum for the month of October and a minimum for the month of December. The null hypothesis states that the fog data series was independent whereas the alternate hypothesis states the existence of a trend. Because the *p*-values were lower than the significance level (0.5), the MK test at the 95% confidence level rejected the null hypothesis and accepted the alternate hypothesis that there exists a trend for all five months under study. Kendall's tau statistics measure correlation and therefore demonstrate the strength of the relationship between the two variables. In this case as shown in (Figure 10.8a and b), the statistics showed that there exists a positive correlation between the time series and number of fog days. The value reaches a maximum of the order of 0.64 for the month of October; followed by 0.54 for the months of November, January, and February; and 0.44 for December.

### TABLE 10.3
### Mann–Kendall Trend Statistics

|  | October | November | December | January | February |
| --- | --- | --- | --- | --- | --- |
| Kendall's Tau | 0.643 | 0.321 | 0.443 | 0.542 | 0.542 |
| S | 470.0 | 378.00 | 320.00 | 393.00 | 390.00 |
| Level of significance | 0.05 | 0.05 | 0.05 | 0.05 | 0.05 |
| Ho | Ho: Rejected | Ho: Rejected | Ho: Rejected | Ho: Rejected | Ho: Rejected |
| Ha | Ha: Accepted | Ha: Accepted | Ha: Accepted | Ha: Accepted | Ha: Accepted |
| P-value | <0.0001 | <0.0001 | <0.0001 | <0.0001 | <0.0001 |
| Sen's slope | 0.50 | 0.47 | 0.30 | 0.39 | 0.37 |

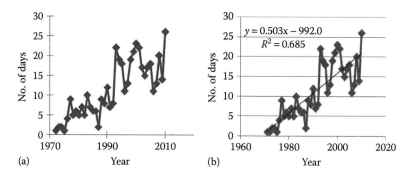

**FIGURE 10.8** Correlation between the time series and number of fog days. (a) Mann–Kendall trend (October) and (b) linear trend (October).

Figure 10.9a and b illustrates the time series analysis of fog frequency over IGP considered independently for the months of December and January over the last 40 years. Fog frequency and interannual variability were generally higher during January when compared to December; however, turnaround points were almost same during both the months. Furthermore, trend analysis was carried out in order to evaluate uncertain patterns of fog in the past and to predict its future pattern. Trend analysis was performed using the number of foggy days for the months of December and January during the period from 1971 to 2010. The result is illustrated in Figure 10.10. The null hypothesis stated that fog data series was independent whereas the alternate hypothesis indicated the existence of a trend. Because the $p$-values were lower than the significant level (0.5), the MK test at 95% confidence level rejected the null hypothesis and accepted the alternate hypothesis that there exists a trend for almost all stations over IGP. During the month of January, the highest MK statistics value of 429 was observed over Ludhiana, followed by Jalandhar and Ghaziabad with values of 402 and 384, respectively. Kendall's tau correlation value of 0.59 (Table 10.4) over Ludhiana was highest over the region followed by Jalandhar and Ghaziabad with values 0.58 and

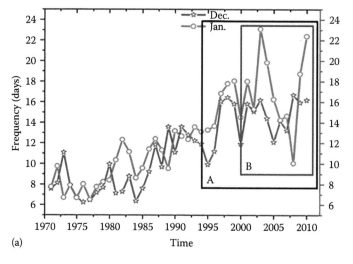

**FIGURE 10.9** Time series analysis of fog frequency over IGP considered independently for the months of December and January over the last 40 years. (a) Time series analysis (January and December). *(Continued)*

# Fog—A Ground Observation-Based Climatology and Forecast over North India

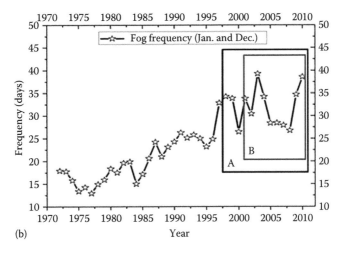

(b)

**FIGURE 10.9 (Continued)** Time series analysis of fog frequency over IGP considered independently for the months of December and January over the last 40 years. (b) Time series analysis (Peak winter).

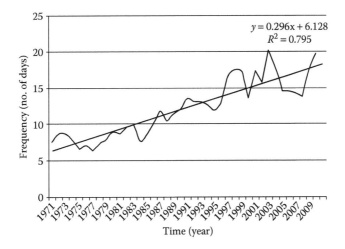

**FIGURE 10.10** Trend analysis of winter fog.

**TABLE 10.4**
**Kendall's Tau Correlation and Sen's Slope Magnitude**

|  | Kendall's Tau (January) | Sen's Slope (December) | Tau's (January) | Sen's Slope (December) |
|---|---|---|---|---|
| Jalandhar | 0.58 | 0.46 | 0.52 | 0.48 |
| Ambala | 0.52 | 0.46 | 0.59 | 0.58 |
| Bhatinda | 0.41 | 0.30 | 0.53 | 0.42 |
| Ludhiana | 0.59 | 0.57 | 0.59 | 0.64 |
| Ghaziabad | 0.56 | 0.40 | −0.25 | 0.31 |
| Saharanpur | −0.09 | 0.00 | 0.52 | 0.00 |
| Bareilly | 0.51 | 0.39 | 0.53 | 0.41 |
| Kanpur | 0.52 | 0.41 | 0.57 | 0.44 |
| Allahabad | 0.57 | 0.47 | 0.13 | 0.47 |
| Jorhat | 0.13 | 0.06 | 0.32 | 0.06 |

0.56, respectively. This signifies that there exists a strong correlation between time in years and fog frequency over the station with high values, that is, as the time progress the fog frequency also tended to increase accordingly.

### 10.4.5 DECADAL TIME SERIES AND TREND ANALYSIS OVER IGP

To have a better appreciation of various characteristics of fog and to know precisely the most significant period of change, the 40 years of data were divided into four decades (1971–1980, 1981–1990, 1991–2000, 2001–2010). Figure 10.11 and Table 10.5 illustrate monthly mean frequency per decade of fog for the peak winter months of January and December. It can be observed from the figure that there was an overall increase over IGP of approximately 99% in fog frequency during the 40-year period. Furthermore, the annual variability increased after 1997 and was much more significant in the decade 2001–2010. It was observed that the mean frequency of fog has increased continuously in each decade from 1971–1980 to 2000–2010. Over the high fog cluster zone, the average fog frequency over IGP has been enhanced by approximately 99% during the months of December and January in the four decades studied. Over the high fog cluster zone during December, the maximum increase of the order of ~42% was observed during 1990–2000. The increase of the orders of 40% and 54% was seen for the moderate and low fog cluster zones, respectively. However, during the month of January, the increase of the orders of 44% and 74% was observed for the high and moderate fog-prone zones, respectively, during 1998–1990 whereas the increase of 97% was seen for the low fog-prone zone during 1990–2000.

### 10.4.6 SPATIAL VARIABILITY

IGP has a spatial coverage of about 867,000 km² and being such a vast area, IGP experiences spatial variability of fog across the entire region. The monthly spatial variability of days having visibility <1000 m indicates the presence of fog there on an average of 7 to 10 days in December, and >8 days during January experience fog (Sawaisarje et al., 2014). Figure 10.12 illustrates the

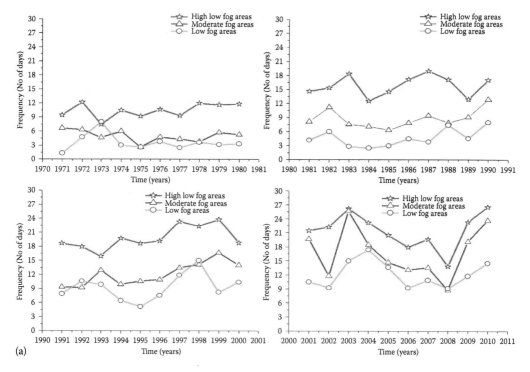

**FIGURE 10.11** Decadewise trend analysis. *(Continued)*

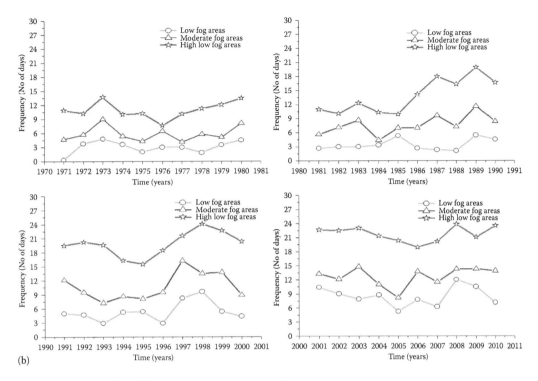

**FIGURE 10.11 (Continued)** Decadewise trend analysis.

**TABLE 10.5**
**Change in Fog Frequency by Decade**

| Month | 1971–1980 (A) | 1981–1990 (B) | 1991–2000 (C) | 2001–2010 (D) | % Change A–B | % Change B–C | % Change C–D | % Change A–D |
|---|---|---|---|---|---|---|---|---|
| December (H) | 10.9 | 13.9 | 19.8 | 21.7 | 27.5 | 42.4 | 9.6 | 99.1 |
| December (M) | 5.9 | 7.7 | 10.8 | 12.7 | 30.5 | 40.3 | 17.6 | 115.3 |
| December (L) | 3.0 | 3.5 | 5.4 | 8.5 | 16.7 | 54.3 | 57.4 | 183.4 |
| January (H) | 10.8 | 15.9 | 19.8 | 21.5 | 44.4 | 26.9 | 8.6 | 99.0 |
| January (M) | 5.0 | 8.7 | 12.1 | 16.8 | 74.0 | 39.1 | 39.2 | 362.8 |
| January (L) | 3.6 | 4.7 | 9.3 | 12.1 | 30.6 | 97.9 | 30.1 | 236.1 |

spatial variability of fog over the entire IGP during the months of December and January. The hot spot analysis tool assesses whether high or low fog frequency values cluster spatially. This analysis calculates GI* statistics, which was a Z-score for fog frequency values for the study period over the study area. Figure 10.12a shows the most statistical significant positive Z-score: 0.966–2.588; Figure 10.12b shows the larger Z-score and high $p$-values: 0.5366–0.8727 for the region of more intense clustering of high fog frequency values. Our analysis revealed that the southeastern parts of Punjab, Haryana, Uttar Pradesh, Bihar, and some eastern parts of northeastern India were the hot spots or high fog-prone zone for occurrence over IGP during peak winter months of December and January. Similarly, the most statistical significant negative Z-scores of −1.66 to −0.30 and the smaller Z-score and small $p$-values of 0.0096–0.2359 were for the region of more intense clustering of low fog frequency values. These were the cold spots. The hilly regions of Jammu and Kashmir, Himachal Pradesh, Uttaranchal, and northeastern India were cold spot areas or low fog-prone zones.

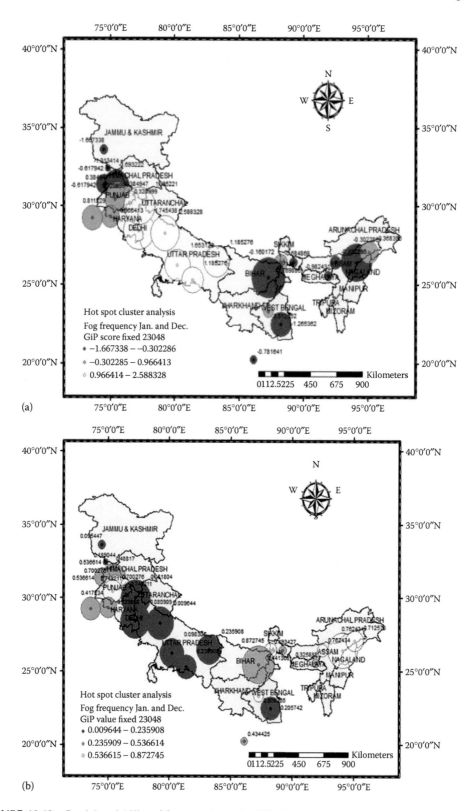

**FIGURE 10.12** Spatial variability of fog over the entire IGP during the months of December and January. (a) Spatial cluster analysis (Z-score) and (b) spatial cluster analysis (*p*-value). *(Continued)*

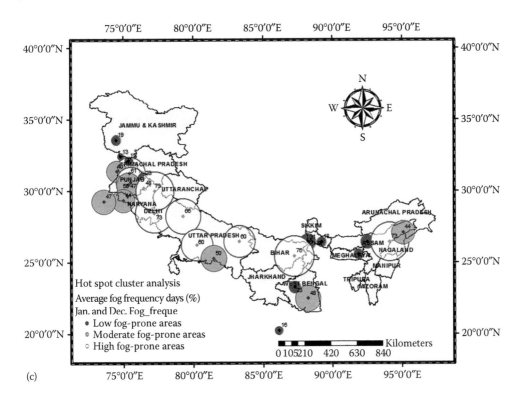

**FIGURE 10.12 (Continued)** Spatial variability of fog over the entire IGP during the months of December and January. (c) Spatial cluster analysis (% average fog frequency).

The most statistical significant Z-score was −0.30 to 0.966, moderate Z-score and moderate p-values were 0.2359–0.5366 in the region of more intense clustering of moderate fog frequency values. These were neither hot spots nor cold spots but were the moderate spots as far as fog occurrence is concerned. West and south Punjab, north Madhya Pradesh, Orissa, parts of west Bengal, and northeastern India were the moderate spots.

Figure 10.12c reveals the percentage frequency of average fog days over IGP during the peak winter months of January and December. The analysis of results reveals that there was an increase in number of fog days in January compared to December over the entire Punjab, south and east Uttar Pradesh, Bihar, West Bengal, and northeastern areas of India. However, a decrease was observed over central and west Uttar Pradesh. Significant increases of up to 17% over Punjab, 21% over West Bengal, and 18% over northeastern India were observed, whereas a small increase of the order of 0%–6% was seen over rest of the IGP area.

### 10.4.7 Autoregressive Integrated Moving Average Modeling

The ARIMA model predicts future values as a product of several past observations and random errors (Yurekli et al., 2007; Narayanan et al., 2013). This model was utilized to forecast the number of foggy days in the months of October to February for the period 2015–2020 over Ghaziabad. The forecast period was restricted to 5 years because it was evident from the trend analysis that fog occurrence did not follow a linear pattern over time. The most parsimonious ARIMA model, which passed the test of randomness in forecasting fog days for the months from October to February, was selected. Figure 10.13 represents the observed and fit values. Model fit statistics $R^2$, root mean square error (RMSE), mean absolute percentage error (MAPE), and normalized Bayesian information

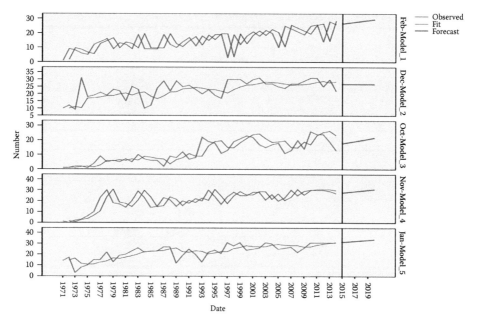

**FIGURE 10.13** ARIMA model forecast.

**TABLE 10.6**
**ARIMA Model Fit Statistics**

| Fit Statistics | Mean | SE | Minimum | Maximum |
| --- | --- | --- | --- | --- |
| Stationary R-squared | 0.118 | 0.117 | 0.000 | 0.236 |
| R-squared | 0.307 | 0.354 | −0.237 | 0.584 |
| RMSE | 5.499 | 0.896 | 4.696 | 6.950 |
| MAPE | 30.963 | 11.752 | 18.748 | 48.119 |
| MaxAPE | 285.220 | 208.328 | 70.026 | 555.713 |
| MAE | 4.160 | 0.759 | 3.490 | 5.347 |
| MaxAE | 15.389 | 3.365 | 12.605 | 20.733 |
| Normalized BIC | 3.494 | 0.296 | 3.181 | 3.965 |

SE: standard errors.

criterion (BIC) are given in Table 10.6. The model in which these statistics were lowest was selected as the most appropriate. Additionally, that model also shows the forecast for the occurrence of fog over Ghaziabad during the months from October to February for the 5 years from 2015 to 2019. The forecast for the study region obtained through the ARIMA model indicates an increasing trend for all months from October to February.

### 10.4.8 RELATIONSHIP BETWEEN FOG OCCURRENCE AND AEROSOL OPTICAL DEPTH

High aerosol concentration of aerosol contents in the lower atmosphere over the study area enhances the formation and persistence of the fog by additional condensation nuclei (Ramanathan et al., 2005; Prasad et al., 2006). Figure 10.14 shows the variability of AOD over the Punjab province from 2000 to 2010 for the months of January and December. A certain increase in the AOD over the years, which might have also influenced the fog occurrence over the region, was noticed. As shown in the

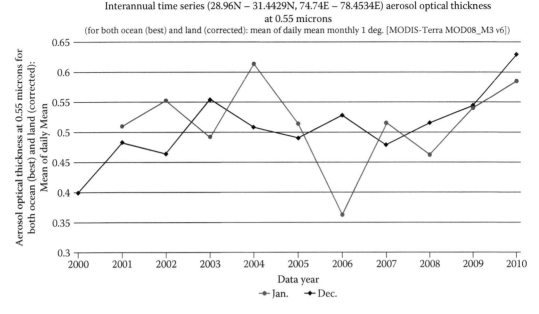

**FIGURE 10.14** AOD for the duration 2000–2010.

Table 10.7, there was a significant positive correlation between the AOD and the fog frequency over the region. Fog frequency for the months of January showed a positive correlation of 0.70, 0.59, 0.43, 0.34, and 0.53 with the AOD values for the same month for Ludhiana, Bhatinda, Amritsar, Ambala, and Jhalandhar, respectively. For December months, all the other cities except Ludhiana showed a significant positive correlation (Table 10.7).

## 10.5 CONCLUSION

Meteorological observations for the period 1971–2010 (40 years) of the NCR of Ghaziabad and the IGP were analyzed. Time series and trend analyses were performed to understand the climatology of fog over the region. Hot spot analysis showing the spatial variability of fog and further ARIMA modeling was performed to predict the future trend. Results suggest the following:

- The frequency of occurrence of fog over Ghaziabad increased significantly in the months from October to February during the last four decades (1971–2010)
- Time series analyses confirmed with a 95% confidence level that there existed a positive (increasing) trend for all months between October and February.
- Data suggests that more than 80% of times the visibility over IGP was poor due to fog between 0600–0900 h during the peak winter months of January and December.
- In the study region, the average duration of fog in a day was about 55% more than the average fog duration as compared to the west IGP.
- Spatial analysis reveals that southeastern parts of Punjab, Haryana, Uttar Pradesh, Bihar, and some eastern parts of northeastern India were the hot spots or high fog-prone zone during the months of December and January.
- The forecast values obtained from the ARIMA model indicate that the number of fog days was expected to increase in future.
- There has been a positive correlation between the AOD and the fog frequency over Punjab region.

## TABLE 10.7
### Relationship with AOD 550 nm

| Year | Jalandhar | | Ambala | | Amritsar | | Bhatinda | | Ludhiana | | Punjab (AOD) | |
|---|---|---|---|---|---|---|---|---|---|---|---|---|
| | January | December | January | December | January | December | January | December | January | December | January | December |
| 2000 | 18 | 26 | 15 | 15 | 14 | 13 | 20 | 20 | 22 | 24 | | 0.39864 |
| 2001 | 23 | 28 | 18 | 24 | 17 | 11 | 23 | 23 | 23 | 27 | 0.513061 | 0.483001 |
| 2002 | 25 | 21 | 22 | 22 | 22 | 15 | 22 | 22 | 26 | 22 | 0.55237 | 0.464341 |
| 2003 | 26 | 24 | 29 | 23 | 27 | 16 | 28 | 28 | 26 | 22 | 0.491773 | 0.55511 |
| 2004 | 25 | 26 | 25 | 26 | 8 | 18 | 27 | 27 | 26 | 23 | 0.614047 | 0.508335 |
| 2005 | 26 | 25 | 24 | 24 | 13 | 16 | 26 | 26 | 17 | 22 | 0.514083 | 0.490194 |
| 2006 | 20 | 24 | 22 | 16 | 5 | 14 | 17 | 17 | 14 | 23 | 0.362599 | 0.528245 |
| 2007 | 23 | 21 | 15 | 21 | 15 | 8 | 18 | 18 | 24 | 18 | 0.515329 | 0.478683 |
| 2008 | 10 | 28 | 10 | 22 | 4 | 15 | 6 | 6 | 7 | 27 | 0.462287 | 0.515681 |
| 2009 | 29 | 20 | 23 | 24 | 16 | 10 | 24 | 24 | 24 | 18 | 0.540341 | 0.54418 |
| 2010 | 27 | 28 | 29 | 28 | 26 | 24 | 31 | 31 | 30 | 24 | 0.585205 | 0.629286 |
| Corr. value | 0.53473 | 0.1446 | 0.34581 | 0.63126 | 0.43211 | 0.57474 | 0.5908 | 0.36658 | 0.7099 | −0.044 | | |

## REFERENCES

Badrinath KVS, Kharol SK, Sharma AR et al. 2009. A study using multi-satellite data and ground observations. *IEEE J. Appl. Earth Obs. Rem. Sens.* 2(3): 185–195.

Bayazit M, Onoz B. 2007. To prewhiten or not to prewhiten in trend analysis. *Hydrol. Sci. J.* 52(4): 611–624.

Boochabun K, Tych W, Chappell NA et al. 2004. Statistical modelling of rainfall and river flow in Thailand. *J. Geol. Soc. India* 64: 503–515.

Box GEP, Jenkins GM, Reinsel GC. 2007. *Time Series Analysis: Forecasting and Control*, 3rd ed. Pearson Education: New York, pp. 8–12.

Chattopadhyay S, Chattopadhyay G. 2010. Univariate modelling of summer monsoon rainfall time series: Comparison between ARIMA and ARNN. *C. R. Geosci.* 342: 100–107.

Chattopadhyay S, Jhajharia D, Chattopadhyay G. 2011. Univariate modelling of monthly maximum temperaturetime series over northeast India: Neural network versus Yule–Walker equation based approach. *Meteorol. Appl.* 18: 70–82.

Collier CG. 1970. Fog at manchester. *Weather* 25(1): 25–29.

Das SK, Jayaraman A, Misra A. 2008. Fog induced variation in aerosol optical and physical properties over the Indo-Gangetic Basin and impact to aerosol radioactive forecasting. *Ann. Geophy.* 26: 1345–1354.

Dickey DA, Fuller WA. 1979. Estimators for autoregressive time series with a unit root. *J. Am. Stat. Assoc.* 74: 427–431.

Dutta HN, Singh B, Kaushik A. 2004. Role of western disturbances in the development of fog over Northern India. *XIII National Space Science Symposium (NSSS)*, Koteyam, India.

Glickman TS. 2000. *Glossary of Meteorology*, 2nd ed. American Meteorological Society: Boston, MA. http://glossary.ametsoc.org/ (accessed April 10, 2016).

Gultepe I, Tardif R, Michaelides SC. 2007. Fog research: A review of past achievements and future perspectives. *Pure Appl. Geophys.* 164: 1121–1159.

Hamed KH, Rao AR. 1998. A modified Mann–Kendall trend test for autocorrelated data. *J. Hydrol.* 204: 182–196.

Hansen JM, Sato P, Kharecha G et al. 2007. Climate change and trace gases. *Philos. Trans. R. Soc. A* 365:1925–1954.

Hyndman RJ, Athanousopoulos G. 2015. Forecasting: Principles and practice. 8.9. Seasonal ARMA models. https://www.otexts.org/fpp/8/9 (accessed June 21, 2015).

India Meteorological Department. 1982. Weather Codes. IMD: Pune, India.

Intergovernmental Panel on Climate Change (IPCC). 2007. The physical science basis. Contribution of Working Group I to the Fourth Assessment Report of the Intergovernmental Panel on Climate Change. Cambridge University Press: New York, pp. 589–663.

Intergovernmental Panel on Climate Change (IPCC). 2014. The physical science basis. Contribution of Working Group I to the Fourth Assessment Report of the Intergovernmental Panel on Climate Change. Cambridge University Press: New York, pp. 589–663.

Jenamani RK, Dash SK, Panda SK. 2007. Some evidence of climate change in twentieth-century India. *Clim. Change.* 85(3): 299–321.

Jenamani RK, Kalsi AR. 2012. Microclimatic study and trend analysis of fog characteristics at IGI airport New Delhi using hourly data (1981–2005). *Mausam* 63(2): 203–218.

Karmeshu N. 2012. Trend detection in annual temperature & precipitation using the Mann Kendall test–A case study to assess climate change on select states in the northeastern United States. http://repository.upenn.edu/mes_capstones/47 (accessed April 10, 2016).

Kendall MG. 1975. *Rank Correlation Methods*, 4th ed. Charles Griffin: London, UK.

Laskar SI, Bhowmik SKR, Sinha V. 2013. Some statistical characteristics of fog over Patna airport. *Mausam* 64(2): 345–350.

Lewis JM, Koracin D, Redmond KT. 2004. Sea fog research in the United Kingdom and United States: A historical essay including outlook. *Bull. Am. Meteorol. Soc.* 85: 395–408.

Mathur V, Singh K, Chawhan MD. 2012. Zero-visibility navigation for the Indian Railways. *International Journal of Computer Applicances* (IJCA),Bangalore,India. pp. 28–31.

Meteorological Office. 1960. *Handbook of Aviation Meteorology*. HMSO: London, UK.

Michaelides S, Gultepe I. 2008. Short range forecasting methods for fog, visibility and low clouds. Office for Official Publications of the European Communities: Luxembourg.

Mishra S, Mohapatra M. 2004. Some climatological characteristics of fog over Bhubaneswar airport. *Mausam* 55(4): 695–698.

Mohan M, Payra S. 2009. Influence of aerosol spectrum and air pollutants on fog formation in urban environment of megacity Delhi, India. *Environ. Monit. Assess.* 151(1–4): 265–277.

Narayanan P, Basistha A, Sachdeva K. 2016a. Trend analysis and forecast of pre-monsoon rainfall over India. *Weather* 71(4): 94–99.

Narayanan P, Basistha A, Sachdeva K. 2016b. Understanding trends and shifts in rainfall in parts of northwestern India based on global climatic indices. *Weather* 71(8): 198–203.

Narayanan P, Basistha A, Sarkar S et al. 2013. Trend analysis & ARIMA modelling of pre-monsoon rainfall data for western India. *C. R. Geosci.* 345: 22–27.

Prasad AK, Sing RP, Kafatos M. 2006. Influence of coal based thermal power plants on aerosol optical properties in the Indo-Gangetic basin. *Geophys. Res.* 118: 2956–2965.

Ram S, Mohapatra M. 2008. Some characteristics of fog over Guwahati airport. *Mausam* 59(2): 159–166.

Ramanathan V, Chung C et al. 2005. Atmospheric brown clouds: Impacts on south Asian climate and hydrological cycle. *Proc. Natl. Acad. Sci. USA*. 102: 5326–5333.

Saraf AK, Bora AK, Das J, Rawat V, Sharma K, Jain SK. 2011. Winter fog over the Indo-Gangetic plains: Mapping and modelling using remote sensing and GIS. *Nat. Hazards*. 58(1): 199–220. doi:10.1007/s11069-010-9660-0.

Singh A, Dey S. 2012. Influence of aerosol composition on visibility in megacity Delhi. *Atmos. Environ.* 62: 367–373.

Singh J, Giri RK, Kant S. 2007. Radiation fog viewed by INSAT-1 D and Kalpana Geo Stationary satellite. *Mausam* 58(2): 251–260.

Singh J, Kant S. 2006. Radiation fog over north India during winter from 1989–2004. *Mausam* 57(2): 271–290.

Srivastava SK, Sharma AR, Sachdeva K. 2016. A ground observation based climatology of winter fog: Study over the Indo-Gangetic Plains, India. *Int. J. Environ. Chem. Ecol. Geol. Geophys. Eng.* 10(7): 734–745. http://waset.org/Publication/a-ground-observation-based-climatology-of-winter-fog-study-over-theindo-gangetic-plains-india/10004884.

Srivastava SK, Sharma AR, Sachdeva K. 2017. An observation-based climatology and forecasts of winter fog in Ghaziabad, India. *Weather*. doi:10.1002/wea.2743.

Sun DZ, Yu Y, Zhang T. 2009. Tropical water vapor and cloud feedbacks in climate models: A further assessment using coupled simulations. *J. Clim.* 22(5): 1287–1304.

Suresh R, Janakiramayya MV, Sukumar E. 2007. An account of fog over Chennai. *Mausam* 58(4): 501–512.

Sawaisarje GK, Khare P, Shirve S, Deepa kumar S, Narkhede NM, 2014. Study of winter fog over Indian sub continent: Climatological perspective. *Mausam* 65:19–28.

Tardif R, Rasmussen RM. 2007. Event-Based climatology and typology of fog in the New York City region. *J. Appl. Meteorol. Climatol.* 46(8): 1141–1168.

Theil H. 1950. A rank-invariant method of linear and polynomial regression analysis. I, II, III. *Nederl. Akad. Wetensch., Proc.* 53: 386–392, 521–525, 1397–1412.

Yasmeen Z, Rasul G, Zahid M. 2012. Impact of aerosols on winter fog of Pakistan. *Pak. J. Meteorol.* 8(16): 21–30.

Yue S, Hashino M. 2003. Long term trends of annual and monthly precipitation in Japan. *J. Am. Water Resour. Assoc.* 39(3): 587–596.

Yue S, Pilon P, Cavadias G. 2002. Power of the Mann–Kendall and Spearman's rho test for detecting monotonictrends in hydrological series. *J. Hydrol.* 259: 254–271.

Yurekli K, Simsek H, Cemek B et al. 2007. Simulating climatic variables by using a stochastic approach. *Build. Environ.* 42: 3493–3499.

# 11 Estimation of Evapotranspiration through Open Access Earth Observation Data Sets and Its Validation with Ground Observation

*Kishan Singh Rawat, Sudhir Kumar Singh, and Anju Bala*

## CONTENTS

| | |
|---|---|
| 11.1 Introduction | 174 |
| 11.2 Methodology | 175 |
|     11.2.1 Study Area | 175 |
|     11.2.2 Method of Evapotranspiration Comparisons | 175 |
|         11.2.2.1 Comparison with Lysimeter | 175 |
|     11.2.3 Remote Sensing Observation | 176 |
|     11.2.4 Surface Energy Balance Algorithm for Land Model | 177 |
|         11.2.4.1 Estimation of Land Surface Variables Using Surface Energy Balance Algorithm for Land | 177 |
|         11.2.4.2 Solving the Surface Radiation Balance Equation for $R_n$ | 177 |
|         11.2.4.3 Soil Heat Flux | 177 |
|         11.2.4.4 Sensible Heat Flux | 178 |
|         11.2.4.5 Evaporative Fraction | 179 |
|         11.2.4.6 Daily Evapotranspiration | 179 |
| 11.3 Results and Discussions | 183 |
|     11.3.1 LAI and Normalized Difference Vegetation Index Relationship | 184 |
|     11.3.2 Result from Comparison with Conventional Method | 185 |
| 11.4 Summary and Conclusions | 186 |
| Acknowledgment | 187 |
| References | 187 |

**ABSTRACT** In this study, the surface energy balance algorithm for land (SEBAL) model is used for estimation of evapotranspiration (ET) by the application of satellite RS. These estimates are validated against a lysimeter at the Water Technology Centre (WTC) experimental field within the Indian Agricultural Research Institute (IARI) farmland, New Delhi. SEBAL model is based on the evaporative fraction concept, and it has been applied to LANDSAT

7-ETM+ (30 m resolution) data acquired over the IARI farmland. Daily ET from SEBAL was compared with observed ET.

Comparison with a lysimeter showed an RMSE of 0.14 mm d$^{-1}$, indicating a good performance using the present approach. The crop-growing period from (the SEBAL model exhibited an $R^2$ of 0.91, R-RMSE of 0.063 mm d$^{-1}$, PBIAS of 5.60, Mean Absolute Error (MAE) of 0.11), almost negligible Normalized Root Mean Square Error (NRMSE) and NSE, and $d$ of 1. These validation studies indicate that remote sensing observations of ET provide a good comparison with ground-based stations for a nonstress wheat crop in the WTC field. The methodology described in the present study can be applied to remote areas of agricultural land where ground-based data are not possible.

## 11.1 INTRODUCTION

Information about the transfers of mass and energy at a surface is essential for hydrological and agricultural (or crop water) management. It is also crucial for a better understanding and prediction of climatic and hydrological systems. Remote sensing (RS) is an excellent and feasible tool for this purpose as it provides information related to mass and energy transfers, particularly to ET (Kumar et al., 2013).

ET is one of the basic processes controlling the equilibrium of our globe. It constitutes the link between the hydrological and energetic equilibrium at the soil vegetation atmosphere interface, and knowledge of it is vital for climatic and meteorological (especially agricultural meteorology) studies. Monitoring ET at large scales is essential for assessing the climatic and anthropogenic impacts on natural and agricultural ecosystems.

At field scales, ET can be measured over a homogenous surface using conventional techniques such as the Bowen ratio, eddy covariance, water balance, and lysimeter systems; however, these systems cannot provide spatial trends at the regional scale, especially in heterogeneous landscapes. Furthermore, the estimation of ET using visible and infrared satellite RS data has been at the center of numerous methodological approaches during the past few decades (Li et al., 2009). The deterministic models based on complex submodels such as soil vegetation atmosphere transfer (SVAT) models (Boulet et al., 2007; Li et al., 2009) are mainly applicable for calculating ET, surface energy exchanges, and water balance. Most of the transfer mechanisms (water transfers and turbulence) and some physiological processes (stomata regulation and photosynthesis) are previously described models. Their time resolution is less than 1 hour in concord with the dynamics of surface and atmospheric processes. However, these models are more complex due to their use of numerous parameters that are difficult to understand, making them not feasible for spatial integration (Jacob, 1999).

Over the last few years, researchers have been involved in estimating ET from RS. For this reason, different methods have been developed to derive surface fluxes from RS observations, such as SEBAL (Bastiaanssen, 2000; Bastiaanssen et al., 1998; Jacob et al., 2002), S-SEBI (Roerink et al., 2000), SEBS (Jia et al., 2003; Su, 2002), and TSEB (French et al., 2003; Kustas and Norman, 1999).

Optical satellite RS is a strong tool for assessing instantaneous and daily ET at regional and global scales with surface energy budget closure. The methods proposed in the literature range from simple and empirical approaches to difficult and data-consuming ones (Glenn et al., 2007). Among the difficult ones are SVAT models, which describe the diurnal course of mass and heat transfers provided that micrometeorological conditions and water/energy balance parameters are recognized (Mahfouf et al., 1995; Olioso et al., 1996, 2005; Calvet et al., 1998; Coudert et al., 2006; Gentine et al., 2007). Among the liner approaches is the basic (normal) relationship, which relates daily ET ($ET_{24h}$) to midday near-surface temperature ($T_s$) gradient (Jackson et al., 1977). Between difficult and empirical models, compromising solutions are energy balance (EB) models. These models compute instantaneous ET as the residual term of the energy budget, net radiation ($R_n$), soil heat flux ($G$), and sensible heat flux ($H$) based on observations from satellite overpasses (Allen et al., 2007;

# Estimation of Evapotranspiration through Open Access Earth Observation

Bastiaanssen et al., 1998; Cleugh et al., 2007; Crow and Kustas, 2005; Caparrini et al., 2003, 2004; French et al., 2007; Mu et al., 2007; Norman et al., 2003; Su, 2002).

Instantaneous values of ET at satellite overpass can be utilized as diagnostic tools for surface status (Chandrapala and Wimalasuriya, 2003) or as controls for hydrological models through integration (Schuurmans et al., 2003). $ET_{24h}$ can be inferred from Food and Agricultural Organization (FAO)-56, but difficulties come (or arise) when extrapolating outside the environmental conditions considered for calibration. The ET diurnal course can be inferred by assimilating sun synchronous observations into the SVAT model, but this is limited by uncertainties when estimating SVAT parameters and primary input variables. The ET diurnal course can also be retrieved from geostationary satellites data sets, but the poor resolutions (kilometric) severely limit water management at the field scale. Perhaps the most practical solution is to assess instantaneous values from energy balance models combined with sun-synchronous satellite data sets and then extrapolating at the daily scale by presuming generic trends for the diurnal courses of ET and related variables.

SEBAL is one of the residual methods (models) of the energy budget (Kustas et al., 1994; Kumar et al., 2013). It consists of empirical models and physical parameterization. These inputs include local weather data (wind speed, resistance to turbulent transfer, air temperature, etc.) and satellite radiance. From the input data, the $R_n$, normalized difference vegetation index (NDVI), albedo ($\alpha$), roughness length, and $G$ are calculated. $H$ is calculated by measuring two contrasting points (as cold and hot pixel points). Then, the ET is calculated as the residual of the energy budget.

The major objectives of this study are as follows:

1. To derive ET from the SEBAL model using high-resolution LANDSAT 7 Enhanced Thematic Mapper (ETM) on nonstressed wheat crop land (WTC experimental field, IARI farmland) conditions in the IARI, New Delhi.
2. To compare and validate ET derived from the SEBAL model using the RS technique with that from a ground-based lysimeter.

## 11.2 METHODOLOGY

### 11.2.1 Study Area

A field experiment was carried out on the field of the WTC, farm of IARI, New Delhi, located at 28° 38′ 23″ and 77° 09′ 27″ with an altitude of 228.6 m above mean sea level; see Figure 11.1a and b.

For validation purposes, a lysimeter in Figure 11.1a was installed near 28° 37′ 54.54″ N 77° 09′ 35.24″ E, in the state of New Delhi (India), which covers the entire 500 m² farm of IARI during which wheat (in winter) is the major crop. The map of the lysimeter location in the WTC field is shown in Figure 11.1a and b.

### 11.2.2 Method of Evapotranspiration Comparisons

The IARI WTC field was selected for validation of ET from SEBAL. SEBAL was applied to four high-resolution (30 m) satellite images (ETM+ images; Table 11.1) and validated against accurate lysimeter observations. The lysimeter used for validation purposes was kept inside the wheat crop field situated in the WTC experimental field.

#### 11.2.2.1 Comparison with Lysimeter

Three lysimeters were located along the WTC field from which daily ET was derived. Using the SEBAL model, daily ET was estimated for the years 2010 and 2011, and these values were compared with the lysimetric data. Two internal calibration points (cold and hot pixels) of the SEBAL were located near the lysimeter location.

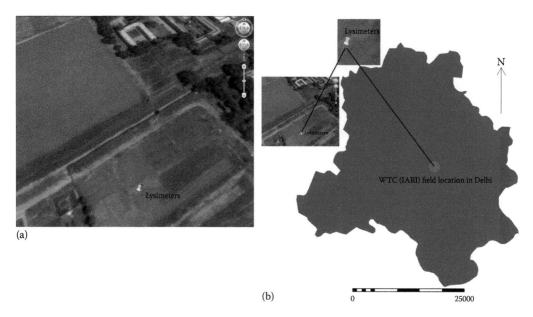

**FIGURE 11.1** Location of study area. (a) Lysimeter location in WTC experimental field and (b) WTC experimental field location over New Delhi.

**TABLE 11.1**
**Detail Remote Sensing (RS) Data Description**

| Date | Row/Path | Time Data Acquired | Sensor |
|---|---|---|---|
| December 23, 2010 | 146/40 | 10:11:55 a.m. | ETM+ |
| February 9, 2011 | 146/40 | 10:12:16 a.m. | ETM+ |
| March 14, 2011 | 146/40 | 10:12:27 a.m. | ETM+ |
| April 29, 2011 | 146/40 | 10:12:26 a.m. | ETM+ |

### 11.2.3 REMOTE SENSING OBSERVATION

Remotely sensed data used in this study were LANDSAT 7-ETM+ measurements. The ETM+ measurements from four different cases during December 2010–April 2011 were downloaded from http://daac.gsfc.nasa.gov/. The descending mode images from ETM+ data during daytime having spatial resolution of 30 m² at satellite nadir were utilized. In this study, four LANDSAT satellite images (path 146/row 40), acquired by LANDSAT 7, were processed. Images with 0% cloud cover were considered for the processing. The LANDSAT overpass time was at approximately 10:12:16 a.m. local time (LT) of India. All images (LANDSAT 7) were geometrically and radiometrically corrected and georegistered.

Due to instrument failure of LANDSAT 7 on May 31, 2003, there was an approximate 22% loss of the image data per scene. The problem was due to the failure of the Scan Line Corrector (SCL), which accounts for the forward motion of the satellite. Without an operational SCL, the ETM+ line of sight traced a zigzag along the satellite ground track, resulting in missing data across most of the image with scan gaps varying in size from 2 pixels near the center of the image to 14 pixels along the east and west edges. These impacts were pronounced along the edge of the scene and gradually diminished toward the center of the scene. The middle of the scene should be similar to previous LANDSAT image data acquired prior to the failure of SCL. In this study, the analysis was applied to the middle part of the scene using four LANDSAT 7-ETM+ measurements (over the lysimeter location) during 2010–2011 seasons. The gap can be filled by using well-defined different techniques such as box averaging, Bessel interpolation, and so on. However, since the lysimeter location always fell inside the undistorted part of ETM+ data set, we avoided such interpolation.

### 11.2.4 Surface Energy Balance Algorithm for Land Model

SEBAL utilizes the generally applied residual approaches of surface energy balance to estimate ET on different spatial and temporal scales. The net energy from the sun and atmosphere in the form of short- and long wave radiation is transformed and used for (1) heating the soil, (2) heating the surface environment, and (3) transforming water into vapor. All the energy involved in the SVAT interface can be expressed as the EB equation:

$$R_n = G + H + \lambda ET$$

In computation of ET, the surface EB equation is the initial boundary condition to be satisfied:

$$R_n - G - H - \lambda ET = 0 \quad (11.1)$$

where:
$R_n$ is the net radiation
$G$ is the soil heat flux
$H$ is the sensible heat flux
$\lambda ET$ is the latent heat flux associated with ET, all are in same unit (Wm$^{-2}$)

#### 11.2.4.1 Estimation of Land Surface Variables Using Surface Energy Balance Algorithm for Land

*11.2.4.1.1 Net Radiation*

Net radiation ($R_n$) is the most important term in the SEBAL since it represents the source of energy that must be balanced by the thermodynamic equilibrium of the remaining terms. The net radiation may also be expressed as an electromagnetic balance of all incoming and outgoing fluxes reaching and leaving a flat horizontal and homogeneous surface as:

$$R_n = (1-\alpha).S\downarrow + (L\downarrow - L\uparrow) \quad (11.2)$$

where:
$\alpha$ is albedo (unit less)
$S\downarrow$ is incoming shortwave (0.3–3 µm) radiation
$L\downarrow$ is incoming long wave (3–100 µm) radiation
$L\uparrow$ is outgoing long wave (3–100 µm) radiation; all are in same unit (Wm$^{-2}$)

#### 11.2.4.2 Solving the Surface Radiation Balance Equation for $R_n$

$R_n$ is computed using Equation 11.2. $R_n$ thus computed is expressed in Figure 11.2.

#### 11.2.4.3 Soil Heat Flux

The following equation was used by Bastiaanssen (2000) to calculate $G$.

$$G = R_n \times (T_s - 273.15)(0.0038 + 0.0074\alpha)(1 - 0.98 \times NDVI^4) \quad (11.3)$$

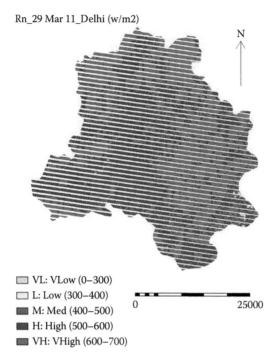

**FIGURE 11.2** Net radiation over New Delhi.

where:
$T_s$ is surface temperature in K (from LANDSAT 7 imagery, Appendix figure f)
NDVI is normalized vegetation index

Calculated $G$ is shown in Figure 11.3.

### 11.2.4.4 Sensible Heat Flux

$H$ was calculated using the bulk aerodynamic resistance model and a procedure that assumed a linear relationship among the aerodynamic surface temperature air temperature difference ($dT$) and $T_s$ calculated from extreme pixels (highest value pixels of $T_s$). Extreme pixels showing hot/dry and cold/wet spots were selected to develop a linear relationship between $dT$ and $T_s$. At the cold pixel in the satellite imagery, $H$ was assumed to be nonexistent, that is, $H_{\text{cold}} = 0$ and at the hot pixel, latent heat flux ($\lambda E$) was set to zero, which in turn allowed setting $H_{\text{hot}} = (R_n - G)_{\text{hot}}$. Then, $dT_{\text{hot}}$ and $dT_{\text{cold}} = 0$ can be obtained by solving the bulk aerodynamic resistance equation for the hot pixel as:

$$H = \rho_a C_p \frac{T_o - T_a}{r_{ah}} \tag{11.4}$$

where:
$\rho_a$ is air density (kg/m³)
$C_p$ is specific heat of air (1004 J kg⁻¹ K⁻¹)
$r_{ah}$ is the aerodynamic resistance to heat transport (s m⁻¹)

After calculating $dT$ at both hot and cold pixels, a linear relationship between $T_s$ and $dT$ was developed to estimate $H$ iteratively correcting $r_{ah}$ for atmospheric stability by using *Calculations_for_05_H_Automatic_GMD_model.xls* (developed by Allen et al., 2007) as presented in *Advanced*

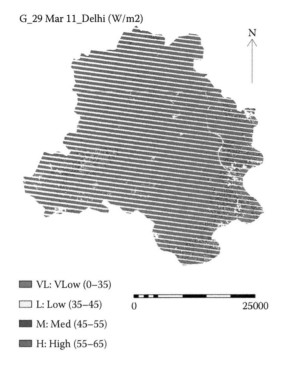

**FIGURE 11.3** Soil heat flux over New Delhi.

*Training and Users Manual*, Version 1.0 (2002). This was implemented by the application of the Monin Obhukov Similarity (MOS) theory (Foken, 2006). This step required $T_s$ and horizontal wind speed (u, m s$^{-1}$) that were assessed at a nearby automatic weather station (see the appendix) and a mechanism that extrapolates wind speed to a blending height of 100–150 m. In this study, a height of 100 m was used for the estimation of distributed friction velocity, a term utilized in the estimation of $H$ (Figure 11.4).

### 11.2.4.5 Evaporative Fraction

$\lambda E$ is the residual term of the energy budget and is used to calculate the instantaneous evaporative fraction ($\Lambda_{ins}$), so Equation 11.1 can be rewritten and expressed as $H$ by considering $\Lambda_{ins}$ (Figure 11.5) and net available energy ($R_n - G$).

$$\Lambda_{ins}(R_n - G) = \lambda ET \tag{11.5}$$

or

$$\Lambda_{ins} = \frac{\lambda ET}{R_n - G} = \frac{R_n - G - H}{R_n - G} \tag{11.6}$$

### 11.2.4.6 Daily Evapotranspiration

$\Lambda_{ins}$ expresses the ratio of the actual to the crop evaporative demand when the atmospheric moisture conditions are in equilibrium with the soil moisture (SM) conditions. The evaporative fraction tends to be constant during the daytime; the $H$ and $\lambda E$ fluxes, to the contrary, vary considerably. The difference between $\Lambda_{ins}$ at the moment of satellite overpass and the evaporative fraction derived from the 24-hour integrated energy balance is marginal ($\Lambda_{ins} \approx \Lambda_{24}$) and may be neglected (Crago, 1996:

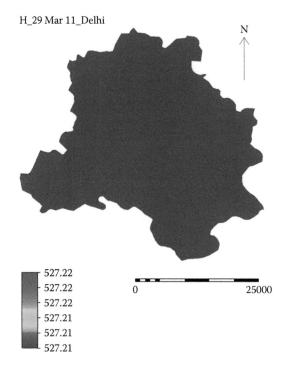

**FIGURE 11.4** Sensible heat flux over New Delhi.

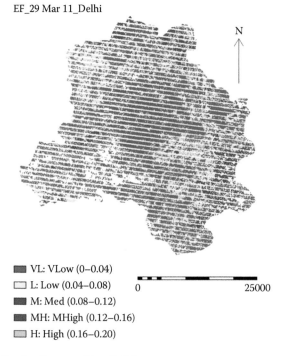

**FIGURE 11.5** Evaporative fraction over New Delhi.

Brutsaert and Sugita, 1992). For timescales of 1 day (24-h) or longer, $G$ can be ignored ($G = 0$) and net available energy ($R_n - G$) reduces to $R_n$. For the daily timescale, $ET_{24}$ (mm d$^{-1}$) can be calculated as follows:

$$\lambda ET_{24} = \Lambda_{24}(R_{n24} - G) = \Lambda_{ins}(R_{n24} - G)$$

$$ET_{24} = \frac{\lambda ET_{24}}{\lambda} = \frac{\Lambda_{ins} R_{n24}}{\lambda}$$

The final equation that is used to estimate $ET_{24}$ (Figure 11.6) is dependent on $\Lambda_{ins}$:

$$ET_{24} = \frac{8.64 \times 10^7 \Lambda_{ins} R_{n24}}{\lambda . \rho_w} \tag{11.7}$$

In Equation 11.7, $R_{n24}$ (24-h averaged $R_n$ = 1 day averaged $R_n$) is given in Wm$^{-2}$; $\lambda$ is $2.47 \times 10^6$ JKg$^{-1}$; and $\rho_w$ is 1000 kgm$^{-3}$. The final form of equation is expressed in Equation 11.8:

$$ET_{24} = \frac{\Lambda_{ins} R_{n24}}{28.588} \tag{11.8}$$

This is $ET_{24}$ (December 23, 2010 [Figure 11.6], February 9, 2011 [Figure 11.7], March 29, 2011 [Figure 11.8], and April 14, 2011 [Figure 11.9]) from SEBAL for the wheat crop field of the Water Technology Centre of IARI, New Delhi. This daily $ET_{24}$ (from SEBAL) was validated from the lysimeter method ET.

Finally, we validate SEBAL algorithm with lysimeter data using different statistical tests.

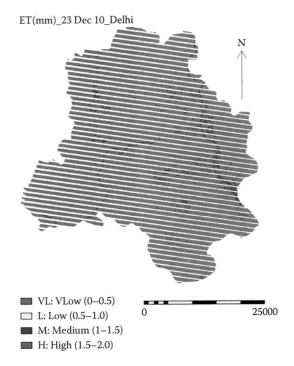

**FIGURE 11.6** $ET_{24h}$ during December 23, 2011, over New Delhi.

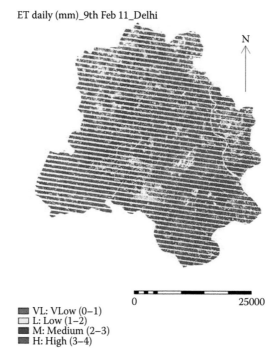

**FIGURE 11.7** $ET_{24h}$ during February 9, 2011, over New Delhi.

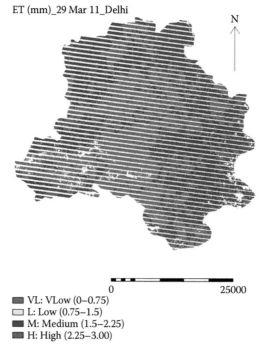

**FIGURE 11.8** $ET_{24h}$ during March 29, 2011, over New Delhi.

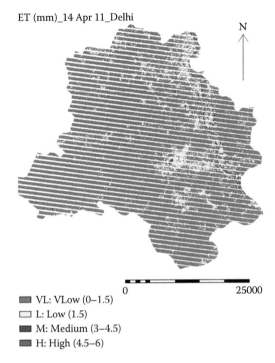

**FIGURE 11.9** $ET_{24h}$ during April 14, 2011, over New Delhi.

## 11.3 RESULTS AND DISCUSSIONS

Distribution maps were generated (as Rawat et al., 2017) for $\alpha$, emissivity, Leaf Area Index (LAI), NDVI, SAVI, $T_s$, and $R_n$, respectively during December 23, 2010, over Delhi. While point values LAI, NDVI, SAVI, $\alpha$, $T_s$, and $R_n$ at lysimeter location (in WTC field) on December 23, 2010, are given in Table 11.2. LAI and NDVI at lysimeter gave an increasing pattern during December 2010 (young crop stages) and February 2011 (exhibited by butting or flowering) at growing states of wheat crop. During March 29, 2011, to April 14, 2011, a decreasing pattern (crop toward the harvesting stages) appeared over the growing period. Furthermore, SAVI showed an inverse pattern with LAI and NDVI because during growing season, there is almost no gap among crops due to leaf density. During December 2010 to February 2011, soil was not exposed, so SAVI was low, but when crops grew, soil started to be exposed, and SAVI increased gradually (because of areas where vegetative cover was low and the soil surface was exposed, the reflectance of light in the red and near-infrared spectra can influence vegetation index values). Table 11.2 indicates that the cultivated wheat crop with high LAI and NDVI (at lysimeter footprint) showed low $\alpha$ and low $T_s$ due to the strong negative correlation between $T_s$ and LAI and NDVI.

**TABLE 11.2**
**Quantified Surfaces Parameter Value over Lysimeter Point**

| Surface Variables | December 23, 2010 | February 9, 2011 | March 29, 2011 | April 14, 2011 |
|---|---|---|---|---|
| NDVI | 0.530 | 0.580 | 0.340 | 0.290 |
| LAI | 2.80 | 3.015 | 1.320 | 1.030 |
| SAVI | 0.413 | 0.474 | 0.200 | 0.297 |
| albedo | 0.102 | 0.112 | 0.130 | 0.145 |
| $T_s$ (K) | 288.80 | 292.10 | 301.00 | 307.80 |
| $T_s$ (°C) | 20.80 | 21.40 | 28.00 | 35.70 |

## 11.3.1 LAI AND NORMALIZED DIFFERENCE VEGETATION INDEX RELATIONSHIP

Figure 11.10 shows the time courses of LAI and NDVI for wheat crop. The two variables (LAI and NDVI) exhibited comparable seasonal patterns, primarily following the dynamics of the density and greenness of leaves. The footprint area of lysimeter showed high values of NDVI (>0.5) and LAI (>3), which reflects acceptable growth conditions. The NDVI deep curve shape (Figure 11.10) at midseason with reduced standard deviation as the NDVI high for high values of LAI, when the soil was entirely covered by the canopy. At the footprint area of lysimeter, the contrast between low NDVI with high SD and low LAI with high SD at the end of the season was due to the fact that NDVI measurement data were selected on crops of dry condition while the falling LAI was sharper than NDVI because of LAI's dependence on NDVI, but under these conditions (sharpness in falling of LAI), it was not possible to quantify a correlation ($r^2 = 0.86$) between NDVI and LAI from Figure 11.11. LAI showed a linear response to NDVI (Figure 11.11) in agreement with the results obtained by previous studies (Wang et al., 2004; Potithep et al., 2010). The comparison clearly shows the high LAI values at saturation of NDVI.

Table 11.3 shows the land surfaces heat fluxes: $R_n$ (Figure 11.2); $G$ (Figure 11.3); $H$ (Figure 11.4); $\Lambda_{ins}$ (Figure 11.5); and daily $ET_{SEBA}$ (Figures 11.6 through 11.9). The result indicates that $\Lambda_{ins}$ and $ET_{24h}$ peaked on March 29, 2011, and February 9, 2011, respectively, as shown by the green full stages of the crop. $G$ of April 14, 2011, was high during the growing period because the crop became

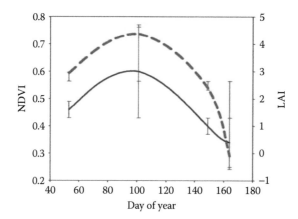

**FIGURE 11.10** Variation of LAI and NDVI with cropping period.

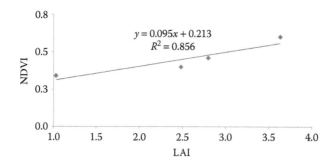

**FIGURE 11.11** LAI–NDVI relationship.

## TABLE 11.3
### Land Surface Heat, Heat Fluxes, and Daily ET over Lysimeter Point

| Flux | December 23, 2010 | February 9, 2011 | March 29, 2011 | April 14, 2011 |
|---|---|---|---|---|
| $R_n$ (W/m$^{-2}$) | 346.49 | 419.00 | 578.00 | 590.00 |
| $H$ (W/m$^{-2}$) | 311.44 | 306.06 | 426.00 | 502.50 |
| $G_0$ (W/m$^{-2}$) | 23.45 | 29.00 | 61.70 | 74.50 |
| $\Lambda_{ins} \approx \Lambda_{24}$ (dimensionless) | 0.036 | 0.215 | 0.175 | 0.025 |
| $ET_{24h}$ (mm d$^{-1}$) | 0.44 | 3.15 | 3.54 | 0.52 |

dry or was in the harvest stage with exposed soil. The ET ($ET_{24h}$) of March 29, 2011, was relatively high because during this period, the crop was in grain-filling states, which required more energy by the plants to produce grain, so more evaporation from the plants in this process only top leaf (flag leaf of wheat) of the plant resulted in a major role (other leafs of plants were in states of dryness); therefore NDVI and LAI (leaf density of crop) were low in March 29, 2011. The $R_n$ during the crop's growing period showed an increasing pattern, which may have been due to clear conditions (clear sky).

### 11.3.2 Result from Comparison with Conventional Method

Because of the small dimensions of the lysimeter fields at WTC (20 × 20 m on average), it was not possible to find a band of more than 1 pixel of the LANDSAT 7-ETM+ (30 × 30 m) completely inside a lysimeter field. Often portions of each band of 1 pixel fell onto a lysimeter field. Since resolution of LANDSAT 7-ETM+ and lysimeter is almost similar, we did not interpolate these observations to compare them.

The ET ($ET_{Lysimeter}$) was estimated by lysimeter using daily data sets of lysimeter units of the WTC (IARI, New Delhi) meteorological station. Considering that the lysimeter unit is exact inside or middle of field and surrounding land condition is same. Table 11.4 gives ET ($ET_{Lysimeter}$, $ET_{SEBAL}$) and the results of the comparison for 10 statistical tests. The statistical analysis provides evidence of the accuracy level of the selected model ET ($ET_{SEBAL}$) with respect to observed ET ($ET_{Lysimeter}$).

## TABLE 11.4
### Comparison of ET (from Wheat Crop) from SEBAL with Lysimeter

| | December 23, 2010 | February 9, 2011 | March 29, 2011 | April 14, 2011 |
|---|---|---|---|---|
| $ET_{Lysimeter}$ | 0.48 | 3.33 | 3.75 | 0.54 |
| $ET_{SEBAL}$ | 0.44 | 3.15 | 3.54 | 0.52 |
| **Methods** | | | | |
| RMSE | | | 0.140 | |
| R-RMSE | | | 0.063 | |
| MAE | | | 0.113 | |
| NRMSE | | | 0.069 | |
| NSE | | | 00.00 | |
| MBE | | | 0.113 | |
| PBIAS | | | 5.601 | |
| D | | | 1.000 | |
| RMSE% | | | 1.734 | |
| $R^2$ | | | 0.910 | |

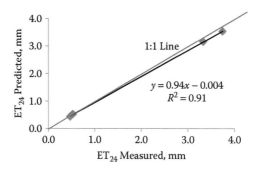

**FIGURE 11.12** Correlation between measured and predicted $ET_{24h}$.

We observed a relative root mean square error (RMSE) of 0.140 mm d$^{-1}$ and R-RMSE of 0.63 mm d$^{-1}$ between SEBAL and lysimeter. R-RMSE standardizes the RMSE to the unperturbed derivative value for each cell location (Kroll and Stedinger, 1996; Rawat et al., 2013). The resulting R-RMSE value is a percentage and reflects the standard variation of the derivative. These results indicate that SEBAL ET data sets are accurate as compared to observed data sets within the resolution of the 30 m resolution images. We also observed MAE and NRMSE values of 0.113 and 0.069 (Table 11.4), respectively, which shows that prediction of our model is accurate with respect to observed values (Figure 11.12). It can be concluded that the model can be used effectively with RS data (LANDSAT 7-ETM+) for prediction of ET at a point scale.

## 11.4 SUMMARY AND CONCLUSIONS

The RS radiances in the visible, near infrared, and thermal infrared ranges are capable tools for determining α, vegetation index, and $T_s$. These parameters can be used to estimate ET and surface energy fluxes.

The results just outlined confirm the opportunities presented by LANDSAT 7-ETM+ satellite data to solve the equation of EB (SEBAL) and to compare it with ground-measured ET using lysimeter. However, the ET and surface-energy flux estimates cannot be regarded as very accurate compared to data points. Despite these inaccuracies, the approach is suitable for use of satellite data to estimate a number of parameters at the soil-plant-atmosphere interface. These parameters have the advantage of being spatialized and offer a spatiotemporal coverage more satisfactory than the data points measured operationally.

We show that land surface EB models using RS data from satellite platforms at different spatial resolution are promising for estimating daily and seasonal ET at a point scale. The spatial and temporal RS data from an existing set of earth observation satellite sensors are not adequate to use their ET products for ET estimation.

In this study, a SEBAL model was applied to WTC field in IARI farmland, Pusa, New Delhi, using four images of LANDSAT 7-ETM+ on December 23, 2010; February 9, 2011; April 14, 2011; and March 29, 2011, and various land surface parameters and fluxes, including ET, were calculated. The main results are summarized as follows:

- ET estimated by SEBAL at the point of lysimeter indicates that the cultivated wheat with high NDVI and low albedo showed high ET.
- The ET for wheat derived by SEBAL for December 23, February 9, March 29, and April 14 were 0.44, 3.15, 3.75, and 0.52 mm d$^{-1}$, respectively. The seasonal pattern of both estimations (from lysimeter) was similar (Table 11.4, Figure 11.13).

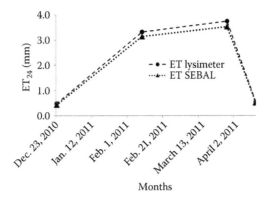

**FIGURE 11.13** Seasonal variation of $ET_{24h}$ over the study area.

This study results indicates that SEBAL has the potential to estimate ET distribution whereas the conventional method provides only real reference ET. The estimated ET of wheat using the SEBAL method is very useful for improving water management on a regional scale. Therefore, SEBAL is a useful and important tool for determining ET on a regional scale. Remote sensing-based ET estimates can be applied over the region where there is lack of ground-based lysimeter.

## ACKNOWLEDGMENT

The authors are very grateful to the project director (WTC) and the director of the Indian Agricultural Research Institute, New Delhi, for kindly providing the necessary lysimeter experimental data.

## REFERENCES

*Advanced Training and Users Manual,* Surface Energy Balance Algorithms for Land, Idaho Implementation, August, 2002, version 1.0. Funded by a NASA EOSDIS/Synergy grant from the Raytheon Company through the Idaho Department of water Resources, 2002, pp. 80–81.

Allen, R.G., Tasumi, M. and Trezza, R., 2007, Satellite-based energy balance for mapping evapotranspiration with internalized calibration (METRIC): Model. *Journal of Irrigation and Drainage Engineering,* 133, 380–394.

Bastiaanssen, W.G.M., 2000, SEBAL-based sensible and latent heat fluxes in the irrigated Gediz Basin, Turkey. *Journal of Hydrology,* 229, 87–100.

Bastiaanssen, W.G.M., Pelgrum, H., Wang, J., Ma, Y., Moreno, J.F., Roerink, G.J. and Vander, W.T., 1998, A surface energy balance algorithm for land (SEBAL): Part 2 validation. *Journal of Hydrology,* 212–229.

Boulet, G., Chehbouni, A., Gentine, P., Duchemin, B., Ezzahar, J. and Hadria, R., 2007, Monitoring water stress using time series of observed to unstressed surface temperature difference. *Agricultural and Forest Meteorology,* 146, 159–172.

Brutsaert, W. and Sugita, M., 1992, Application of self-preservation in the diurnal evolution of the surface energy budget to determine daily evaporation. *Journal of Geophysical Research,* 97, 18288–18294.

Calvet, J.C., Noilhan, J., Roujean, J.L., Bessemoulin, P., Cabelguenne, M., Olioso, A. and Wigneron, J.P., 1998. An interactive vegetation SVAT model tested against data from six contrasting sites. *Agricultural and Forest Meteorology,* 92, 73–95.

Caparrini, F., Castelli, F. and Entekhabi, D., 2003, Variational estimation of soil and vegetation turbulent transfer and heat flux parameters from sequences of multi-sensor imagery. *Water Resources Research,* 4, 1–15.

Caparrini, F., Castelli, F. and Entekhabi, D., 2004, Estimation of surface turbulent fluxes through assimilation of radiometric surface temperature sequences. *Journal of Hydrometeorology,* 5, 145–159.

Chandrapala, L. and Wimalasuriya, M., 2003, Satellite measurements supplemented with meteorological data to operationally estimate evaporation in Sri Lanka. *Agricultural Water Management,* 58, 89–107.

Cleugh, H.A., Leuning, R., Mu, Q. and Running, S.W., 2007, Regional evaporation estimates from flux tower and MODIS satellite data. *Remote Sensing of Environment*, 106, 285–304.

Coudert, B., Ottle, C., Boudevillain, B., Demarty, J. and Guillevic, P., 2006, Contribution of thermal infrared remote sensing data in multiobjective calibration of a dual source SVAT model. *Journal of Hydrometeorology*, 7, 404–420.

Crago, R.D., 1996, Comparison of the evaporative fraction and the Priestley-Taylor α for parameterizing daytime evaporation. *Water Resources Research*, 32, 1403–1409.

Crow, W.T. and Kustas, W.P., 2005, Utility of assimilating surface radiometric temperature observations for evaporative fraction and heat transfer coefficient retrieval. *Bound-Layer Meteorology*, 115, 105–130.

Foken, T., 2006, 50 years of the Monin–Obukhov similarity theory. *Boundary-Layer Meteorolgy*, 119, 431–447.

French, A.N., Hunsaker, D.J., Clarke, T.R., Fitzgerald, G.J., Luckett, W.E. and Pinter, P.J., 2007, Energy balance estimation of evapotranspiration for wheat grown under variable management practices in Central Arizona. *American Society of Agricultural and Biological Engineers*, 50, 2059–2071.

French, A.N., Schmugge, T.J., Kustas, W.P., Brubaker, K.L. and Prueger, J., 2003, Surface energy fluxes over El Reno, Oklahoma, using high-resolution remotely sensed data. *Water Resources Research*, 39, 1–12.

Gentine, P., Entekhabi, D., Chehbouni, G., Boulet, G. and Duchemin, B., 2007, Analysis of evaporative fraction diurnal behavior. *Agricultural and Forest Meteorology*, 143, 13–29.

Glenn, E.P., Huete, A.R., Nagler, P.L., Hirschboeck, K.K. and Brown, P., 2007, Integrating remote sensing and ground methods to estimate evapotranspiration. *Critical Reviews in Plant Sciences*, 26, 139–168.

Jackson, R.D., Reginato, R.J. and Idso, S.B., 1977, Wheat canopy temperature: A practical tool for evaluating water requirements. *Water Resources Research*, 13, 651–656.

Jacob, F., 1999, Utilisation de la télédétection courtes longueurs d'onde et infrarouge thermique à haute résolution spatiale pour l'estimation des flux d'énergie à l'échelle de la parcelle agricole. Thèse de l'université Paul Sabatier, Toulouse, 3.

Jacob, F., Olioso, A., Gu, X. F., Su, Z. and Seguin, B., 2002, Mapping surface fluxes using airborne visible, near infrared, thermal infrared remote sensing data and a spatialized surface energy balance model. *Agronomie*, 22, 669–680.

Jia, L., Su, Z., Van der Hurk, B., Menenti, M., Moene, A., De Bruin, H.A.R., 2003. Estimation of sensible heat flux using the Surface Energy Balance System (SEBS) and ATSR measurements. *Physics and Chemistry of the Earth*, 28(1–3), 75–88.

Kroll, C. and Stedinger, J., 1996, Estimation of moments and quantiles using censored data. *Water Resources Research*, 32, 1005–1012.

Kumar, R., Shambhavi, S., Kumar, R., Singh Y.K. and Rawat, K.S., 2013, Evapotranspiration mapping for agricultural water management: An overview. *Journal of Applied and Natural Science*, 5, 522–534.

Kustas, W.P., Blanford, J.H., Stannard, D.I., Daughtry, C.S.T., Nichols, W.D. and Weltz, M.A., 1994, Local energy flux estimates for unstable conditions using variance data in semiarid rangelands. *Water Resources Research*, 30, 1351–1361.

Kustas, W.P. and Norman, J.M., 1999, A two-source approach for estimating turbulent fluxes using multiple angle thermal infrared observations. *Water Resources Research*, 33, 1495–1508.

Li, Z. L., Tang, R., Wan, Z., Bi, Y., Zhou, C., Tang, B., Yan, G. and Zhang, X., 2009, A review of current methodologies for regional evapotranspiration estimation from remotely sensed data. *Sensors*, 9, 3801–3853.

Mahfouf, J.F., Manzi, A.O., Noilhan, J., Giordani, H. and Deque, M., 1995, The land surface scheme ISBA within the M6 to France Climate model ARPEGE. Part I: Implementation and preliminary results. *Journal of Climate*, 8, 2039–2057.

Mu, Q., Heinsch, F.A., Zhao, M. and Running, S.W., 2007, Development of a global evapotranspiration algorithm based on MODIS and global meteorology data. *Remote Sensing of Environment*, 111, 519–536.

Norman, J.M., Anderson, M.C., Kustas, W.P., French, A.N., Mecikalski, J., Torn, R., Diak, G.R., Schmugge, T.J. and Tanner, B.C.W., 2003, Remote sensing of surface energy fluxes at 101-m pixel resolutions. *Water Resources Research*, 39, 1–8.

Olioso, A., Carlson, T.N. and Brisson, N., 1996, Simulation of diurnal transpiration and photosynthesis of a water stressed soybean crop. *Agricultural and Forest Meteorology*, 81, 41–59.

Olioso, A., Inoue, Y., Ortega-Farias, S., Demarty, J., Wigneron, J.P., Braud, I., Jacob, F. et al., 2005, Future directions for advanced evapotranspiration modeling: Assimilation of remote sensing data into crop simulation models and SVAT models. *Irrigation and Drainage Systems*, 19, 377–412.

Potithep S, Nasahara NK, Muraoka H et al., 2010. What is the actual relationship between LAI and VI in a deciduous broadleaf forest. *International Archives of the Photogrammetry, Remote Sensing and Spatial Information Science*. Kyoto International Conference Center (ICC Kyoto), August 9–12, 2010 38(8): 609–614.

Rawat, K.S., Mishra, A.K., Sehgal, V.K., Ahmed, N. and Tripathid, V.K., 2013, Comparative evaluation of horizontal accuracy of elevations of selected ground control points from ASTER and SRTM DEM with respect to CARTOSAT-1 DEM: A case study of district Shahjahanpur (Uttar Pradesh), India. *Geocarto International*, 28, 439–452.

Roerink, G.J., Su, Z. and Menenti, M., 2000. S-SEBI: A simple remote sensing algorithm to estimate the surface energy balance. *Physics and Chemistry of the Earth. Part B: Hydrology, Oceans and Atmosphere*, 25(2), 147–157.

Schuurmans, J.M., Troch P.A., Veldhuizen A.A., Bastiaanssen W.G.M. and Bierkens M.F.P., 2003, Assimilation of remotely sensed latent heat flux in a distributed hydrological model. *Advances in Water Resources*, 26, 151–159.

Su, Z., 2002, The Surface Energy Balance System (SEBS) for estimation of turbulent heat fluxes. *Hydrology and Earth System Sciences*, 6, 85–100.

Wang, Y., Woodcock, C.E., Buermann, W., Stenberg, P., Voipioc, P., Smolander, H., Hame, T. et al. 2004. Evaluation of the MODIS LAI algorithm at a coniferous forest site in Finland. *Remote Sensing of Environment*, 91, 114–127.

# 12 Use of Hydrological Modeling Coupled with Geographical Information System for Plotting Sustainable Management Framework

*Pankaj Kumar and Chander Kumar Singh*

## CONTENTS

12.1 Introduction ........................................................................................................ 192
12.2 Study Area .......................................................................................................... 193
    12.2.1 Hydrogeologic Setting and Aquifer Systems of the Area ...................... 194
12.3 Methodology ...................................................................................................... 195
    12.3.1 Data Collection and Processing ............................................................. 195
    12.3.2 Ground Water Simulation Model ........................................................... 195
    12.3.3 Spatial and Temporal Discretization of the Model Domain .................. 195
    12.3.4 Boundary Conditions ............................................................................. 197
        12.3.4.1 Land Use Land Cover ............................................................. 198
        12.3.4.2 Soil Types ................................................................................ 199
12.4 Results and Discussion ....................................................................................... 201
12.5 Conclusion ......................................................................................................... 204
References .................................................................................................................... 204

**ABSTRACT** The continual replenishment of groundwater storage and its sustainable utilization is indispensable to meet the ever-increasing water needs of the Killinochi district of northern Sri Lanka. To ascertain the sustainable exploitation of this resource, a modeling study was conducted there. A widely accepted 3D modular groundwater flow model was employed to simulate the hydrodynamics and stresses of the Killinochi basin for an initial stabilization period of 1 year, followed by a 3-year period simulation. Spatially distributed recharge was considered using the recharge package and the multiplier function to incorporate the effects of variable infiltration rates in the recharge estimation. A literature review on the availability of data and prior groundwater investigations in the area shows a huge gap in soil and aquifer hydraulic parameters, water-level information, and related aquifer and confining bed geometry data. Therefore, it was not possible to calibrate the model following the standard calibration techniques. However, in the conceptualization process, all efforts were made to make sure that the model domain fairly followed the actual field conditions of the Killinochi basin by incorporating all available data. In addition, a comparison of the simulated zonal water budgets with the zonal budgets calculated based on observed data from International Water Management Institute (IWMI) and water resources board (WRB) reports were used to validate the results of the model. The recharge pattern from the model results follows a close resemblance to the rainfall pattern, which can be explained by the more or less flat topography of the basin. We believe the flat topography has limited any significant contribution of the lateral inflows to the groundwater dynamics. Hence, the simulated heads and the recharge and flow patterns obtained at the end of the simulations were to the most part influenced by the interaction between percolated rainwater and the groundwater aquifer as a function of land use and soil physical properties.

## 12.1 INTRODUCTION

In arid and semiarid areas characterized by short periods of heavy rainfall and prolonged dry periods, the continual replenishment of groundwater storage, and its sustainable utilization are indispensable to address the water needs of the community. Water availability and demand in Sri Lanka shows a highly spatial and temporal variability. The water scarcity is prominent, especially in the dry areas of the country where there is a high demand for usable water to irrigate 85% of the land, which aims to increase agricultural productivity (Hettiarachchi, 2008). Previous studies, especially in dry land areas of Bangladesh, have proven the ability of groundwater to facilitate the spread of the green revolution in the dry areas through the use of tube wells to maximize the agricultural development and hence improved living conditions through income distribution (Fujita and Hossain, 1995).

The fact that groundwater is the major source of fresh water, especially among the rural communities in the dry parts of Sri Lanka, will have an invaluable role in the development of the economy and living standards of the region. Economic empowerment of the locals through improved productivity will greatly curb environmental stressors related to urbanization and influx of people from less developed areas to urban areas. However, the groundwater's flow dynamics and hence its availability is known to be highly dependent on a multitude of factors including the hydrogeological formations through which the flow passes (Freeze and Cherry, 1979).

In the dry zones of the Killinochi basin in northern Sri Lanka, the monsoon and intermonsoonal periods play the major role in replenishing these groundwater resources. However, the decreasing trend in the rainfall regime of the dry zones and the increase in population size calls for a sustainable use of the groundwater resources in the region.

In general, three types of recharge are recognized: potential, actual, and rejected recharge where potential recharge is the recharge as a result of net precipitation. Actual recharge may be much less than the potential recharge because of low aquifer storability while the rejected recharge is the amount of total recharge that overflows as a result of aquifer saturation. In principle, the potential recharge is the sum of actual recharge and rejected recharge (Orehova et al., 2012).

The task of estimating the volume of groundwater entering the aquifer system and its spatial and temporal distribution over the domain is very difficult (Gupta, 2008). Hence, the task is highly demanding both in terms of workforce and resources to obtain reliable information on its availability and distribution in space.

Previous studies have shown that the seasonal available water resource per unit area in the Killinochi district is less than 0.1 m for both monsoon seasons whereas the per capita water withdrawals for the Killinochi district are among the highest in Sri Lanka. According to a projection by the same study considering the current level of irrigation efficiency, it is expected that by 2025, more regions will fall into the water-scarce category (Amarasinghe et al., 1999). Hence, sustainable development of the groundwater resources can be considered as the only viable alternative to support the community in the dry areas without depleting the groundwater reserve, mainly to improve members' livelihood by maximizing agricultural output. The Asian Development Bank report (2011) indicates that the poverty rate in the region is considerably high as compared to the national average.

Therefore, in the Sri Lankan context, a sustainable development of the groundwater resources especially in the dry zones is very important because it is widely used for domestic, commercial, and industrial purposes, small-scale irrigation, water supply schemes, and so on. However, the amount of research done on issues related to the availability and quality of groundwater, especially in the dry zones of the country, is very limited (Hettiarachchi, 2008).

The main objective of this study is therefore to evaluate the groundwater recharge potentials of the area for a sustainable and more efficient use of the groundwater resources for agricultural, industrial, and other domestic uses on a regional scale. In the reconstruction of the war-ravaged region, knowledge of the potential of these resources is important in designing a long-term sustainable development strategy.

## 12.2 STUDY AREA

The Killinochi area is situated in the dry zones of northern Sri Lanka on 9°22′39.54″ N and 80°22′33.54″ E coordinates with an annual average temperature of 28°C with a total land area of approximately 1205 km². Sri Lanka has four seasons, namely the southwest monsoon (*swm*) from May to September, the northeast monsoon (*nem*) from November to February, the first intermonsoon (*fim*) from March to April, and the second intermonsoon (*sim*) from October to November. The rice-growing seasons are defined based on these rains with the Yala season being defined as the combination of *fim* and *swm* and the Maha season as the combination of *sim* and *nem*. These rains are not equally distributed throughout the island. Over the Kilinochchi area, *swm* is not effective but Yala is from March to early May whereas the major growing season is the Maha from October to February. The area receives an average rainfall of 1250 mm (Figure 12.1).

The area is characterized by two major rainfall seasons, namely Maha and Yala, the wet and dry seasons, respectively. The poverty level in this area is as high as 64%, which is more than double the national average (Asian Development Bank, 2011). The Killinochi area at the center of the 26-year long devastating civil war (1983–2009) has seen the major destruction of its irrigation facilities and water reservoirs. Hence, the poverty in the area can be mainly attributed to the war that resulted in the destruction of most of the infrastructure in the district. As war-displaced residents return to their villages, competition for the already scarce water resource is expected to grow sharply. According to the national census data, the population quadrupled from 23,625 in 2009 to 112,875 in 2012 (Wikipedia) and is expected to stress the already scarce fresh water reserves in this semiarid region of Sri Lanka.

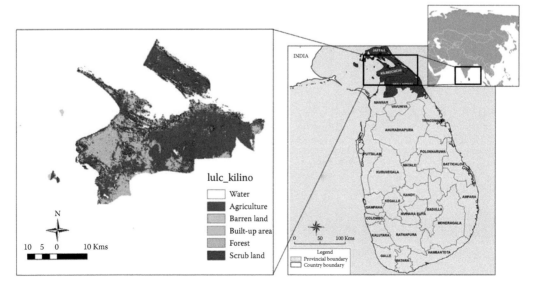

**FIGURE 12.1** Location map of the study area.

### 12.2.1 HYDROGEOLOGIC SETTING AND AQUIFER SYSTEMS OF THE AREA

According to a report by the WRB, six major categories of groundwater aquifers are recognized in Sri Lanka. The shallow karstic aquifers of Jaffna peninsula, deep confined aquifers, coastal sand aquifers, alluvial aquifers, shallow regolith aquifers of the hard rock region, and the southwestern lateritic (cabook) aquifers (Panabokke et al., 2005). The Killinochi basin constituted by the shallow coastal sand aquifers, deep confined aquifer, and alluvial aquifer (Figure 12.2).

The hydrostratigraphic units of the area are divided into a two-layer system consisting of an unconfined upper layer and a confined bottom layer separated by a thin confining bed.

An alluvial aquifer and a strip of coastal sand aquifer near the coastal beach areas dominate the unconfined upper layer. The lower confined layer is composed of limestone and sandstone formations of the northwest coastal plains (Panabokke et al., 2005).

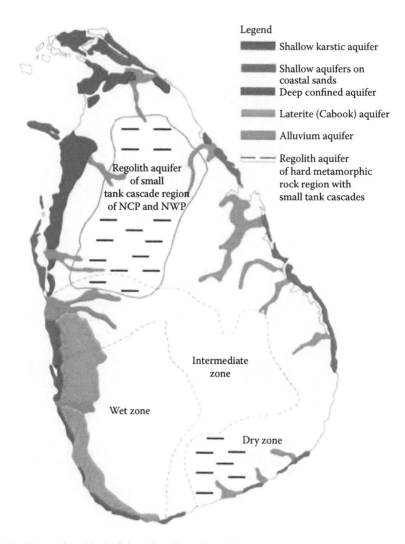

**FIGURE 12.2** Types of aquifer in Sri Lanka. (From Panabokke CR, Perera APGRL, 2005.)

## 12.3 METHODOLOGY

### 12.3.1 Data Collection and Processing

Because a holistic reconstruction of the groundwater flow system as it exists in nature is not feasible, mainly because of the lack of detailed information, a simplification of the modeling domain was attempted in this study. The various data on the Sri Lankan aquifer systems documented mainly on the WRB (Mawatha) and IWMI reports were summarized and integrated to help build the conceptual model of the Killinochi groundwater system. The land use and land cover (LULC), soil type distributions, hydraulic properties and parameters, and water demand of the area both for domestic and agricultural purposes (Amarasinghe et al. 1999) were among the major variables incorporated into the database. The database and literature reviews played a major role in ensuring the considered premodeling conceptualization of the groundwater flow system to be a valid representation of the vital field hydraulic properties and conditions of the modeled domain.

Major sources of data include WRB data (Mawatha), data from the joint project of the Soil Science Society of Sri Lanka and Canadian Society of Soil Science (SRICANSOL; 2009), IWMI (Amarasinghe et al., 1999; Kodituwakku and Pathirana, 2003; Dassanayake et al., 2005; Hydrosult Inc., 2010).

Hydrometric variables such as hydraulic conductivity and specific storage were drawn mainly from the SRICANSOL report (2009). Moreover, preparation of the geospatial data from readily available sources was conducted, including base maps, land use maps, soil maps, and the groundwater aquifer system. Main sources of geospatial data include, HydroSHEDS (United States Geological Service [USGS], 2010), and ArcGIS.

### 12.3.2 Ground Water Simulation Model

The widely used USGS modular 3D finite difference groundwater flow model (MODFLOW) was selected to numerically represent the simplified dynamics of the groundwater flow system as a function of hydrogeomorphologic, climatic, and environmental variables (Harbaugh 2005). MODFLOW is considered to be the most reliable, verified, and utilized groundwater flow computer program available (Kresic, 2007).

The MODFLOW model is organized in such a way that it can be easily modified without changing the entire code to solve a particular problem by incorporating a new package to the modular structure of the program.

The MODFLOW model input files were imported to the USGS open sources Model-Viewer and GW_chart software programs for a better visualization and postprocessing of the results.

### 12.3.3 Spatial and Temporal Discretization of the Model Domain

The entire model domain is spatially discretized into two simulated layers (unconfined upper layer and confined bottom layer) with a confining bed in between. The confining bed is a nonsimulated layer because it has no significant groundwater potential due to its very low conductivity. The entire model domain was spatially discretized into two layers of a 1 km grid cell constituting 55 rows and 64 columns. The top elevation of the model domain was determined from the Shuttle Radar Topography Mission (SRTM) elevation data, which is equivalent to the land surface altitudes. The intermittent streams in the modeled area were assigned an elevation value of 1 m less than the elevation of the surrounding land area. The bottom elevations for the top unconfined layer were assigned a constant depth of 10 m from the elevation of the top layer. The bottom elevations of the confining bed were assigned three different depth values based on their respective location with reference to the Mulankavil basin, the basin west of Killinochi, and the Paranthan basin (Figure 12.3 and Table 12.1). Temporally, the system was discretized into 12 stress periods of daily time steps for a

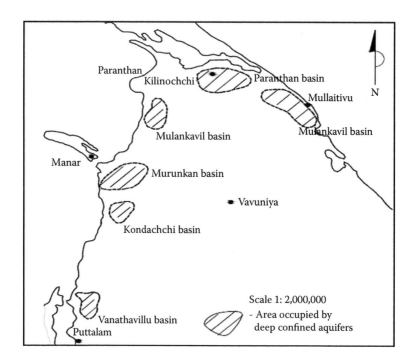

**FIGURE 12.3** Distribution of the confined aquifers in the various basins. (Adopted from Panabokke CR, Perera APGRL, 2005.)

**TABLE 12.1**
**Major Land Use Classes in the Basins of Killinochi**

| Basin No. | Basin | Major Land Type | R.R (Maha) | R.R (Yala) | EVT Rate |
|---|---|---|---|---|---|
| 78 | Thoravil Aru | Forest (59.3%) | 0.20 | 0.04 | 0.65 |
| 79 | Piramenthal Aru | Forest (33.3%), homesteads (30%) | 0.15 | 0.04 | 0.71 |
| 80 | Nathali Aru | Paddy (46.8%) | 0.10 | 0.02 | 0.71 |
| 81 | Kanakarayan Aru | Paddy (50.6%) | 0.10 | 0.02 | 0.72 |
| 82 | Kalavalappu Aru | Homesteads (38.2), paddy (32.7%) | 0.15 | 0.04 | 0.72 |
| 83 | Akkarayan Aru | Paddy (35.5%), forest (29.5%) | 0.15 | 0.04 | 0.71 |
| 84 | Mandekal Aru | Forest (64.9%) | 0.20 | 0.05 | 0.70 |
| 85 | Pallavarayankadd Aru | Forest (55.0%) | 0.20 | 0.05 | 0.69 |
| 86 | Pali Aru | Forest (63.2%) | 0.20 | 0.05 | 0.70 |
| N1 | – | Paddy (43.8%) | 0.10 | 0.02 | 0.70 |
| N2 | – | Coconut (24.1%), homestead (21.9%) | 0.05 | 0.01 | 0.70 |

N1, N2—no name for these basins; R.R—runoff coefficient.

1-year period of transient state modeling to simulate the seasonal variation in head and the recharge volume into the saturated zone. After obtaining the initial condition from the result of the 1-year simulation results, we run the model for a period of three consecutive years (1984–1986). The stress periods were designed mainly to reflect the impact of the monthly rainfall variation on the groundwater recharge process for the year simulated (Figure 12.4). The results of change in head at the end of the simulation were used to fix the initial conditions for subsequent runs under different rainfall patterns.

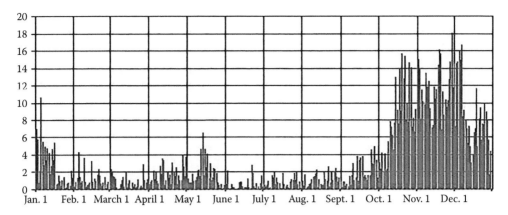

**FIGURE 12.4** Rainfall time series used in the model (in m/day).

### 12.3.4 BOUNDARY CONDITIONS

Because the sea surrounds the active parts of the model layer, we assigned a constant head for the entire outer boundaries of the flow domain. Similarly, the inland water bodies and streams were given a constant head status. The rest of the ocean area was considered a no-flow zone. All the land area was given a variable head status, and inland water bodies were assigned a constant head status assuming that there was a strong source of flow from the expanding sea toward the inland areas.

Based on the detailed conceptual representation of the saturated porous media, a numerical representation of the system that approximates a simplified representation of its responses and stresses was considered. The model grid cells of the top layer were characterized as unconfined, and the grid cells of the bottom layer were classified as confined. The hydraulic conductivities of the soil in the grid cells ranged from 0.65 to 10.3 m/d. These values are based on the different soil type classes considered (Figure 12.5). For the confined layer, a constant value of 0.0086 m/d was used for all the soils in the grid cells of the active model domain. The specific yield values were again based on soil type classes based on the data from SRICANSOL (2009). A constant value of specific storage, $1 \times 10^{-3}$, was used for all the grid cells of the confined layer active domain. Furthermore, an evapotranspiration (Figure 12.6) extinction depth was assigned based on the LULC type of each basin within the Killinochi district (Amarasinghe et al., 1999). The

**FIGURE 12.5** Horizontal hydraulic conductivity distribution map of the area—From the groundwater flow model (m/day).

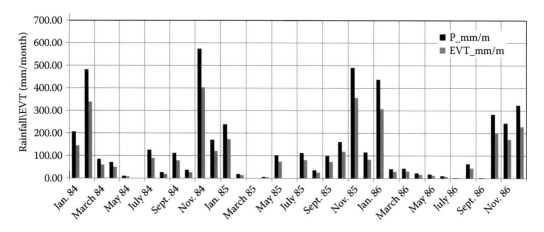

**FIGURE 12.6** Rainfall and EVT for the simulated years (1984–1986).

**TABLE 12.2**
**Extinction Depth Estimates Based on Land Use Classes**

| Type of Land Use | Extinction Depth (in meters from land surface) |
|---|---|
| Water body | 1.000 |
| Forest land (sandy loam) | 3.000 |
| Paddy land | 0.875 |
| Grassland | 2.000 |
| Homestead | 1.000 |
| Crop land | 1.000 |
| Coconut land | 0.500 |
| Sparse crop land | 1.000 |
| Barren land | 1.000 |
| Built-up land | 1.000 |

initial water content of the unsaturated zone (THTI) was assumed to be equivalent to the saturated water content because the model starts right after the wet season. The extinction water content for the unsaturated zone was set based on the experimental data of soil moisture content at 15 bar of applied pressure (SRICANSOL 2009; Table 12.2). Evapotranspiration demand rates were specified for stress period 1 and for the rest of the stress periods the values from the previous stress period were used. Although in nature the physical soil properties are highly heterogeneous, in this study, we assumed an isotropic condition for the horizontal and vertical conductivities of the cells in the model domain.

### 12.3.4.1 Land Use Land Cover

LULC attributes (Figure 12.7) are known to influence the surface groundwater interaction and hence the groundwater recharge rate. Therefore, for each basin in the Killinochi district, extinction depths of the different land use classes were determined based on the rooting depth of the

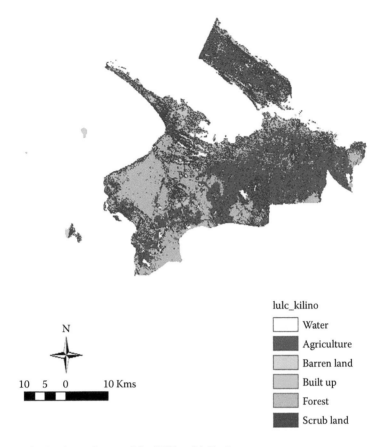

**FIGURE 12.7** Major land use classes of the Killinochi district.

specific crop type in the land use classes to characterize the depth below which moisture cannot be removed by the evapotranspiration processes (Table 12.2). The values are based on a study conducted on the extinction depth and evapotranspiration from groundwater under selected land covers (Shah et al., 2007). Preliminary water balance analysis based on field measured data (Amarasinghe et al., 1999) showed that approximately 65% of the precipitation was lost to evapotranspiration (Figure 12.6) and Table 12.3.

### 12.3.4.2 Soil Types

According to the (United Nations Development Programme (UNDP), 2011) soil map (Figure 12.8), the soils of the Killinochi area are classified into six major groups. The classification of the soil groups in Sri Lanka is based on the morphological analysis of the soil units both in the field and in the laboratory (Moormann and Panabokke. 1961). Hence, major features such as the nature and organization of the soil horizons, texture color, structure, and consistence are considered in the classification process. In this study, the different soil class groups were considered in assigning the different hydraulic conductivity values cell by cell in the model domain (Table 12.4). Accordingly, the data from the integrated strategic environmental assessment in the northern province was employed.

## TABLE 12.3
### Evapotranspiration Rate Estimated Based on the Data from WRB (mm/Month; Killinochi Basin)

| Basin No. | Basin | January | February | March | April | May | June | July | August | September | October | November | December |
|---|---|---|---|---|---|---|---|---|---|---|---|---|---|
| 78 | Thoravil Aru | 53.82 | 21.35 | 18.02 | 34.27 | 35.44 | 11.00 | 15.33 | 22.87 | 36.96 | 164.23 | 225.77 | 165.02 |
| 79 | Piramenthal Aru | 59.16 | 23.47 | 19.81 | 37.67 | 38.96 | 12.10 | 16.85 | 25.14 | 40.63 | 180.53 | 248.17 | 181.40 |
| 80 | Nathali Aru | 59.46 | 23.59 | 19.91 | 37.86 | 39.16 | 12.16 | 16.94 | 25.27 | 40.83 | 181.46 | 249.44 | 182.33 |
| 81 | Kanakarayan Aru | 60.08 | 23.84 | 20.12 | 38.25 | 39.57 | 12.29 | 17.11 | 25.54 | 41.26 | 183.35 | 252.04 | 184.23 |
| 82 | Kalavalappu Aru | 59.85 | 23.75 | 20.04 | 38.11 | 39.42 | 12.24 | 17.05 | 25.44 | 41.10 | 182.65 | 251.09 | 183.53 |
| 83 | Akkarayan Aru | 59.09 | 23.44 | 19.79 | 37.63 | 38.92 | 12.08 | 16.83 | 25.12 | 40.58 | 180.34 | 247.90 | 181.20 |
| 84 | Mandekal Aru | 58.39 | 23.16 | 19.55 | 37.18 | 38.45 | 11.94 | 16.63 | 24.82 | 40.10 | 178.18 | 244.94 | 179.04 |
| 85 | Pallavarayankadd Aru | 57.55 | 22.83 | 19.27 | 36.64 | 37.90 | 11.77 | 16.39 | 24.46 | 39.52 | 175.63 | 241.43 | 176.47 |
| 86 | Pali Aru | 58.04 | 23.03 | 19.44 | 36.95 | 38.22 | 11.87 | 16.53 | 24.67 | 39.86 | 177.12 | 243.48 | 177.97 |
| N1 | — | 58.52 | 23.22 | 19.60 | 37.26 | 38.54 | 11.97 | 16.67 | 24.87 | 40.19 | 178.59 | 245.51 | 179.45 |
| N2 | — | 58.52 | 23.22 | 19.60 | 37.26 | 38.54 | 11.97 | 16.67 | 24.87 | 40.19 | 178.59 | 245.51 | 179.45 |

N1, N2—No names for these basins.

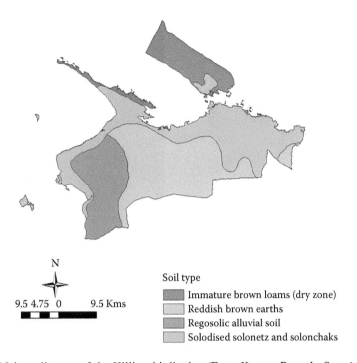

**FIGURE 12.8** Major soil types of the Killinochi district. (From Kumar, P, et al., *Sustain. Water Resour. Manage.*, 2, 419–430, 2016.)

**TABLE 12.4**
**Saturated Hydraulic Conductivity Values of the Various Soil Groups**

| Type of Soil | Hydraulic Conductivity (m/d) | Specific Storage |
| --- | --- | --- |
| 1. Red yellow latosols | 3.00 | 0.19 |
| 2. Alluvial soils of variable texture and drainage | 2.00 | 0.21 |
| 3. Solodized solonetz and solonchaks | 2.00 | 0.18 |
| 4. Regosols on raised beaches, spits, and dune sands | 10.30 | 0.25 |
| 5. Grumusols | 2.00 | 0.14 |
| 6. Eroded land | 0.65 | 0.17 |

*Source:* Soil Science Society of Sri Lanka and Canadian Society of Soil Science (SRICANSOL), Benchmark soils of the dry zones of Sri Lanka—Fact sheets for the major soil series, Peradeniya, Sri Lanka, 2009.

## 12.4 RESULTS AND DISCUSSION

The head contour values at the start and end of the initial simulation year were analyzed to properly assign initial conditions for the consecutive runs between 1984 and 1986 to study the temporal behavior of the groundwater recharge volumes; see Figure 12.9(a) and (b). The recharge into the groundwater system was considered to be entirely from rainfall percolation because the more or less flat topography inhibits the possibility of any significant contribution from lateral inflows. The recharge regime observed from the 3-year simulation results (Figure 12.10) also corroborates the assumption that the groundwater fluctuation was synchronized with the fall and rise of the rainfall regime (Figures 12.11 and 12.12). The lack of the observed data imposed a limitation on the calibration and validation of the results of the simulation. However, in the conceptualization process, all efforts are made to make sure that the model represents fairly the actual field conditions of the

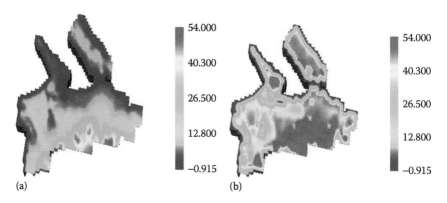

**FIGURE 12.9** Head contours in meters. (a) At the beginning of the simulation (SP-1, TS-1) and (b) at the end of the simulation (SP-12, TS-365). SP = Stress period, TS = Time step.

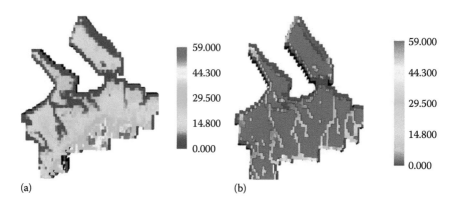

**FIGURE 12.10** Head contours in meters. (a) At the beginning of the simulation (SP-1, TS-1) and (b) at the end of the simulation (SP-36, TS-1095). SP = Stress period, TS = Time step.

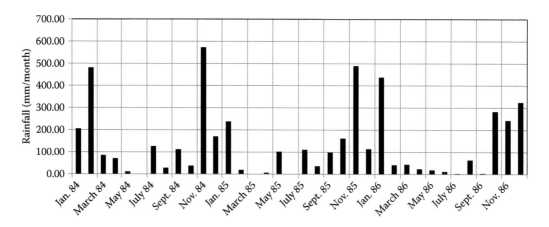

**FIGURE 12.11** Rainfall time-series (in mm/month).

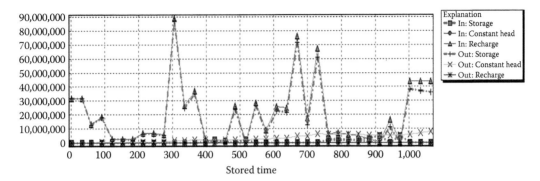

**FIGURE 12.12** Recharge rate for the simulated consecutive years in cubic meter (1984–1986).

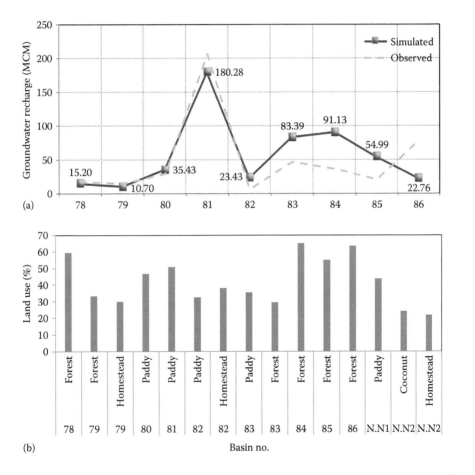

**FIGURE 12.13** (a) Zonal budget recharge rates for the different basins, simulated, and observed (MCM) and (b) major basin land use distribution. N.N1 and N.N2 are the two basins with no name.

Killinochi basin by incorporating all available data. In addition, a rough evaluation of the model performance in estimating the groundwater aquifer recharge volumes was attempted through a comparison of the simulated zonal water budgets with the zonal budgets calculated based on limited observed data; refer to Figure 12.13(a). The difference in cumulative recharge volumes between the observed and simulated results was mainly due to the interannual variations in the rainfall regime. Moreover, the basinwise spatiotemporal groundwater recharge distribution shows that, in the paddy- and forest-dominated zone areas where runoff was less, the recharge appeared to be much larger as compared to the recharge in the other land use classes; see Figure 12.13(a) and (b).

## 12.5 CONCLUSION

The study was conducted based on a groundwater modeling analysis at a regional scale with a 1 km grid resolution. Hence, the current groundwater model can be used largely as a framework to refine future local, more detailed groundwater modeling studies to better predict with a high level of confidence. The classical model calibration techniques were not applied in this study mainly due to the lack of observed hydrometric time series data. However, in the conceptualization process, all efforts were made to make sure that the model domain mimicked fairly the actual field conditions of the Killinochi basin by incorporating all available data. In addition, a rough evaluation of the model performance in estimating the groundwater aquifer recharge volumes was attempted through a comparison of the simulated zonal water budgets with the zonal budgets calculated based on observed data from IWMI and WRB reports. The modeling results at the end of the 1-year simulation period were used as an initial condition for further analysis of the recharge patterns of the Killinochi area for various temporal scales. Based on the 3-year simulation results, we have concluded that the recharge into the groundwater system was considered entirely to be from rainfall percolation because the more or less flat topography inhibits the possibility of any significant contribution from lateral inflows. The recharge regime observed from the modeling results also corroborated the assumption that the groundwater fluctuation was synchronized with the fall and rise of the rainfall regime. Although this study gives a positive insight into the groundwater dynamics and potential of the area, we believe that well-established knowledge of the availability and its spatial and temporal distribution is needed to accurately quantify and evaluate the groundwater resource and its sustainable exploitation with a greater confidence.

## REFERENCES

Amarasinghe UA, Mutuwatta L, Sakthivadivel R, 1999. Water scarcity variations within a country: A case study of Sri Lanka—Research report IWMI, Vol. 32. Colombo, Sri Lanka: International Water Management Institute.
Asian Development Bank, 2011. Jaffna and Killinochi water supply and sanitation project: Summary poverty reduction and social strategy. ADB, BRP SRI 37378.
Dassanayake AR, De Silva GGR, Mapa RB, Kumaragame D, 2005. *Benchmark Soils of the Dry Zone of Sri Lanka*. Peradeniya, Sri Lanka: Soil Science Society of Sri Lanka.
Hettiarachchi I, 2008. A review on ground water management issues in the dry zone of Sri Lanka. *BALWOIS 2008*, Ohrid, Republic of Macedonia, May 27–31.
Freeze RA, Cherry JA, 1979. *Groundwater*. Englewood Cliffs, NJ: Prentice Hall.
Fujita K, Hossain F, 1995. Role of the groundwater market in agricultural development and income distribution: A case study in a northwest Bangladesh village. *The Developing Economies*, 33, 460–463.
Gupta RS, 2008. *Hydrology and Hydraulic Systems*, 3rd ed. Long Grove, IL: Waveland Press.
Harbaugh AW, 2005. MODFLOW-2005, The U.S. Geological Survey modular ground-water model—The ground water flow process. U.S. Geological Survey Techniques and Methods 6-A16.
Hydrosult Inc., 2010. Dam safety and water resource planning project, DSWRPP-2/CS/QCBS/08.
Kodituwakku KAW, Pathirana SRK, 2003. North eastern coastal sand aquifer in Trincomalee district. Colombo, Sri Lanka: Integrated Food Security Programme. www.ifsp-srilanka.org.
Kresic N, 2007. *Hydrogeology and Groundwater Modeling*, 2nd ed. Boca Raton, FL: CRC Press, Taylor & Francis Group.
Kumar P, Herath S, Avtar R, Takeuchi K, 2016. Mapping of groundwater potential zones in Killinochi area, Sri Lanka, using GIS and remote sensing techniques. *Sustainable Water Resource Management*, 2(4), 419–430.

Mawatha HK. Water resources in Northern Province. Water Resources Board. http://www.isea.lk/dl_gal/198/239.pdf. Accessed on July 15, 2015.

Moormann FR, Panabokke CR, 1961. A new approach to the identification and classification of the most important soil groups of Ceylon. *Tropical Agriculture*, 117, 3–65.

Orehova T, Zektser I, Benderev A, Karimova O, 2012. Evaluation of the potential groundwater recharge. Example of the Ogosta River basin, NW Bulgaria. *Comptes rendus de l'Acade'mie bulgare des Sciences*, 65(10), 1387–1394.

Panabokke CR, Perera APGRL, 2005. Groundwater resources of Sri Lanka (WRB). Colombo, Sri Lanka: Water Resources Board.

Shah N, Nachabe M, Ross M, 2007. Extinction depth and evapotranspiration from ground water under selected land covers. *Ground Water*, 45(3), 329–338.

Soil Science Society of Sri Lanka and Canadian Society of Soil Science (SRICANSOL), 2009. Benchmark soils of the dry zones of Sri Lanka—Fact sheets for the major soil series. Peradeniya, Sri Lanka.

UNDP, 2011. Integrated strategic environmental assessments, northern province of Sri Lanka. www.isea.lk. Accessed on September 20, 2016.

United States Geological Service, 2010. HydroSHEDS. www.hydrosheds.cr.usgs.gov. Accessed on July 16, 2017.

Wikipedia, http://en.wikipedia.org/wiki/Kilinochchi_District.

# 13 CERES-Rice Model to Define Management Strategies for Rice Production; Soil Moisture and Evapotranspiration Estimation during Drought Years—A Study over Parts of Madhya Pradesh, India

*Sourabh Shrivastava, S.C. Kar, and Anu Rani Sharma*

## CONTENTS

| | | |
|---|---|---|
| 13.1 | Introduction | 208 |
| 13.2 | Material and Method | 209 |
| | 13.2.1 Study Region | 209 |
| | 13.2.2 Data Used | 209 |
| 13.3 | Result and Discussion | 210 |
| | 13.3.1 Observed Rainfall and Temperature Variability | 210 |
| | 13.3.2 Rainfall Anomalies and Madden–Julian Oscillation Indices | 212 |
| | 13.3.3 Variability in Crop Yield | 214 |
| | 13.3.4 Sensitivity Analysis | 215 |
| | 13.3.5 Detrend Analysis of the Simulated Yield | 217 |
| | 13.3.6 Simulated and Remotely Sensed Evapotranspiration | 218 |
| | 13.3.7 Simulated and Remotely Sensed Soil Moisture | 219 |
| 13.4 | Conclusion | 220 |
| References | | 220 |

**ABSTRACT** The performance of the Decision Support System for Agrometeorology Transfer (DSSAT) v4.6 model was assessed for two major rice-growing varieties (IR 36 and Swarna) over selected stations, viz. Balaghat, Jabalpur, Narsinghpur, and Seoni in Madhya Pradesh, India. Drought years in these regions were identified using India Meteorological Department (IMD) rainfall data sets during monsoon season of June, July, August, and September (JJAS), 1990–2011 and different sowing dates were chosen, respectively. It was found that the DSSAT model predicted crop yield higher than the observed yield. Furthermore, a bias correction and detrended yield anomaly analysis was carried out; it suggested that the model followed an observed pattern during most of the drought or deficit phases. Additionally, a sensitivity analysis was carried out to examine the model sensitivity with respect to variations in daily rainfall and genetic coefficients of the crop, affecting crop yield. Remote sensing data have also been used to compare the model, which estimates that daily soil moisture is closer to European Space Agency (ESA)–derived soil moisture. The model-simulated evapotranspiration (ET) reflects a minor gap with the remote sensing observations. The relationship of weak phases of rainfall over central India to real-time multivariate (RMM) indices of Madden–Julian oscillation (MJO) has been examined. RMM-6, RMM-7, RMM-1, and RMM-2 describe the drought conditions over central India. However, the frequency of drought occurrence over Madhya Pradesh is more during RMM-7 phase.

## 13.1 INTRODUCTION

The prediction of crop yield plays a vital role in developing and applying policies in extreme conditions. Crop simulation models are mathematical equations describing crop growth, yields, climate, and management practices (Motha, 2011). Numerous researchers worldwide used simulated model approach in various crops (rice, wheat, maize, etc.) with satisfactory results (Mourice et al., 2014; Nyang'au et al., 2014). The performance of crop prediction models during drought and non-drought years has been assessed over various stations around the world. The DSSAT v4.6 model is a computer-based system that has combination of different crop growth models that is, CERES-Barley, CERES-Maize, CERES-Rice, CERES-Sorghum, CERES-Sunflower, CERES-Wheat, Cropgro-Drybean, Crogro-Soyabean and so on, and it is an important tool to use to predict crop yield (Jones et al., 2003). DSSAT simulates crop growth, development, and yield through the effects of weather, management, genetics, soil water, carbon, and nitrogen (Timsina and Humphreys, 2006). The DSSAT model, however, does not account for incidences of pest, diseases, and extreme weather conditions, hence hampering the precision of predicted output (Jones et al., 2003). Singh et al. (2016) found the inaccuracies of the DSSAT soil water balance module and suggested modifications to improve the model forecast. It is anticipated that the cropping system will be very complex and that the crop production varies spatially as well as temporally (Motha, 2011). Nyang'au et al. (2014) carried out sensitivity analysis of CERES-Rice and suggested that maximum, mean, and minimum temperature or rainfall directly correlates with observed crop yield. These researchers have documented the impact of increased temperature over crop yield.

The direct relationship between crop yield, soil moisture (SM), and ET has been known (Cemek et al., 2011). The ability of the crop simulation model is to simulate ET, SM, and various parameters thereby influencing crop output levels (Jones et al., 2003). Dallacort et al. (2011) found an error in model-simulated ET and simulated SM. Olioso et al. (2003) concluded that the crop simulation model may be used in combination with remote sensing data for monitoring ET of irrigated crops. These researchers carried out a continuous estimation of ET and SM over the crop cycle and found that the model procedures make it possible to monitor ET, SM, plant growth, and, in some cases, irrigation practices. Dallacort et al. (2011) noticed a variability in the model's simulated-SM and ET.

Indian agriculture mainly depends on summer monsoon rainfall to account for 75–90% of India's annual rainfall (Varikodan et al., 2014). Rice is one of the important food crops in the Asian subcontinent, primarily India. In Madhya Pradesh, the majority of the rice area has rainfed conditions (78.4%) of which 20.7% in the area is sown in Jabalpur, Seoni, Balaghat, and Narsinghpur districts (Commissioner, Land Records, 2012). This suggests that the meteorological parameters largely vary interannually on the rice yield in Madhya Pradesh. There have been no studies on the use of the DSSAT model for forecasting rice yield, variations in SM, and ET in India during drought conditions. The aim of this study is to document the model performance on the estimates of rice yield under rainfed conditions. The model's estimated ET and SM were compared with the remote sensing data sets. In Section 13.2, the material and method of analysis is described. The results of this study are described in Section 13.3. Section 13.4 discusses the conclusion of the study.

## 13.2 MATERIAL AND METHOD

### 13.2.1 STUDY REGION

The study region is located in eastern Madhya Pradesh, India (Figure 13.1). According to the selection of the highest rice yield–producing district in Madhya Pradesh and data availability for the experiment, four stations selected for the experiments were Balaghat (21.86 N, 80.36 E), Jabalpur (23.18 N, 79.98 E), Narsinghpur (22.91 N, 79.10 E), and Seoni (22.08 N, 79.54 E). The experiments were conducted during the 1990 to 2013 growing season of *kharif* rice crop. The soil textures are silty clay loam, sandy clay loam, and black cotton soil of the selected region (Rao, 2013).

### 13.2.2 DATA USED

The daily weather data (rainfall, maximum and minimum temperature, solar radiation) for all the four stations in Madhya Pradesh were collected from the IMD, New Delhi. The IMD's gridded rainfall (0.25° × 0.25°) and mean temperature (1.0° × 1.0°) were used. The soil of the region is classified as a rice soil according to the Jawaharlal Nehru Agricultural University (JNKVV), Jabalpur, Madhya Pradesh, India under Forecasting Agricultural output using Space, Agrometeorology and Land based observations (FASAL) scheme (Table 13.1). The Moderate Resolution Imaging Spectroradiometer (MODIS) derived to be 0.05° × 0.05° gridded monthly ET data (Mu et al., 2011) was used during 2000–2014. The ET algorithm is based on the Penman–Monteith equation

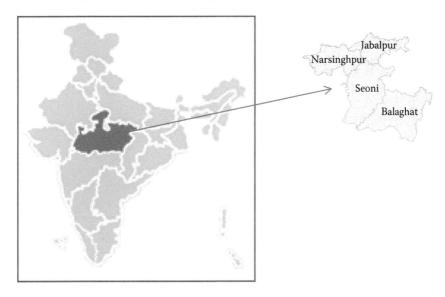

**FIGURE 13.1** Study area Balaghat, Jabalpur, Natshinghpur, and Seoni districts.

**TABLE 13.1**
**Soil Physical Characteristics of 0–60 cm Soil Depth at the Experimental Area**

| Layer (cm) | Saturation | Upper Limit | Lower Limit | Bulk Density (g/cm³) | Sand (%) | Silt (%) | Clay (%) |
|---|---|---|---|---|---|---|---|
| 0–5 | 0.60 | 0.30 | 0.17 | 1.36 | 12 | 44.8 | 43.2 |
| 5–15 | 0.60 | 0.30 | 0.17 | 1.36 | 11.2 | 46.0 | 42.8 |
| 15–30 | 0.60 | 0.30 | 0.17 | 1.36 | 11.3 | 46.3 | 42.4 |
| 30–45 | 0.60 | 0.30 | 0.17 | 1.36 | 10.8 | 47.5 | 41.7 |
| 45–60 | 0.60 | 0.30 | 0.17 | 1.36 | 11.90 | 45.20 | 42.90 |

## TABLE 13.2
### Description of Data Sets

| Parameter | Data Source | Resolution | Data Used | Type |
|---|---|---|---|---|
| Rainfall | IMD | 0.25° | 1990–2011 | Observed |
| Rainfall | IMD | Station | 1990–2011 | Observed |
| Soil moisture | ESA–CCI | 0.25° | 1990–2011 | Remotely sensed |
| Evapotranspiration | MODIS | 0.05° | 1990–2011 | Remotely sensed |

## TABLE 13.3
### Calibrated Genetic Coefficients for Different Varieties of Rice

| Varity | P1 | P2R | P5 | P2o | G1 | G2 | G3 | References |
|---|---|---|---|---|---|---|---|---|
| IR 36 | 450.0 | 149.0 | 350.0 | 11.7 | 45.0 | 0.0230 | 1.0 | Singh et al. (2005) |
| Swarna | 800.0 | 052.0 | 550.0 | 11.1 | 45.0 | 0.0280 | 1.0 | Jain and Sastri (2015) |
| IR 36 | 500.0 | 149.0 | 450.0 | 11.5 | 30.0 | 0.0230 | 1.0 | DSSAT default |

(Monteith, 1965). The ESA–CCI merged product of 0.25° × 0.25° daily SM data sets were developed following the procedure described by Christopher et al. (2011), Parinussa et al. (2012), Liu et al. (2012), and Wagner et al. (2012). The crop yield was collected from the Farmer Welfare and Agriculture Development Department of Madhya Pradesh (available at www.mpkrishi.mp.gov.in) during the years 1990–2011. The DSSAT input such as soil profile, fertilizer, and other management practices followed the recommendations collected from IMD and JNKVV under the FASAL scheme. Data used in this study are presented in the Table 13.2. Similarly, genetic coefficients of different rice varieties (IR 36 and Swarna) derived for the Raipur station is presented in the Table 13.3.

The RMM MJO indices (RMM1 & RMM2) of Wheeler and Hendon (2004) were used for defining the various MJO phases. The data were obtained from http://www.bom.gov.au/bmrc/clfor/cfstaff/matw/maproom/RMM/.

The CERES-Rice model was run with different management practices with different dates of sowing for each station. The sowing window in Balaghat, Jabalpur, Narsinghpur, and Seoni was from June 5–July 10. Therefore, the first week of the southwest monsoon with 50–110 mm was chosen for the direct seed sowing of rice under rainfed conditions. Plot management followed the local standard practices (weed control and fertilizer application) for rice production in all four districts.

The sensitivity of the model was tested by making changes (one by one) in the genetic coefficients and the rainfall. In the genetic coefficients, the value of the basic vegetative phase of the plant (P1), the critical photoperiod, or the longest day at which the plant development occurs at a maximum rate (P2O), the extent to which phasic development leading to panicle initiation was delayed (P2R), and the potential spikelet number (G1) were changed to those of the test model. Each variety coefficient was changed, and the remaining coefficients were fixed on their normal values following Singh et al. (2005). Six different genetic coefficients for sensitivity analysis were tested. The model's sensitivity to rainfall was also evaluated. The daily rainfall values were reduced and increased by 20% and 50% in the experiments for all stations.

## 13.3 RESULT AND DISCUSSION

### 13.3.1 Observed Rainfall and Temperature Variability

The seasonal mean climatology of rainfall, SM, and ET over India during June, July, and August (JJA) is presented in Figure 13.2. The Western Ghats receive highest rainfall during JJAS, but the core zone of the monsoon is located in central India (Madhya Pradesh and the surrounding region).

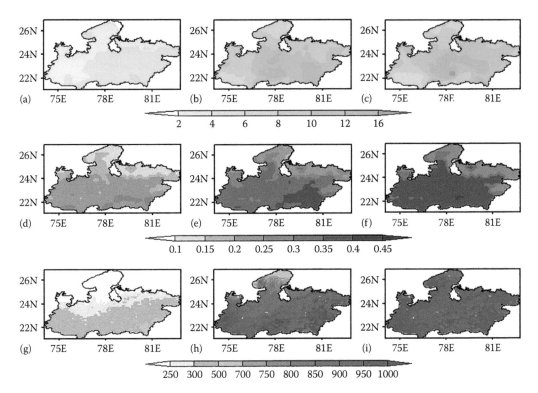

**FIGURE 13.2** Rainfall (mm/day), soil moisture (m³/m³), and evapotranspiration over Madhya Pradesh. (a) Rain June climatology, (b) rain July climatology, (c) rain August climatology, (d) SM June climatology, (e) SM July climatology, (f) SM August climatology, (g) ET June climatology, (h) ET July climatology, and (i) ET August climatology.

The mean rainfall over Madhya Pradesh is 7–9 mm/day according to the observed rainfall data from IMD. The ESA's SM climatology for JJA is shown in Figure 13.2d–f. The figures indicate that during June, SM ranges from 0.1 to 0.3 m³/m³. In July, the SM values over Madhya Pradesh increase, ranging from 0.2 to 0.4 m³/m³. August is the wettest month as far as soil is concerned with SM ranging from 0.3 to 0.45 m³/m³. Consistent with rainfall and SM, ET values also have the same pattern with June having the least ET (200–500 mm) and August (>850 mm) having the maximum. However, this rainfall has considerable intraseasonal and interannual variability.

Figure 13.3a shows the observed rainfall and mean temperature for the Balaghat, Jabalpur, Narsinghpur, and Seoni stations. The rainfall amount over all four stations varied from 0 to 14 mm/day from January to December. The peak months of rainfall were JJAS when the rainfall reached 14 mm/day or more. The maximum amount of rainfall received in Madhya Pradesh is during July and August. June and September are the primary months that contribute to the total rainfall over these districts. Figure 13.3b shows the time series of seasonal rainfall anomaly for JJAS during 1990–2011 at each station. The JJAS average anomaly of rainfall for Balaghat, Jabalpur, Narsinghpur, and Seoni was plotted. The year with the deficit and the year with excess rainfall for all stations were identified. A year with rainfall of more than 2 mm/day is considered as an excess and one with less than 2 mm/day as a deficit year. Table 13.4 shows the excess and deficit years for all four Madhya Pradesh stations. Regarding the climatology of temperature of these districts, April, May, and June were the hottest months in every year. The maximum mean temperature reached 30° during the hottest month of May, and the minimum mean temperature reached 14° in January. As shown in Figure 13.3a, November to March are the coldest months in Madhya Pradesh. During the peak monsoon months (July and August), the temperature decreases, but after August, it increases by more than 3°–4°.

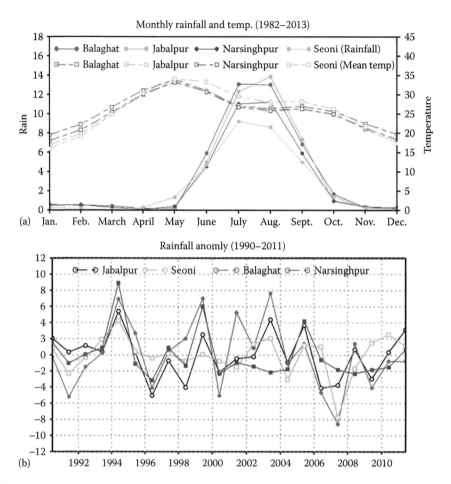

**FIGURE 13.3** Observed rainfall and mean temperature for the four stations in Madhya Pradesh, India. (a) Climatology of rainfall and mean temperature during January to December, (b) seasonal mean (JJAS) rainfall anomaly for the Balaghat, Jabalpur, Narsinghpur, and Seoni districts.

**TABLE 13.4**
**Excess and Deficit Rainfall Years for the Four Districts of Madhya Pradesh**

| District | Excess | Deficit |
| --- | --- | --- |
| Balaghat | 1994, 1995, 1999, 2001, 2003 | 1991, 1996, 2000, 2006, 2007, 2009 |
| Jabalpur | 1994, 1999, 2003, 2005 | 1996, 1998, 2006, 2007, 2009 |
| Narsinghpur | 1994, 1999, 2005 | 1996, 2000, 2003, 2004, 2007, 2008, 2009 |
| Seoni | 1993, 1994, 2003, 2010 | 1991, 2004, 2007, 2008, 2009 |

### 13.3.2 Rainfall Anomalies and Madden–Julian Oscillation Indices

As discussed in the previous section, monsoon rainfall over Madhya Pradesh exhibits strong intraseasonal variability that causes droughts or floods; see Figure 13.3b. MJO is one of the most influential factors of the intraseasonal variability of monsoon rainfall over India. Using IMD daily gridded rainfall data and Wheeler–Hendon MJO indices, Pai et al. (2011) examined the intraseasonal variation of daily rainfall distribution over India associated with various phases of the eastward-propagating MJO life cycle. These researchers found that during MJO phases of 1 and 2, the formation of positive convective anomaly over the equatorial Indian causes break monsoon type rainfall distribution over India. As the MJO propagates eastward to the west equatorial Pacific through the maritime continent, a gradual northward shift of the convective activity over the Indian Ocean is observed. During phase 4, the northward propagating convective zone merges with the monsoon trough and enhances rainfall over the region. During phases 5 and 6, the patterns are reversed compared to those during phases 1 and 2, and India experiences active monsoon conditions. During the subsequent phases (7 and 8), the convective anomaly patterns are very similar to those during phases 1 and 2. A general decrease in rainfall is also observed over most parts of the country. However, it is not clear from the Pai et al. (2011) study and similar past studies whether the real-time MJO indices represent drought conditions over Madhya Pradesh and if so, which MJO phases need to be monitored or forecasted.

The relationship of weak phases of rainfall over central India with RMM indices (Wheeler and Hendon, 2004) for the intraseasonal time scale has been studied. Figure 13.4a–e shows the composite of weak phases over central India during different RMM phases. Maximum weak phases over central India occur during positive phases of RMM-6 and RMM-7 rather than RMM-1 or RMM-2. Longer weak phases of rainfall are seen more during RMM-7 than in other modes. RMM phase 1 experiences a deficiency in rainfall (10–15 mmday$^{-1}$) over central India and excess rainfall (8–10 mmday$^{-1}$) over northeastern India. During the RMM phase 2, the deficiency in rainfall is 12–15 mmday$^{-1}$ over central India, and the excess rainfall over northeastern India is 8–10 mmday$^{-1}$. In phases 6 and 7, the whole of central India and surrounding regions have a 10–15 mmday$^{-1}$ deficiency in rainfall. Therefore, during RMM phase 7, long week phases of rainfall occur in the study region, and they should be monitored very closely for drought monitoring and prediction.

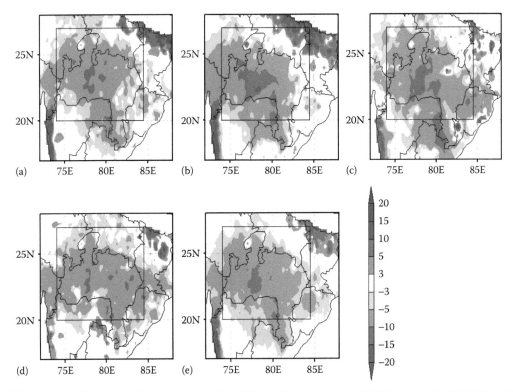

**FIGURE 13.4** Composite of break phases during different RMM phases. (a) RMM phase 1, (b) RMM phase 2, (c) RMM phase 6, (d) RMM phase 7, and (e) composite of all phases (1, 2, 6, and 7).

### 13.3.3 Variability in Crop Yield

Figure 13.5a shows the year-to-year variations in the observed rice yield for the Balaghat, Jabalpur, Narsinghpur, and Seoni stations during 1990–2011. The observed yield varied from 826 to 1547 kg/ha in Balaghat, 578 to 1302 kg/ha in Jabalpur, 749 to 1971 kg/ha in Narsinghpur, and 416 to 1532 in Seoni. Among these four stations, Jabalpur had the lowest yield, especially during 1991–1998. In 2000, Seoni and Jabalpur had the minimum rice yield and in 2003 and 2005, Balaghat had the maximum yield during the study period. An increasing trend was noticed in the observed rice yield for all the stations (not shown in the figure). The DSSAT model was used to simulate rice yield for these stations using observed meteorological and other parameters as mentioned in Section 13.2. Figure 13.5b shows the model-simulated yield over Balaghat, Jabalpur, Narsinghpur, and Seoni during 1990–2011. Large variations in the simulated field were noticed, which was not seen in the observed yield. The interannual standard deviation (SD) of observed yield was 224.78 kg/ha, and the simulated yield was 1217.19 for Balaghat; for Jabalpur, the SD observed and simulated yields were 237.64 and 1037.48 kg/ha, respectively. For Narsinghpur, the SD of the observed yield was 253.06 kg/ha, and the simulated yield was 1499.90 kg/ha. For Seoni, the SD of observed yield was 315.46 kg/ha and the simulated yield was 1323.38 kg/ha. In addition, a large yield gap between observed and simulated results was also noticed for all four districts. Figure 13.6a shows the difference in observed and simulated yields for all four stations. The average bias in the model simulations for Balaghat was −2389.45 kg/ha and for Jabalpur was −2475.92 kg/ha. For Narsinghpur and Seoni, the bias values were −2789.13 and −2397.59 kg/ha, respectively. As a result of this large bias between the model and observed yields, the root mean square errors (RMSE) were also very large

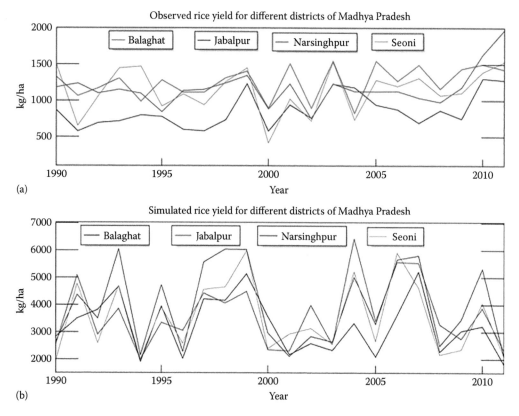

**FIGURE 13.5** Year-to-year variations in the observed rice yield. (a) Observed yield and (b) simulated yield from 1990 to 2011 for the Balaghat, Jabalpur, Narsinghpur, and Seoni districts.

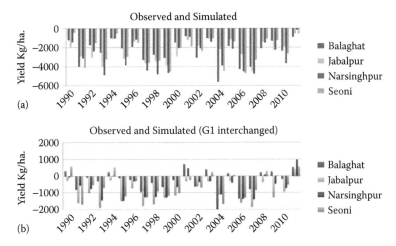

**FIGURE 13.6** Difference between the observed and simulated yield for (a) control runs (b) by changing the genetic coefficients for IR36 rice crop.

for all four stations. The RMSE were 2690.21, 2624.74, 3158.96, and 2752.71 kg/ha for the Balaghat, Jabalpur, Narsinghpur, and Seoni stations, respectively.

### 13.3.4 SENSITIVITY ANALYSIS

Sensitivity of the CERES-Rice model was studied using the changes in weather data and genetic coefficient of the crop. Table 13.3 shows the six different genetic coefficients used in this study. The genetic coefficients of Singh et al. (2005) have been used by changing the default DSSAT genetic coefficients for the rice crop. Coefficients P2R, G2, and G3 values are found to be similar in both studies. Remaining values have been changed one by one in the RICER.CUL file of the DSSAT model. Only minor differences have been seen in the simulated yield when the values of P1, P5, and P2O are changed. Therefore, the model is not very sensitive to these parameters in the study region. The main difference is seen in the simulated yield when the value of G1 was interchanged from the default values provided by the DSSAT model.

The simulated yield in the G1 interchanged experiment is reduced (maximum up to 3000 kg/ha or more) in the Balaghat, Jabalpur, Narsinghpur, and Seoni stations. The simulated yield range in the CERES-Rice for Balaghat was from 800 to 2600 kg/ha, for Jabalpur was from 800 to 2500 kg/ha, for Narsinghpur was from 600 to 2500 kg/ha, and for Seoni was from 900 to 2500 kg/ha. The bias, RMS error, and yield gap in simulated crop yield were found to be much less for the four stations. Figure 13.6b shows the difference between the observed yield and the simulated yield from the interchanged genetic coefficient. It has seen that, the model performance is improved when modified GC have used. Table 13.5 shows that the RMS error for Balaghat was 724.92 kg/ha, for Jabalpur

**TABLE 13.5**
**RMSE of Simulated Yield for Both Genetic Coefficients (GC)**

| Stations | RMSE (kg/ha) (Interchanged GC) | RMSE (kg/ha) (Singh et al., 2005) |
|---|---|---|
| Balaghat | 724.92 | 2577.92 |
| Jabalpur | 1150.88 | 3061.71 |
| Narsinghpur | 901.10 | 3150.66 |
| Seoni | 875.23 | 2770.35 |

was 1150.88 kg/ha, for Narsinghpur was 901.10 kg/ha, and for Seoni was 875.23 kg/ha. Therefore, it is found that the potential spikelet number (G1) is one of the most crucial parameters in the DSSAT model for rice crop as far as the Madhya Pradesh region is concerned.

For the second set of sensitivity studies, experiments were carried out by reducing the rainfall by 20% and 50% and increasing the rainfall by 20% and 50%. These experiments show how the model responds to changes in rainfall in the selected districts of Madhya Pradesh while all other parameters were the same. The change was made in the weather data (.WTH) file of the DSSAT model for each station. The impact of rainfall over the simulated yield was clearly seen during normal and excess years when the rainfall amount increased and the yield amount reduced. However, during drought years, when the rainfall amount increased, an increase in rice yield was noticed. Similarly, rice yield increased when that rainfall amount was reduced in flood years. This feature is seen for all the stations for which experiments were conducted. These experiments show that the model is sensitive to rainfall parameters and that proper amount of rainfall values is required to simulate the model's correct trend in rice. The sensitivity analysis found that the model is sensitive to both rainfall and genetic coefficients. If a change is made in the genetic coefficient, the model will respond in terms of increasing or decreasing the yield. It was also found that in the model, the simulated yield is not totally dependent on rainfall. However, in all cases, the model-simulated yield was larger than the observed yield when much less rainfall (−50%) was used.

To further understand the role of the genetic coefficients, five different ones have been used for the rice crop, and sensitivity experiments were carried out for the four stations in Madhya Pradesh. These genetic coefficients are derived for IR 36 variety for Asia, specifically for different locations in India. The source of the genetic coefficients and their values are shown in Table 13.6. The model was run with all the genetic coefficients for different districts of Madhya Pradesh during the period 1990–2011. The comparison was made by examining the model performance with all genetic coefficients of rice crop. The experiments have shown huge differences between observed and simulated yield for IR 36. The variations in the yield for Balaghat were from 1225.01 to 1966.59 kg/ha, for Jabalpur were from 1427.89 to 2392.13 kg/ha, for Narsinghpur were from 1394.25 to 2533.74 kg/ha, and for Seoni were from 1324.75 to 2089.24 kg/ha. The RMSEs obtained from these experiments are summarized in Table 13.7. The RMSE for Balaghat was from 2577.92 to 4423.93 kg/ha, for

**TABLE 13.6**
**Genetic Coefficients for Different Sensitivity Experiments**

| Experiment No. | Source | P1 | P2O | P2R | P5 | G1 | G2 | G3 |
|---|---|---|---|---|---|---|---|---|
| 1 | Singh et al. (2005) | 450 | 11.7 | 149 | 350 | 45 | 0.023 | 1.0 |
| 2 | Hoogenboom et al. (1997) | 450 | 11.7 | 149 | 350 | 68 | 0.023 | 1.0 |
| 3 | Rao (2008) | 550 | 11.7 | 149 | 550 | 70 | 0.023 | 1.0 |
| 4 | Satapathy et al. (2013) | 470 | 11.5 | 50 | 350 | 65 | 0.020 | 1.0 |
| 5 | Swain et al. (2007) | 475 | 11.7 | 90 | 385 | 68 | 0.023 | 1.0 |

**TABLE 13.7**
**RMSE of Simulated Yield for All Genetic Coefficients (IR36) Experiments**

| Stations | Singh et al. (2005) | Hoogenboom et al. (1997) | Rao (2008) | Satapathy et al. (2013) | Swain et al. (2007) | Interchanged (G1) |
|---|---|---|---|---|---|---|
| Balaghat | 2577.92 | 4268.46 | 4423.93 | 3262.45 | 4258.25 | 724.92 |
| Jabalpur | 3016.71 | 5248.79 | 5224.93 | 4373.67 | 5270.59 | 1150.88 |
| Narsinghpur | 3150.65 | 4899.93 | 4766.70 | 4214.04 | 4947.30 | 901.10 |
| Seoni | 2770.35 | 4514.11 | 4289.56 | 3738.63 | 4568.72 | 875.23 |

Jabalpur was from 3061.71 to 5270.59 kg/ha, for Narsinghpur was from 3150.65 to 4947.30 kg/ha, and for Seoni was from 2770.35 to 4568.72 kg/ha. Although the amount of simulated yield in the experiments changed, the pattern of simulated yield did not.

### 13.3.5 DETREND ANALYSIS OF THE SIMULATED YIELD

Due to the large yield gap noticed between the observed and the model-simulated yield, direct use of the model-simulated yield values could lead to improper policy decisions. In order to know the model performance and the ability of the model to capture the trend, a detrend technique was used. The simulated yield was detrended by removing the tendency in the simulated yield as a function of time. Figure 13.7 shows the detrended anomalies of the model-simulated yield along with observed yield anomalies. After detrending, the model was able to capture the trend as well as the fall in yield during specific years. For Balaghat, the observed anomaly varied from −1000 to 1500 kg/ha and the simulated detrend anomaly varied from −3000 to 2000 kg/ha. The model captured the observed anomaly pattern during 1997–2011 very well and was able to capture the pattern of the observed yield, especially in drought years. For Jabalpur, the observed anomaly varied between −400 and

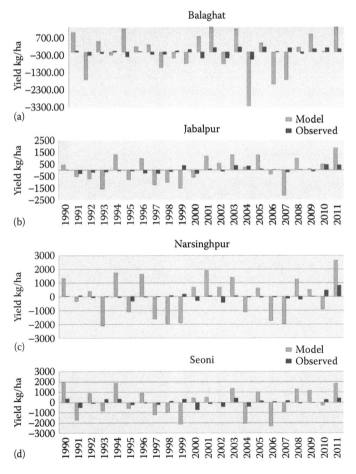

**FIGURE 13.7** Observed anomaly and detrend anomaly of rice yield the (a) Balaghat, (b) Jabalpur, (c) Narsinghpur, and (d) Seoni districts.

500 kg/ha, and the simulated detrend anomaly ranged from −2000 to 2000 kg/ha. From 1990 to 1996, the pattern did not match the observation well. In the drought years, the simulated detrend yield pattern was very similar to the observed yield anomaly. In nondrought years, the simulated yield matched the observed anomaly well. For Narsinghpur, the range of the observed yield anomaly was between −500 and 1000 kg/ha, and the simulated detrend anomaly varied from −2000 to 2000 kg/ha. The simulated detrend model yield was also able to capture the observed anomaly pattern, especially in drought years. For Seoni, the observed anomaly varied between −700 and 300 kg/ha, and the simulated detrend anomaly varied from −3000 to 2000 kg/ha. The simulated detrend anomaly captured the observed anomaly pattern very well.

### 13.3.6 SIMULATED AND REMOTELY SENSED EVAPOTRANSPIRATION

Figure 13.8 shows that the model-simulated ET and MODIS-derived ET during the period 2000–2011. It is found that the model underestimated the ET for Balaghat district whereas it overestimated it for the Jabalpur, Narsinghpur, and Seoni stations. The average MODIS ET and simulated ET for Balaghat was 501 mm and 416 mm, respectively; for Jabalpur, they were 385 mm and 419 mm, respectively; for Narsinghpur, they were 366 mm and 415 mm; for Seoni, they were 355 mm and 419 mm, respectively. The model-simulated ET for all the stations was almost similar whereas the remotely sensed ET had major differences. Table 13.8 shows the differences in model-simulated ET and MODIS ET. The model simulated ET was more than MODIS ET except for the Balaghat station because the model simulates average ET (nearly 400 mm) for all the stations, and there were no variations among the simulated ET. Table 13.8 shows the RMS error, SD, and difference of ET for the model and MODIS during the excess and deficit years. The RMS error for Balaghat station was 97.25 mm, for Jabalpur was 62.7 mm, for Narsinghpur was 70.62 mm, and for Seoni was 79.80 mm. The SD of MODIS varied from 17 to 35 mm, and the simulated ET variation was from 34 to 41 mm. The model overestimated the ET, and the variation in model output was seen to be much higher than the

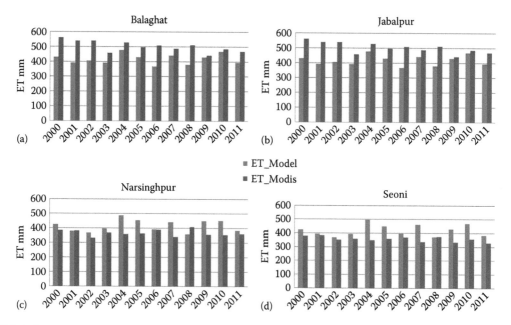

**FIGURE 13.8** Model-simulated and MODIS evapotranspiration from 2000 to 2011 (per data availability) for (a) Balaghat, (b) Jabalpur, (c) Narsinghpur, and (d) Seoni districts.

## TABLE 13.8
### RMSE and SD of MODIS and Simulated Evapotranspiration: Difference of ET between MODIS and Simulated Values during Excess and Deficit Years

| District | RMSE | MODIS (SD) | Simulated (SD) | MODIS (Excess-Deficit) | Model (Excess-Deficit) |
|---|---|---|---|---|---|
| Balaghat | 97.25 | 35.91 | 34.38 | −0.75 | −24.75 |
| Jabalpur | 62.70 | 21.88 | 41.79 | 15.88 | −8.67 |
| Narsinghpur | 70.62 | 21.42 | 40.83 | −5.78 | 27.33 |
| Seoni | 79.80 | 17.97 | 41.79 | 10.27 | −8.75 |

remote sensing data. The difference was calculated for excess and deficit years (Table 13.4), and the remote sensing data difference was found to be positive or near to zero, and the simulated difference was negative and high except for the Narsinghpur station. The model overestimated the ET during deficit and drought years.

### 13.3.7 SIMULATED AND REMOTELY SENSED SOIL MOISTURE

The daily SM variation over each district occurred during drought years only. The daily remote sensing SM variation over Balaghat was similar to that of the simulated SM, which was very close to the remote sensing data in 1996, 1998, and 2007, but there was more difference in 2006 and 2009. In Jabalpur, the simulated SM followed the remote sensing time series of SM. In 2000, 2007, and 2009, both (remotely sensed and simulate SM) are closer but large differences were seen in 1996 and 2006. In Narsinghpur, there were six deficit years in the last decade and, the model-simulated SM was not closer to the observation in any of the years. The simulated SM and remote sensing SM difference was very high during these years. In Seoni, the simulated SM values were less than the remote sensing data. Simulated SM was underestimated by the model during drought years. Figure 13.9 shows the RMSEs of SM simulations during drought years. The RMSE varies from 0.15 to 0.35 m$^3$/m$^3$ for all four stations. The maximum error of 0.32 was noticed in 1996 for the Balaghat and Narsinghpur stations. The minimum RMSE (0.15 m$^3$/m$^3$) was noticed in 2007, 2008, and 2009 for Jabalpur and Seoni.

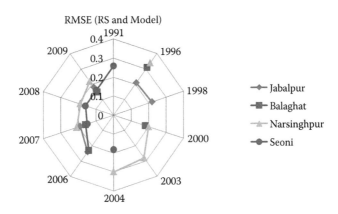

**FIGURE 13.9** RMSE of SM from remote sensing and model simulations during drought years.

## 13.4 CONCLUSION

The model discussed here overestimated the yield in every year for all four stations; however, the simulated yield followed the observed pattern, especially in the drought years. The results for the simulated yield pattern, ET, and the SM relationship were comparable with the remote sensing data sets. The lack of experimental data on locally adopted rice varieties made a difference in observed and simulated yields. In addition, many forces work on crop growth and development in the field, and it is not possible to provide all information in a model because of the lack of observation. The sensitivity of the model shows large variations with genetic coefficient parameters, particularly G1 coefficients. The yield gap between the observed and the simulated yields was reduced by choosing a proper G1 coefficient. Overall, the study shows that with proper refinements, the DSSAT model can be used to estimate crop yield, SM, and ET for the Madhya Pradesh region during drought years.

## REFERENCES

Cemek, B., Unlukara, A., Karamen, S. and Gokalp, Z. 2011. Effect of evapotranspiration and soil salinity on some growth parameters and yield of lettuce. *Agriculture*, 98(2), 139–148.

Christopher, R.H., Crow, W.T., Mecikalski, J.R., Anderson, M.C. and Holmes, T. 2011. An inter comparison of available soil moisture estimates from thermal infrared and passive microwave remote sensing and land surface modelling. *Journal of Geophysical Research*, 116(D15107), 1–18.

Commissioner, Land Records. 2012. Available at Commissioner, Land Records, Gwalior at www.landrecords.mp.gov.in/agr_Stat.htm.

Dallacort, R., de-Freitas, P., Faria, R.T., Goncalves, A.C.A., Rezende, R. and Guimarães, R.M.L. 2011. Simulation of bean crop growth, evapotranspiration and yield in Paraná State by the CROPGRO-Drybean model. *Acta Scientiarum Agronomy*, 33(3), 429–436.

Hoogenboom, G., Georgiev, G. and Singh, U. 1997. Performance of CERES–Rice model for three location in the Asia region. Report of the workshop Climate variability, agricultural productivity and food security in the Asian monsoon region. *A Joint START/WCRP/IGBP-GCTE Workshop*, Bogor, Indonesia, February 19–22. START Report 1997/2, pp. 35–42.

Jain, S. and Sastri, A.S. 2015. Comparison of ceres and infocrop dynamic crop simulation models for assessing the production potential of rice in irrigated and rainfed condition. Indira Gandhi Agriculture University, Raipur, India.

Jones, J., Hoogenboom, G., Porter, C.H., Boote, K.J., Batchelor, W.D., Hunt, L.A., Wilkens, P.W., Singh, U., Gijsman, A.J. and Ritchie, J.T. 2003. The DSSAT cropping system model. *European Journal of Agronomy*, 18(3–4), 235–265.

Liu, Y.Y., Dorigo, W.A., Parinussa, R.M., Jeu, de R.A.M., Wagner, W., McCabe, M.F., Evans, J.P. and Dijk, van, A.I.J.M. 2012. Trend-preserving blending of passive and active microwave soil moisture retrievals. *Remote Sensing and Environment*, 123, 280–297.

Monteith, J.L. 1965. Evaporation and environment. In: B.D. Fogg (Ed.), *The State and Movement of Water in Living Organism, Symposium of the Society of Experimental Biology*, 19, Cambridge University Press, Cambridge, UK, pp. 205–234.

Motha, R.P. 2011. Use of crop models for drought analysis-drought mitigation center Faculty publications. Paper 58. http://digitalcommons.unl.edu/droughtfacpub/58.

Mourice, S.K., Rweyemamu, C.L., Tumbo, S.D. and Amuri, N. 2014. Maize cultivar specific parameters for Decision Support System for Agrotechnology Transfer (DSSAT) application in Tanzania. *American Journal of Plant Sciences*, 5(6), 821–833.

Mu, Q., Zhao, M. and Running, S.W. 2011. Improvements to a MODIS global terrestrial evapotranspiration algorithm. *Remote Sensing of Environment*, 115(8), 1781–1800.

Nyang'au, O.W., Mati, M.B., Kalamwa, K., Wanjogu, K.R. and Kiplagat, K.L. 2014. Estimating rice yield under changing weather conditions in Kenya using CERES rice model. *International Journal of Agronomy*. Article ID 849496.

Olioso, A., Inoue, Y., Demarty, J., Wigneron, J.P., Braud, I. and Ortega-Farias, S. 2003. Assimilation of remote sensing data into crop simulation models and SVAT models. In: J.A. Sobrino(Ed.), *Proceedings of 1st International Symposium on Recent Advances in Quantitative Remote Sensing*, Valencia, September 16–18, 2002, pp. 329–338.

Pai, D.S., Bhate, J., Sreejith, O.P. and Hatwar, H.R. 2011. Impact of MJO on the intraseasonal variation of summer monsoon rainfall over India. *Climate Dynamics*, 36(1–6), 41–55.

Parinussa, R.M., Holmes, T.R.H. and de Jeu, R.A.M. 2012. Soil moisture retrievals from the WindSat space borne polarimetric microwave radiometer. *IEEE Transactions on Geoscience and Remote Sensing*, 50(7), 2683–2694.

Rao, P. 2008. *Agriculture Meteorology*. Delhi, India: PHI Learning Limited.

Rao, S.K. 2013. *Status Paper on Rice in Madhya Pradesh*. Hyderabad, India: Rice Knowledge Management Portal (RKMP).

Satapathy, S.S., Swain, D.K. and Ghosh, M. 2013. Effect of climate change on growth, phenology and yield of rice crop grown in open top chamber in Eastern India. *Global Journal of Applied Agricultural Research*, 3(1), 19–30.

Singh, K.K, Baxla, A.K., Chaudhary, J.L., Kaushik, S. and Gupta, A. 2005. Exploring the possibility of second crop in Bastar Plateau region of Chhattishgarh using DSSAT crop simulation model. *Agrometeorology Journal*, 7(2), 149–160.

Singh, P.K., Singh, K.K., Rathore, L.S., Baxla, A.K., Bhan, S.C., Gupta, A., Gohain, G.B., Balasubramanian, R., Singh, R.S. and Mall, R.K. 2016. Rice (*Oryza sativa* L.) yield gap using the CERES-rice model of climate variability for different agroclimatic zones of India. *Current Science*, 110(3), 405–413.

Swain, D., Srikantha, H., Saha, S. and Rabindra, D.N. 2007. CERES-Rice model: Calibration, evaluation and application for solar radiation stress assessment on rice production. *Association of Agromet*, 9(2), 198–148.

Timsina, J. and Humphreys, E. 2006. Performance of CERES-rice and CERES-wheat models in rice-wheat systems: A review. *Agriculture Systems*, 90(1–3), 5–31.

Varikoden, H., Kumar, M.R.R. and Babu, A.C. 2014. Indian summer monsoon rainfall characteristics during contrasting monsoon years. *Pure an Applied Geophysics*, 171(7), 1461–1472.

Wagner, W., Dorigo, W., de Jeu R., Fernandez, D., Benveniste, J., Haas, E. and Ertl, M. 2012. Fusion of active and passive microwave observations to create an essential climate variable data record on soil moisture, ISPRS annals of the photogrammetry, remote sensing and spatial information sciences (ISPRS Annals). *XXII ISPRS Congress*, Melbourne, Australia, August 25–September 1, 1–7, pp. 315–321.

Wheeler, M.C. and Hendon, H.H. 2004. An all-season real-time multivariate MJO index: Development of an index for monitoring and prediction. *Monthly Weather Review*, 132(8), 1917–1932.

# 14 Simulation of Hydrologic Processes through Calibration of SWAT Model with MODIS Evapotranspiration Data for an Ungauged Basin in Western Himalaya, India

*Pratik Dash*

## CONTENTS

| | | |
|---|---|---|
| 14.1 | Introduction | 224 |
| 14.2 | Study Site | 226 |
| 14.3 | Material and Methods | 227 |
| | 14.3.1 The Soil and Water Assessment Tool Model | 227 |
| | 14.3.2 Data Preparation | 227 |
| | 14.3.3 Model Setup | 228 |
| | 14.3.4 Calibration and Validation | 228 |
| 14.4 | Results and Discussions | 230 |
| | 14.4.1 Sensitivity Analysis | 230 |
| | 14.4.2 Groundwater Parameters | 230 |
| |     14.4.2.1 Threshold Depth of Water in the Shallow Aquifer Required for Return Flow | 230 |
| |     14.4.2.2 Threshold Water Level in Shallow Aquifer for Re-evaporation (REVAP) | 230 |
| |     14.4.2.3 Groundwater "Revap" Coefficient | 230 |
| |     14.4.2.4 Groundwater Recharge to Deep Aquifer | 231 |
| | 14.4.3 Soil Parameters | 232 |
| |     14.4.3.1 Soil Depth | 232 |
| |     14.4.3.2 Available Water Capacity in Soil | 233 |
| |     14.4.3.3 Saturated Hydraulic Conductivity | 233 |
| | 14.4.4 Hydrologic Response Units Parameters (.hru) | 233 |
| |     14.4.4.1 Plant Uptake Compensation Factor | 233 |
| |     14.4.4.2 Soil Evaporation Compensation Coefficient | 233 |
| |     14.4.4.3 Maximum Canopy Storage | 234 |
| | 14.4.5 Calibration and Validation | 234 |
| | 14.4.6 Hydrologic Simulations (Water Balance) | 236 |

14.5  Limitations ..................................................................................................... 238
14.6  Conclusion ..................................................................................................... 238
Acknowledgments ..................................................................................................... 239
References ................................................................................................................. 239

**ABSTRACT**  Hydrological modeling is widely practiced to understand the hydrological processes and its response to environmental changes. To obtain realistic results, sensitivity analysis and calibration are performed to identify the key parameters and their optimal values. Availability of discharge data makes the calibration process easier for gauged basins whereas for ungauged basins, the process is a challenging but essential task. Although regionalization approaches are popularly applied for ungauged basins, these approaches have some limitations and data availability issues. Satellite-based evapotranspiration can be effective for calibration of physically based models. The present study describes a simple and effective approach to calibrate hydrological model for areas with sparse data. The study used the Soil and Water Assessment Tool (SWAT) model on an ungauged river basin (Sirsa River) in northwest Himalaya, India. This model was parameterized through sensitivity analysis and manual calibration. The model's simulated actual evapotranspiration ($ET_a$) was compared with a Moderate Resolution Imaging Spectroradiometer (MODIS) $ET_a$ data product (MOD16A2) at daily (8-day composite) and monthly time-step for calibrations (2004–2006) and validations (2007–2008). The sensitivity of key parameters on selected five components (stream flow, surface runoff, base flow, deep aquifer recharge, and $ET_a$) was tested by manually changing parameters one at a time. The simulated $ET_a$ on iteration was verified with four statistical parameters to choose the optimal parameter value. After optimization, the model was run to simulate hydrological components for the period 2003–2008. The calibration and validation results showed "good" and "very good" performance of the model for daily and monthly comparisons, respectively. Overall, the model overestimated $ET_a$ in premonsoon and monsoon periods and underestimated it in postmonsoon and winter periods. This study showed that satellite-based evapotranspiration data can be effectively used for calibration of distributed hydrological model in data-limited regions worldwide.

## 14.1  INTRODUCTION

Quantitative assessment of hydrological components helps to understand the governing processes of and to manage water resources under changing environmental conditions. Among all hydrological components, streamflow is commonly measured at several points on major streams. Although stream discharge data are readily available in the majority of river basins in developed countries, they are sparsely available and poorly maintained in developing countries. So, quantification of rainfall–runoff relation and other hydrological components is essential but is a challenging task for ungauged catchments. Several hydrological and environmental models have been recently developed to quantify hydrological components and to probe the hydrologic response to human activity (Wu and Xu, 2006; Wu and Liu, 2012).

Hydrological models, especially rainfall–runoff models, are "simplifications of the real-world system under investigation" (Gupta et al., 2005). Based on the description of hydrological process, models can be either lumped (conceptual) or fully distributed or semidistributed. In the last two decades with the integration of geospatial tools and remotely sensed data, various distributed and semidistributed models have been developed to estimate water quality and quantity (Praskievicz and Chang, 2009). Among various models, the SWAT model has been popularly applied worldwide to various ranges of watersheds with varying topography, climate, soil, and management conditions over long periods of time (Gassman et al., 2007; Zhang et al., 2007; Krysanova and White, 2015). The SWAT model (Arnold et al., 1998) is a physically based semidistributed, basin-scale, and continuous-time model. It is suitably used for estimating (1) water balance components

(Van Liew and Garbrecht, 2003), (2) sediment and nutrition loss (Chu et al., 2004; Behera and Panda, 2006), (3) impact of nonpoint-source pollution and water management (Santhi et al., 2006), (4) impact of land use change (Nie et al., 2011; Li et al., 2013; Yan et al., 2013), and climate change (Wu and Johnston, 2007; Githui et al., 2009; Jha and Gassman, 2014) on water quality and quantity.

Physically based models (like SWAT) incorporate a high number of parameters (model parameter*) of which most are not physically measurable. It can be obtained through automated or manual calibration processes in which parameters are adjusted by comparing selected model output with observed data. The adjustment of such huge parameters is cumbersome and labor intensive (Immerzeel and Droogers, 2008). Therefore, sensitivity analysis is performed to identify the key parameters that have significant influence on model output (Saltelli et al., 2000; Arnold et al., 2012). A combined method of Latin Hypercube (LH) sampling and the one-factor-at-a-time (OAT) approach is commonly used for sensitivity analysis in which each model parameter is changed at predefined intervals while others are kept constant at their nominal value (Holvoct et al., 2005; Van Griensven et al., 2006; Cibin et al., 2010). However, understanding the variations in model output with the change of sensitive parameters value is of utmost importance for manual calibration. Few studies experimented model accuracy by manually varying SWAT parameters. For example, Wu and Johnston (2007) evaluated the effect of the plant uptake compensation factor (EPCO) and the soil evaporation compensation coefficient (ESCO) on the deviation of simulated discharge under dry and average climate conditions. Kannan et al. (2007) tested the effect of four most sensitive parameters on streamflow components by varying the parameter one at a time at low, medium, and high values. Mosbahi et al. (2015) compared the Nash–Sutcliffe coefficients of simulated runoff at various points in a range of sensitive parameter values by varying OAT.

In gauged basins, the availability of observed data makes the calibration process easier for a realistic simulation. But for the ungauged basin, accurately estimating hydrologic variables is a difficult task (Cibin et al., 2013). The International Association of Hydrological Sciences (IAHS) adopted the "Predictions in Ungauged Basins" (PUB) in 2003 to improve research on hydrologic simulation for ungauged basins (Hrachowitz et al., 2013). These studies were based on either physical considerations or other theories, such as Grey information theory, fuzzy theory, and so on (Nayak et al., 2005; Wang et al., 2013). Most of the studies found a regionalization approach suitable for estimation of runoff in ungauged basins (Parajka et al., 2005; Götzinger and Bárdossy, 2007; Samuel et al., 2011). In this approach, hydrologic information, that is, model parameters or model structure, is transferred from gauged (donor) to ungauged (target) catchment based on the similarity in catchment characteristics or spatial proximity (Samuel et al., 2011; Razavi et al., 2013). But this approach is not applicable for an ungauged basin if a donor basin is not available. Additionally, the regionalization approach is limited in hydrologic simulations that occur either due to equifinality problems caused from the optimization with data from few stations (Beven and Freer, 2001) or uncertainty in transfer of parameters (Sellami et al., 2014). Hence, hydrological parameters measured using satellite data could provide a viable solution to calibrate hydrological models for regions for which data are scarce (Immerzeel and Droogers, 2008).

Applications of remote sensing data derived hydrological components such as evapotranspiration (ET) and soil moisture for parameterization of hydrological models in ungauged basin have recently been used more often in hydrological engineering (Boegh et al., 2004; Immerzeel and Droogers, 2008; Stehr et al., 2009; Zhang et al., 2009; Jhorar et al., 2011; Githui et al., 2012). Most of these studies found that MODIS data determined that ET is suitable for calibration of SWAT or other hydrological models in regions for which data are scarce. Most of these studies used surface energy balance algorithm (SEBAL) to estimate ET from MODIS products of the Normalized Differenced Vegetation Index (NDVI) and leaf area index (LAI). However, calculation of long-term ET from NDVI and LAI through the SEBAL algorithm is quite complex and cumbersome. The MODIS product of global actual ET data (MOD16A2; Mu et al., 2007) can also be considered as a research opportunity to calibrate the SWAT model for ungauged basins.

---

* Model parameters are used in designing models to quantify detailed physical processes whereas input parameters involve spatial information that adds to the model for simulation of selected output.

Hydrological information in developing countries such as India are available on a limited basis. In addition, future water availability is a dilemma due to unprecedented changes in industrial and urban growth and the consequent huge water consumption. Hence, hydrological modeling for basins for which data are scarce is required by planners and managers to provide sustainable management of water resources. The present study is mainly focused on developing a simple and efficient approach for calibration of a physically based hydrological model in an ungauged basin. We applied the SWAT model to calculate water balance components of the Sirsa River basin, an ungauged tributary basin of the Satluj River in the northwestern Himalayas, India. The study used ET (MODIS ET) to derive remote sensing data to parameterize the SWAT model through sensitivity analysis and manual calibration.

## 14.2 STUDY SITE

The Sirsa River basin, a downstream tributary channel of the Satluj River, flows through Himachal Pradesh, Haryana, and Punjab states in India. The study site covers an area of approximately 680 km² of which 75% lies in the Solan district of Himachal Pradesh. The basin extends from 30°49′22″ to 31°11′00″ N latitudes and 76°32′48″ to 76°59′22″ E longitudes in the northwestern Himalayas at the fringe of Ganga plain. The basin is an intermontane river system bounded by the outer Siwalik range in the southwest and the Kasauli–Ramshahr Tertiary ranges in the northeast (Figure 14.1). Elevation of the basin varies between 250 and 1900 m, almost half of which is characterized as an intermontane valley (Nalagarh valley). The basin landscape is characterized by ridge and valley topography, eroded undulating surface, alluvial fan, and so on (Dash et al., 2014). The tributaries of the Sirsa River that originate from the Kasauli–Ramshahr ranges are long whereas rivulets developed in the outer Siwalik are short. The drainage morphometry indicates that the basin is elongated and well drained with an average drainage density of 3 km/km² (Dash et al., 2013).

The Sirsa River basin is located in a subtropical monsoon climate with a mean annual temperature of 23.5°C and an annual mean rainfall of 900 mm. About 80% of its annual precipitation is received

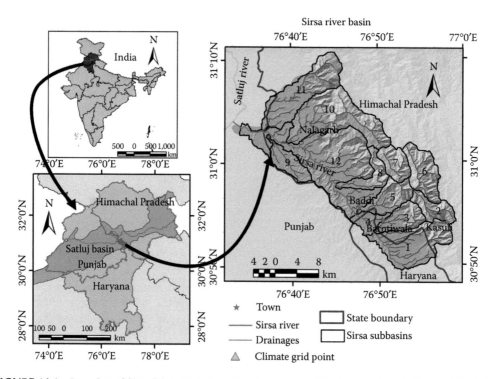

**FIGURE 14.1** Location of Sirsa River basin including towns and subbasins (numbers indicate subbasin ID).

during the summer monsoon (June–September). The major land use and land cover (LULC) classes include dense forests, mixed forests, agricultural lands, barren lands, and built-up areas. Most part of the basin is covered with loamy soil although loamy skeletal soil is also found in the Kasauli–Ramshahr ranges. Soil layers are quite thick in the intermontane valley and outer Himalaya but not in the Kasauli–Ramshahr ranges. Soils are characterized by low to moderate permeability. The major industrial hub of Himachal Pradesh, that is, the Baddi-Barotiwala-Nalagarh corridor, is located in the study basin. The growing urbanization has caused the increase in decadal (2001–2011) population growth by 13.40% and 32.34% in the towns of Nalagarh and Baddi, respectively.

## 14.3 MATERIAL AND METHODS

### 14.3.1 THE SOIL AND WATER ASSESSMENT TOOL MODEL

The 2005 version of SWAT has been used in this study because it accounts for (1) detailed surface and subsurface hydrologic processes, (2) spatial heterogeneity in the input-output model, (3) long-term hydrologic simulation with limited data even for ungauged basins, and (4) a simple and user-friendly platform, good documentation, and feedback solutions from a large number of user communities. The model incorporates large-scale spatial variability of hydrologic processes partitioning a basin into a number of land parcels in two phases. Initially, based on topography, the basin is divided into numerous subbasins, considering drainage area as threshold. Next, subbasins are further segregated into numerous conceptual homogeneous land parcels, known as hydrologic response units (HRUs) combining slope, soil, and land use layers.

The water budget of surface, subsurface, and deep aquifer phases is calculated for each HRU and is routed at basin and subbasin outlets. The SWAT model simulates various hydrological components, including ET, surface runoff, lateral flow, base flow, deep aquifer recharge, and so on, based on the water balance equation expressed as follows:

$$SW_t = SW_0 + \sum_{i=1}^{t} \left( R_{day} - Q_{surf} - E_a - w_{seep} - Q_{gw} \right)$$

where:
$SW_t$ is the final soil water content (mm)
$SW_0$ is the initial soil water content on day $i$ (mm)
$R_{day}$, $Q_{surf}$, $E_a$, $w_{seep}$, and $Q_{gw}$ are precipitation (mm), surface runoff (mm), evapotranspiration (mm), seepage flow (mm), and return flow (mm) on day $i$, respectively

The SWAT model offers two methods for surface runoff simulation, Of them, the SCS curve number (CN) method (Arnold et al., 1998) was chosen for this study. Potential evapotranspiration (PET) was calculated using the Penman–Monteith method. Percolation was estimated by using the storage routing method, and the Muskingum method was used for channel routing. A more detailed description of the model can be found in Neitsch et al. (2005b), and online documentation is available at http://swatmodel.tamu.edu/.

### 14.3.2 DATA PREPARATION

Topographic, land use, soil, and meteorological data are essentially required to set up the SWAT model. The Advanced Spaceborne Thermal Emission and Reflection Radiometer (ASTER) global digital elevation model (GDEM, hereafter DEM) of 1 arc-second resolution was used as topographic input. A soil map of 1:125,000 scale was acquired from the National Bureau of Soil Survey and Land Use Planning (NBSS&LUP), India. Soil classes were reclassified according to SWAT soil database. For LULC information, a LULC map of 2009 was prepared from a Landsat TM image through a supervised classification technique along with ground information collected from the field. Because of the lack of long-term in situ meteorological data, gridded raster climatic data of the National Centers for Environmental Prediction (NCEP)/National Center for Atmospheric Research (NCAR) global reanalysis products of Global Meteorological Forcing Dataset for Land Surface Modeling (ds314) of 1° spatial resolution was used in this study. This global gridded datasets were collected from the Computational and Information Systems Laboratory (CISL) archive (http://rda.ucar.edu/datasets/ds314.0/). The daily meteorological data include minimum and maximum temperatures, precipitation, solar radiation, wind speed, and relative humidity that were extracted for the study domain of two grid locations (Figure 14.1) for the period 2001–2008. Climate data were prepared in a suitable format for SWAT2005 as guided by Neitsch et al. (2005a).

The 8-day composite MODIS actual ET data (MODIS $ET_a$) of 1 km spatial resolution (MOD16A2) was acquired from http://www.ntsg.umt.edu/project/modis/mod16.php for the period of 2004–2008. This global land surface $ET_a$ dataset was prepared by Mu et al. (2007, 2011) coupling MODIS land cover, albedo, LAI data, and daily global meteorological reanalysis data (GMAO) of $1.00° \times 1.25°$ resolution. Based on the Penman–Monteith method, $ET_a$ was calculated considering soil heat flux, evaporation from wet and moist soil, day- and nighttime transpiration, and so on (for details see Mu et al., 2007, 2011). Due to the lack of availability of discharge data, we used MODIS $ET_a$ data as the ground reference for the calibration and validation of the SWAT model.

### 14.3.3 Model Setup

For this study, 12 subbasins were delineated from DEM, considering the 2000 ha as the minimum drainage area and fifth-order streams as thresholds. Subbasins were further divided into 179 HRUs by overlaying LULC, soil, and slope layers. Thereafter, climate input parameters were loaded to run the model for the period 2001–2008 while considering the first three years as a warm-up period. After the initial SWAT run, sensitivity analysis was performed based on the LH-OAT method. The analysis was carried out for 20 model parameters with 10 intervals in the LH sample. To make the calibration process easier, most sensitive parameters were manually varied one at a time within the range as suggested in the SWAT user's manual (Neitsch et al., 2001). The rate of change in the selected hydrological components (model output) with respect to the change in each parameter's value was tested to identify suitable value range of each sensitive parameter.

### 14.3.4 Calibration and Validation

The calibration of SWAT model was performed by comparing SWAT simulated $ET_a$ (hereafter SWAT $ET_a$) with MODIS $ET_a$ because of the lack of measured streamflow data. The daily SWAT $ET_a$ data were assembled at 8-day and monthly interval for comparison with MODIS $ET_a$ during the calibration (2004–2006) and validation (2007–2008). The sensitive parameters were manually adjusted during the calibration for all plausible hydrological components (Nie et al., 2011). During calibration, groundwater parameters (.gw), soil parameters (.sol), and HRU parameters (.hru) (Table 14.1) were iteratively modified until SWAT $ET_a$ closely matched MODIS $ET_a$. Initially, SWAT $ET_a$ and MODIS $ET_a$ were compared for annual values. Afterward, parameters were fine-tuned for monthly and daily (8-day) values until the modeled $ET_a$ was acceptable according to performance ratings proposed by Moriasi et al. (2007). Because parameters were adjusted by only comparing $ET_a$, uncertainty in separation of streamflow components may subsist in simulation. Hence, simulated results were checked with a SWAT Check program (Arnold et al., 2012). Based on the error report of each iteration, surface runoff, lateral flow, base flow, and deep aquifer recharges were adjusted by varying model parameters until satisfactory results were obtained. Finally, model performance for the most suitable set of parameter values was again tested and validated. The four objective functions used for evaluating the model performance were coefficient of determination ($R^2$), Nash–Sutcliffe efficiency ($E_{NS}$), percent bias (PBIAS), and root mean square error-observations standard deviation ratio (RSR) (Table 14.2). Finally, the calibrated model was run for 2003–2008 to simulate basin hydrological behavior.

## TABLE 14.1
### Description and Statistics of Sensitive Parameters of SWAT Including Sensitivity Rank and Optimal Parameterization Values

| Parameters | Description | Ranks | Range | Optimal Value |
|---|---|---|---|---|
| GWQMN | Threshold depth of water in the shallow aquifer required for return flow to occur (mm) | 1 | 0–5000 | 46.44 |
| ALPHA_BF | Base flow alpha factor (days) | 2 | 0–1 | 0.2 |
| REVAPMN | Threshold depth of water in the shallow aquifer for "revap" or percolation to occur (mm) | 3 | 0–500 | 46.5 |
| GW_REVAP | Groundwater "revap" coefficient | 4 | 0.02–0.2 | 0.03 |
| RCHR_DP | Groundwater recharge to deep aquifer (fraction) | 5 | 0–1 | 0.36 |
| CN2 | SCS runoff CN for moisture condition II | 6 | 35–98 | 3.25%[a] |
| CANMX | Maximum canopy storage (mm) | 7 | 0–10 | 5 |
| EPCO | Plant evaporation compensation factor | 8 | 0.01–1 | 0.6 |
| SOL_AWC | Available water capacity of the soil layer (mm $H_2O$/mm soil) | 9 | 0–1 | 0.09 |
| SOL_Z | Soil depth (mm) | 10 | 0–3000 | 480 |
| SOL_K | Saturated hydraulic conductivity of soil (mm/hrs) | 11 | 0–100 | 10 |
| ESCO | Soil evaporation compensation factor | 12 | 0.01–1 | 0.3 |
| GW_DELAY | Groundwater delay (days) | 13 | 0–50 | 18 |

[a] Varies with LULC and soil types.

## TABLE 14.2
### Description and Results of Criteria Used for Testing Statistical Performance of SWAT Model during Calibration

| Name | Formula | Time Step | Performance Calibration | Performance Validation |
|---|---|---|---|---|
| Coefficient of determination | $R^2 = \left[ \dfrac{\sum_{i=1}^{n}(O_i - \bar{O})(S_i - \bar{S})}{\sqrt{\sum_{i=1}^{n}(O_i - \bar{O})^2} \sqrt{\sum_{i=1}^{n}(S_i - \bar{S})^2}} \right]^2$ | Daily | 0.75 | 0.75 |
| | | Monthly | 0.81 | 0.89 |
| Nash–Sutcliffe efficiency | $E_{NS} = \left[ 1 - \dfrac{\sum_{i=1}^{n}(O_i - S_i)^2}{\sum_{i=1}^{n}(O_i - \bar{O})^2} \right]$ | Daily | 0.67 | 0.72 |
| | | Monthly | 0.76 | 0.89 |
| Percent bias | $PBIAS = \dfrac{\sum_{i=1}^{n}(O_i - S_i) * 100}{\sum_{i=1}^{n} O_i}$ | Daily | 2.84 | 2.43 |
| | | Monthly | 2.84 | 2.43 |
| Standardized RMSE | $RSR = \left[ \dfrac{\sqrt{\sum_{i=1}^{n}(O_i - S_i)^2}}{\sqrt{\sum_{i=1}^{n}(O_i - \bar{O})^2}} \right]$ | Daily | 0.57 | 0.53 |
| | | Monthly | 0.48 | 0.33 |

*Note:* $O_i$: Observed data of $i$th day
$\bar{O}$: Mean of observed data for the period being evaluated
$S_i$: Simulated value of $i$th day
$\bar{S}$: Mean of simulated value for the period being evaluated.

## 14.4 RESULTS AND DISCUSSIONS

### 14.4.1 Sensitivity Analysis

LH-OAT–based sensitivity analysis was performed to identify influential parameters by ignoring redundant parameters. The most sensitive parameters found in this study were CN, soil available water capacity (SOL_AWC), soil depth (SOL_Z), ESCO, plant evaporation compensation factor (EPCO), saturated hydraulic conductivity (SOL_K), threshold depth of water in the shallow aquifer required for return flow (GWQMN), groundwater "revap" coefficient (GW_REVAP), threshold water level in shallow aquifer for revap (REVAPMN), groundwater recession factor (ALPHA_BF), groundwater recharge to deep aquifer (RCHR_DP), and maximum canopy storage (CANMX). Table 14.1 lists the rank of sensitive parameters and their final values. The surface runoff was adjusted by changing CN2 (SCS CN for moisture condition II) based on the literature findings. The simulated $ET_a$ was adjusted by changing EPCO, ESCO, CANMX, SOL_Z, and SOL_AWC parameters. Lateral flow was adjusted by varying soil parameters such as SOL_K and SOL_AWC. The sensitivity of groundwater parameters (.gw), soil parameters (.sol), and HRU parameters (.hru) for the simulation of surface runoff, streamflow, base flow, deep aquifer recharge, and $ET_a$ are discussed in this section. During this analysis, each parameter was changed randomly while others were kept as constant.

### 14.4.2 Groundwater Parameters

The sensitivity of groundwater parameters, particularly GWQMN, REVAPMN, GW_REVAP, and RCHR_DP on hydrological components are presented in Figure 14.2. The parameters were found to be most sensitive for simulation of the base flow and consequently of streamflow.

#### 14.4.2.1 Threshold Depth of Water in the Shallow Aquifer Required for Return Flow

The variations in water balance components for changes in GWQMN values are presented in Figure 14.2a. It is clear that with the increase of GWQMN, streamflow and base flow decreased. Initially, with a GWQMN value up to 60, base flow and streamflow decreased at a low to moderate rate with the increase of parameter value. But afterward, base flow and streamflow sharply decreased to a GWQMN value of 175. A high value of GWQMN indicates that a considerable portion of infiltrated water was stored in soil whereas at a low value of GWQMN, the model produced more base flow that, in turn, increased streamflow (Kannan et al., 2007). In this study, a low GWQMN value was found to be suitable for the realistic prediction of daily streamflow as suggested by Kannan et al. (2007).

#### 14.4.2.2 Threshold Water Level in Shallow Aquifer for Re-evaporation (REVAP)

REVAPMN is the threshold depth of water in shallow aquifer that controls water movement to an unsaturated zone for re-evaporation to occur. With the increase of REVAPMN, base flow as well as streamflow increased (Figure 14.2b). But after a certain value (REVAPMN = 60), base flow and streamflow remained constant because at this threshold, no "revap" occurred for the basin. At a low REVAPMN, the contribution of base flow to streamflow is very low due to the high "revap" from soil. In this study, the REVAPN value was finally adjusted close to the GWQMN value.

#### 14.4.2.3 Groundwater "Revap" Coefficient

GW_REVAP controlled the amount of water that will "revap" to upper soil layer. In this study, base flow and streamflow varied inversely with GW_REVAP value in a smooth curvilinear trend Figure 14.2c). For a high value of GW_REVAP, the model returned more water to the root zone for "revap," resulting a decrease in base flow contribution to streamflow. However, GW_REVAP value was finally set at 0.03 for this study.

# Hydrological Modelling of an Ungauged Basin with SWAT Model

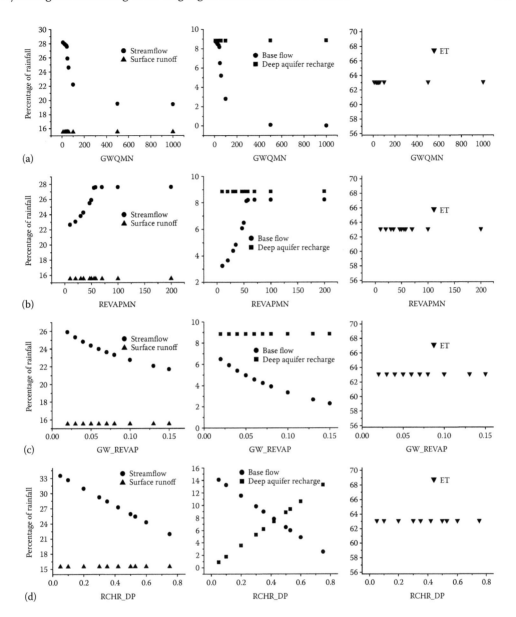

**FIGURE 14.2** Sensitivity of water balance components to groundwater parameters. (a) GWQMN, (b) REVAPMN, (c) GW_REVAP, and (d) RCHR_DP.

### 14.4.2.4 Groundwater Recharge to Deep Aquifer

RCHR_DP controls the amount of water that will move from a shallow aquifer to a deep aquifer. A linear trend is noticed in a decrease of baseflow and streamflow while the parameter values are increased (Figure 14.2d). Likewise, the deep aquifer recharge is changed positively with the parameter values.

However, the adjustment of groundwater configuration parameters was a very challenging task because these parameters do not influence $ET_a$, which was considered for parameterization during calibration. Based on extensive literature findings, these parameters were adjusted using a "trial-and-error" method and were verified repeatedly with the SWAT Check program.

### 14.4.3 Soil Parameters

SOL_Z, SOL_AWC, and SOL_K are the main soil parameters that showed significant control on each water balance component (Figure 14.3). The sensitivity of these three parameters is analyzed next.

#### 14.4.3.1 Soil Depth

The water balance components varied in a curvilinear trend with the change in depth of soil layer (SOL_Z). Surface runoff and $ET_a$ increased with an increase in SOL_Z value whereas streamflow, base flow, and deep aquifer recharge decreased (Figure 14.3a). As soil depth increased, root zone depth and soil profile depth increased. It allowed the model to increase the water-holding capacity as well as water availability in the soil profile that, in turn, increased evaporation from the soil profile and transpiration from plants. Thus, the increase of the vadose zone depth caused a decrease in shallow and deep aquifer recharge and an increase in $ET_a$ generation. According to the SWAT model, the low depth of the soil profile helped in the quick downward movement of water from the lowest soil layer in the vadose zone to the shallow aquifer. Thus, with the increase of soil depth, water movement to shallow aquifer is delayed. Thus, groundwater recharge decreased and $ET_a$ increased. From Figure 14.3a, it was assumed that with the increase of soil water content due to the increase of soil depth, the surface runoff increased, but total streamflow decreased due to the decrease in the

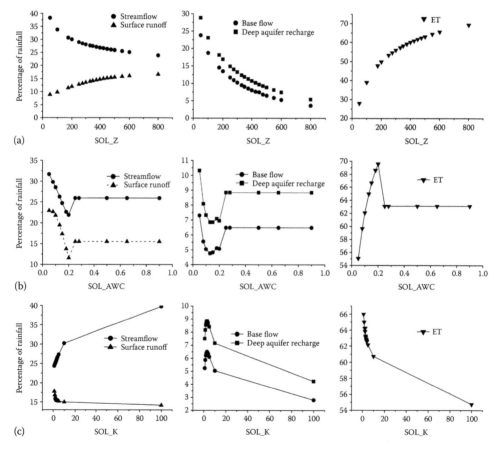

**FIGURE 14.3** Sensitivity of water balance components to soil parameters. (a) SOL_Z, (b) SOL_AWC, and (c) SOL_K.

base flow. However, interaction of this parameter with other soil and groundwater parameters can change its influence on hydrologic components.

### 14.4.3.2  Available Water Capacity in Soil

Available water capacity in soil (SOL_AWC) is one of the crucial parameters that determines the field capacity of soil. In this study, SOL_AWC was found to be sensitive to various water balance components in a similar pattern (Figure 14.3b). At an initial value up to 0.2 within the range of SOL_AWC (0–1), the surface runoff, streamflow, base flow, and deep aquifer recharge decreased, and $ET_a$ increased with the increase in the parameter value. However, between the SOL_AWC value of 0.2 and 0.3, the response of each component was reversed. But beyond 0.3, the SOL_AWC did not show sensitivity to any water balance components. It can be inferred that the increase in SOL_AWC by a fractional amount led to the increase in ET from soil and canopy as soil moisture increased. But beyond a critical value (here 0.2), percolation from the vadose zone to the shallow and deep aquifer increased with the change of SOL_AWC value.

### 14.4.3.3  Saturated Hydraulic Conductivity

The saturated hydraulic conductivity of soil (SOL_K) plays a significant role in hydrological processes. Infiltration and percolation capacity of a soil is directly proportional to the SOL_K according to Neitsch et al., 2005(b). At a low value range (0–10), SOL_K was found very sensitive to all hydrological components (Figure 14.3c). As SOL_K approached from 0 to 10, streamflow, base flow, and deep aquifer recharge increased; however, surface flow and $ET_a$ decreased. But when the value increased from moderate to high range, the value of these components decreased except for streamflow. The results are quite similar to the findings of Kannan et al. (2007). For the SOL_K value below 10, base flow, streamflow, and groundwater recharge increased significantly as infiltration and percolation capacity increased. But beyond the SOL_K value 10, infiltrated water maximally converted to a lateral flow rather than ground water recharge. The substantial increase in lateral flow raised streamflow generation though the contribution of surface flow and reduced base flow (Figure 14.3c).

### 14.4.4  Hydrologic Response Units Parameters (.hru)

EPCO, ESCO, and CANMX were found to be most sensitive to HRU configuration parameters in this study. The sensitivity of parameters to model output is shown in Figure 14.4.

### 14.4.4.1  Plant Uptake Compensation Factor

EPCO controls $ET_a$ by allowing a plant to take water from layers within the rooting zone (Wu and Johnston, 2007). The value of EPCO ranges between 0 and 1. At low EPCO values, the model allows the plant to take water from the top soil layer, but as EPCO approaches 1.0, plant water uptake demand also will be met from the deep soil layer. In this study with the increase in EPCO value, $ET_a$ increased linearly although the rate of increase is less (Figure 14.4a). For a high EPCO value, as model allows more water uptake from the lower soil layer to meet the demand of plants, $ET_a$ increased but at a marginal rate.

### 14.4.4.2  Soil Evaporation Compensation Coefficient

ESCO controls evaporation from soil by modifying depth distribution in soil profile. ESCO is found to be sensitive to all water balance components (Figure 14.4b). While $ET_a$ decreased in curvilinear shape with the increase of the ESCO value, the rest of the hydrologic components increased moderately (Figure 14.4b). As the value of ESCO is reduced, the model is able to extract more of the evaporative demand from the lower soil layer (Neitsch et al., 2005[a]), resulting in an increase

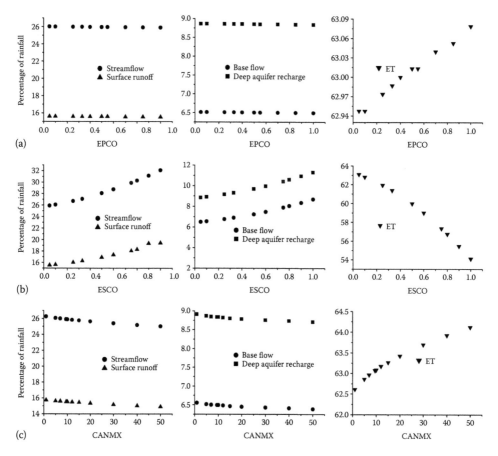

**FIGURE 14.4** Sensitivity of water balance components to HRU parameters. (a) EPCO, (b) ESCO, and (c) CANMX.

in $ET_a$. For high ESCO values, $ET_a$ decreased; consequently, surface runoff, base flow, and other components increase. The sensitivity of this parameter is quite similar to that of a few previous studies (Kannan et al., 2007; Wu and Johnston, 2007).

### 14.4.4.3 Maximum Canopy Storage

CANMX showed low to moderate sensitivity in the simulation of water balance components (Figure 14.4c). With the increase in canopy storage, all components except $ET_a$ decreased. As increase in CANMX value directed the model to increase the holding of intercepted water, the simulated $ET_a$ increased through the increase in the evaporation of intercepted water.

### 14.4.5 CALIBRATION AND VALIDATION

For estimating water balance components, stream discharge data, which are commonly used in the hydrological model for calibration, were simulated and gauged flows were compared. This study tried to calibrate the SWAT model using estimated $ET_a$ and MODIS $ET_a$ instead of gauge flow data. Using Penman–Monteith method, $ET_a$ was estimated from the SWAT model and compared with MODIS $ET_a$ for calibration (2004–2006) and validation (2007–2008) at 8-day and monthly time intervals (Figure 14.5). The statistical performances of the SWAT model for $ET_a$ estimation is presented in Table 14.2. The simulation (8-day and monthly) showed better performance in validation than the calibration period. The overall statistical performance of the model was found to

# Hydrological Modelling of an Ungauged Basin with SWAT Model

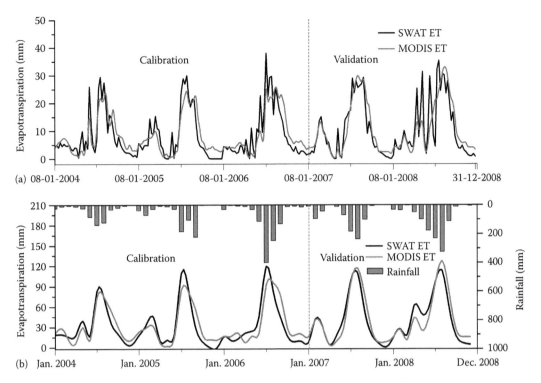

**FIGURE 14.5** Comparison of actual evapotranspiration ($ET_a$) between SWAT simulation and MODIS data at (a) daily (8-day composite) and (b) monthly time step during calibration (2004–2006) and validation (2007–2008) periods.

be "good" for daily (8-day) simulations and "very good" for monthly simulations according to the criteria provided by Moriasi et al. (2007) for streamflow estimation. A comparison of simulations at daily (8-day) intervals indicated that the statistics of $R^2$, $E_{NS}$, PBIAS, and RSR are 0.75, 0.67, 2.84, and 0.57, respectively, for calibration, and 0.75, 0.72, 2.43, and 0.53, respectively, for validation. Similarly, for monthly comparisons, the statistics of $R^2$, $E_{NS}$, PBIAS, and RSR for monthly simulation were 0.81, 0.76, 2.84, and 0.48, respectively, in calibration, and 0.89, 0.89, 2.43, and 0.33, respectively, in validation (Table 14.2).

During the calibration, differences between MODIS $ET_a$ and SWAT $ET_a$ ranged at 0 ~ 11.35 mm/8-day (mean 3.31 mm) and 0.3 ~ 32.42 mm/month (mean 10.81 mm) for daily and monthly comparisons, respectively. During the validation period, the differences varied at 0 ~ 18 mm/8-day (mean 3.31 mm) and 0.44 ~ 24.62 mm/month (mean 9.24 mm) for daily and monthly comparisons, respectively. The deviation of SWAT $ET_a$ from MODIS $ET_a$ for each month during the daily and monthly comparisons for the period of 2004–2008 is shown in box-and-whisker plots (Figure 14.6). In this figure, the height of the box shows the deviation from the first quartile to the third quartile, while the mark on two ends indicated minimum and maximum deviations. The simulated $ET_a$ deviated from MODIS $ET_a$ by much less during August–March whereas the deviation noticeably increased during April–July. During daily comparisons, the minimal deviation was observed in postmonsoon and winter seasons, with a median close to zero (Figure 14.6a). While the deviation was moderate (−8 to 5) during spring, it increased from the summer to the monsoon (wet) period. However, in both daily calibration and validation periods, the difference of SWAT $ET_a$ from MODIS $ET_a$ was less in the late monsoon period.

Similar to the daily simulation, the monthly difference during the postmonsoon and winter period (October–January) was much less and quite constant for all years (Figure 14.6b). For the postmonsoon period, underestimation of the model was marginal and consistent. The average

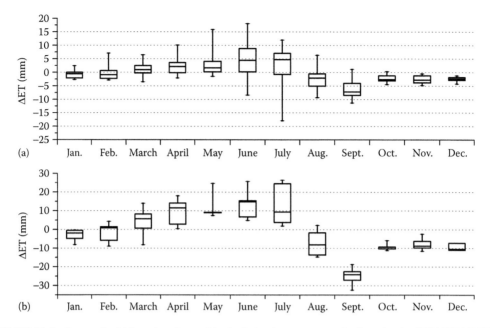

**FIGURE 14.6** Box-and-whisker plot of monthly deviation in evapotranspiration, that is, ΔET (SWAT ET—MODIS ET) for (a) daily (8-day cumulative), and (b) monthly comparisons during the period of 2004–2008. The box-and-whisker plots show the median, first, and third quartiles. The caps at the end of the boxes show the extreme values.

monthly difference during October–December was 8.9 mm/month. In April–July, the median of difference was positive, indicating an overestimation by the model. The deviation was very high in July (the peak rainfall period); the median of difference was about 10 mm. It can be inferred that the differences in the input climatic parameter for both SWAT and MODIS could cause such variation in the water excess period. However, the deviation in the model simulation was minimum in December and maximum in July.

The simulated $ET_a$ was maximally underestimated during August–September, that is, the late monsoon period. However, from Figure 14.6b, it is clear that the model simulated more accurately the during winter season, overestimated during the spring and mid-monsoon period, and underestimated during the postmonsoon and early winter period. This analysis could be verified from Figure 14.7a in which the average monthly $ET_a$ simulated from SWAT and MODIS data for the period of 2004–2008 was plotted. The scatter plot of monthly simulated $ET_a$ against MODIS $ET_a$ indicates that the SWAT simulation reasonably matched the MODIS $ET_a$ with $R^2$ value 0.84 (Figure 14.7b). In terms of the annual comparison, the simulated results closely matched the MODIS $ET_a$ by maximally varied up to 25 mm/yr during the period 2004–2008. From the overall discussion, it can be concluded that satellite-based ET data are very useful for the application of the hydrological model in ungauged basins. The calibration and validation results also indicate that the SWAT model can be effectively applied to the simulation of other hydrologic components.

### 14.4.6 Hydrologic Simulations (Water Balance)

The parameters obtained after the optimal calibration were finally used for the hydrologic simulation of the Sirsa River basin for the period 2003–2008. The simulated average monthly water balance components of the Sirsa River basin are presented in Figure 14.8. On average, 51% and 39% of total rainfall contributed to $ET_a$ and streamflow, respectively. The annual streamflow of the basin

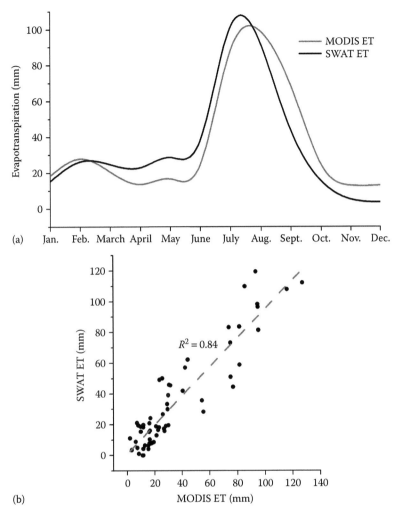

**FIGURE 14.7** Average monthly $ET_a$ simulated from SWAT and MODIS data for the period 2004–2008. (a) Comparison of average monthly ET simulated from SWAT model with MODIS ET and (b) scatter plot of SWAT simulated monthly ET against monthly MODIS ET values.

varied from 105 mm (2004) to 496 mm (2006). The average monthly streamflow (Figure 14.8) varied between less than 1 mm (December) and 90 mm (August). About 78% (240 mm) of the total annual streamflow was yielded during July–September whereas the least amount of streamflow was contributed in the early winter and premonsoon periods. Annually, the streamflow was primarily from surface runoff (60%) followed by base flow and lateral flow of about 27% and 13%, respectively. During the winter (January–February) and monsoon (June–September) periods, about 57% of total streamflow came from surface runoff because of the maximum precipitation. During the postmonsoon period, water in river flows by very negligible amount, of which about 80% came from base flow. Maximum net contribution from the lateral flow and base flow was noticed in July–August and August–September, respectively. The monthly $ET_a$ varied between 4.60 mm (December) and 91.50 mm (July). Only during the monsoon period was the average monthly $ET_a$ above 40 mm whereas during the rest of the period, especially the winter and postmonsoon periods, $ET_a$ was very low (Figure 14.8). March–April (spring) and October–November (postmonsoon) months were water-stressed periods when $ET_a$ exceeded rainfall. The rainfall in these months was insufficient to meet vegetation and crop water demand and was supplied from soil water storage.

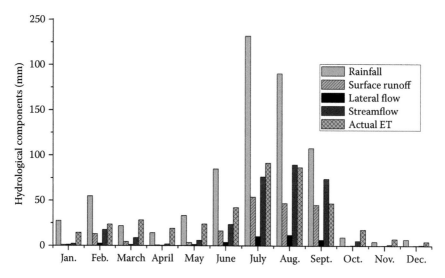

**FIGURE 14.8** Mean monthly value of selected hydrologic components of Sirsa basin for the period of 2003–2008.

## 14.5 LIMITATIONS

Assessing hydrological behavior through hydrological modeling requires identifying the limitations and uncertainties related to data input, model structure, and model parameters (Lindenschmidt et al., 2007). Limitations in the availability of in situ hydro-meteorological data resulted in increased errors in the model output. In this study, the use of used NCEP/NCAR reanalysis data instead of gauge station data increased the uncertainty because of its coarse resolution and inability to capture realistic observed trends. Moreover, the climate data used in this study and for the MODIS ET calculation were collected from two different sources. This fact could cause ambiguity in the parameterization process. However, in regions for which there were sparse data, these limitations should be considered to enhance further research.

## 14.6 CONCLUSION

The SWAT model, a physically based semidistributed hydrological model, was used in this study to simulate water balance components of the Sirsa River basin in the northwestern Himalayas in India. The study adopted an efficient and simple approach to parameterize the SWAT model for this ungauged basin. This study attempted to calibrate the SWAT model data with MODIS $ET_a$ data by identifying sensitivity parameters for the hydrologic simulation. This study sought to test whether remote sensing data can be effective for hydrological modeling in ungauged basins.

The study experimented with the influence of sensitive parameters on the simulation of selected hydrological components to increase the efficiency of the manual calibration process. The results showed that streamflow is most sensitive to SOL_AWC, SOL_K, CN2, and ESCO parameters. Base flow was most sensitive to GWQMN, REVAPMN, GW_REVAP, and SOL_Z parameters whereas ET was most sensitive to ESCO, SOL_Z, and SOL_AWC. Overall, soil parameters were found to be sensitive to all hydrological components. However, the knowledge of the rate of change in hydrologic components with variation in model parameters values increased the efficiency of manual calibration processes.

The sensitive parameters were adjusted by comparing SWAT–simulated $ET_a$ with MODIS $ET_a$ during the calibration period (2004–2006) and the validation period (2007–2008) at monthly and

8-day intervals. The statistical performance of the model for optimal parameters was reasonably "good" and "very good" for daily (8-day) and monthly simulations, respectively. The values of $R^2$, $E_{NS}$, PBIAS, and RSR vary in a range of 0.75–0.89, 0.67–0.89, 2.43–2.84, and 0.33–0.57, respectively, for calibration and validation at monthly and 8-day time intervals. Overall, the model's performance during the validation period was better than during the calibration period. During these two periods, modeled $ET_a$ was overestimated in premonsoon and monsoon periods but underestimated in postmonsoon and winter periods. The performance of the model was more consistent in the month of October to February.

The present study reveals that the SWAT model is a powerful tool for hydrologic simulation of the Sirsa River basin. The application of satellite-based evapotranspiration data for calibration of distributed hydrological model, like SWAT, could be effective for hydrological modeling in data-limited regions worldwide. The present methodology can be tested for gauged basins to increase its reliability.

## ACKNOWLEDGMENTS

The author is grateful to the Council of Scientific and Industrial Research (CSIR), New Delhi for providing financial support for the present work as a doctoral fellowship. NASA- METI, USGS, CSIL, and MODIS are acknowledged for providing freely available data used in this study. Anonymous reviewers are greatly acknowledged for their very fruitful comments and suggestions that improved the quality of the original manuscript.

## REFERENCES

Arnold, J.G., Moriasi, D.N., Gassman, P.W., Abbaspour, K.C., White, M.J., et al. (2012). SWAT: Model use, calibration, and validation. *American Society of Agricultural and Biological Engineers*, 55(4): 1491–1508.

Arnold, J.G., Srinivasan, R., Muttiah, R.S., and William, J.R. (1998). Large area hydrologic modeling and assessment–Part I: Model development. *Journal of American Water Resource Association*, 34(1): 73–89.

Behera, S., and Panda, R.K. (2006). Evaluation of management alternatives for an agricultural watershed in a sub-humid subtropical region using a physical process model. *Agriculture, Ecosystems & Environment*, 113(1–4): 62–72.

Beven, K.J., and Freer, J. (2001). Equifinality, data assimilation, and uncertainty estimation in mechanistic modelling of complex environmental systems. *Journal of Hydrology*, 249: 11–29.

Boegh, E., Thorsen, M., Butts, M.B., Hansen, S., Christiansen, J.S. et al., (2004). Incorporating remote sensing data in physically based distributed agro-hydrological modeling. *Journal of Hydrology*, 287: 279–299.

Chu, T.W., Shirmohammadi, A., Montas, H., and Sadeghi, A. (2004). Evaluation of the SWAT model's sediment and nutrient components in the Piedmont physiographic region of Maryland. *Transactions of ASAE*, 47(5): 1523–1538.

Cibin, R., Athira, P., Sudheer, K.P., and Chaubey, I. (2013). Application of distributed hydrological models for predictions in ungauged basins: A method to quantify predictive uncertainty. *Hydrological Processes*. doi:10.1002/hyp.9721.

Cibin, R., Sudheer, K.P. and Chaubey, I. (2010). Sensitivity and identifiability of streamflow generation parameters of the SWAT model. *Hydrological Processes*. doi:10.1002/hyp.7568.

Dash, P., Aggarwal, S.P., and Verma, N. (2013). Correlation based morphometric analysis to understand drainage basin evolution: A case study of Sirsa river basin, western Himalaya, India. *Scientific Annals of "Alexandru Ioan Cuza" University of Iași- Geography Series*, 59(1): 35–58.

Dash, P., Aggarwal, S.P., Verma, N., and Ghosh, S. (2014). Investigation of scale dependence and geomorphic stages of evolution through hypsometric analysis: A case study of Sirsa basin, western Himalaya, India. *Geocarto International*, 29(7): 758–777.

Gassman, P.W., Reyes, M.R., Green, C.H., and Arnold, J.G. (2007). The soil and water assessment tool: Historical development, applications, and future research directions. *Transactions of the ASABE*, 50(4): 1211–1250.

Githui, F., Gitau, W., Mutua, F., and Bauwens, W. (2009). Climate change impact on SWAT simulated streamflow in western Kenya. *International Journal of Climatology*, 29: 1823–1834.

Githui, F., Selle, B., and Thayalakumaran, T. (2012). Recharge estimation using remotely sensed evapotranspiration in an irrigated catchment in southeast Australia. *Hydrological Processes*, 26: 1379–1389.

Götzinger, J., and Bárdossy, A. (2007). Comparison of four regionalisation methods for a distributed hydrological model. *Journal of Hydrology*, 333: 374–384.

Gupta, H.V., Beven, K.J., and Wagener, T. (2005). Model calibration and uncertainty estimation. In *Encyclopedia of Hydrological Sciences*, Anderson, M. (Ed.). Chichester, UK: John Wiley & Sons.

Holvoet, K., van Griensven, A., Seuntjens, P., and Vanrolleghem, P.A. (2005). Sensitivity analysis for hydrology and pesticide supply towards the river in SWAT. *Physics and Chemistry of the Earth*, 30(8–10): 518–526.

Hrachowitz, M., Savenije, H.H.G., Blöschl, G., McDonnell, J.J., Sivapalan, M. et al. (2013). A decade of predictions in ungauged basins (PUB)—A review. *Hydrological Sciences Journal*, 58(6): 1198–1255.

Immerzeel, W.W., and Droogers, P. (2008). Calibration of a distributed hydrological model based on satellite evapotranspiration. *Journal of Hydrology*, 349: 411–424.

Jha, M.K., and Gassman, P.W. (2014). Changes in hydrology and streamflow as predicted by a modelling experiment forced with climate models. *Hydrological Processes*, 28: 2772–2781.

Jhorar, R.K., Smit, A.A.M.F.R., Bastiaanssen, W.G.M., and Roest, C.W.J. (2011). Calibration of a distributed irrigation water management model using remotely sensed evapotranspiration rates and groundwater heads. *Irrigation and Drainage*, 60: 57–69.

Kannan, N., White, S.M., Worrall, F., and Whelan, M.J. (2007). Sensitivity analysis and identification of the best evapotranspiration and runoff options for hydrological modelling in SWAT-2000. *Journal of Hydrology*, 332: 456–466.

Krysanova, V., and White, M. (2015). Advances in water resources assessment with SWAT—An overview. *Hydrological Sciences Journal*, 60(5): 771–783.

Li, Q., Cai, T., Yu, M., Lu, G., Xie, W., and Bai, X. (2013). Investigation into the impacts of land-use change on runoff generation characteristics in the Upper Huaihe River Basin, China. *Journal of Hydrological Engineering*, 18: 1464–1470.

Lindenschmidt, K.E., Fleischbein, K., and Baborowski, M. (2007). Structural uncertainty in a river water quality modelling system. *Ecological Modeling*, 204: 289–300.

Moriasi, D.N., Arnold, J.G., Van Liew, M.W., Bingner, R.L., Harmel, R.D., and Veith, T.L. (2007). Model evaluation guidelines for systematic quantification of accuracy in watershed simulations. *Transactions of the ASABE*, 50: 885–900.

Mosbahi, M., Benabdallah, S., and Boussema, M.R. (2015). Sensitivity analysis of a GIS-based model: A case study of a large semi-arid catchment. *Earth Science Information*, 8: 569–581.

Mu, Q., Heinsch, F.A., Zhao, M., and Running, S.W. (2007). Development of a global evapotranspiration algorithm based on MODIS and global meteorology data. *Remote Sensing of the Environment*, 111: 519–536.

Mu, Q., Zhao, M., and Running, S.W. (2011). Improvements to a MODIS global terrestrial evapotranspiration algorithm. *Remote Sensing of the Environment*, 115: 1781–1800.

Nayak, P.C., Sudheer, K.P., and Ramasastri, K.S. (2005). Fuzzy computing based rainfall–runoff model for real timeflood forecasting. *Hydrological Processes*, 19(4): 955–968.

Neitsch, S.L., Arnold, J.G., Kiniry, J.R., Srinivasan, R., and Williams, J.R. (2005a). *Soil and Water Assessment Tool Input/Output File Documentation, Version 2005*. Temple, TX: Grassland, Soil and Water Research Laboratory, Agriculture Research Service.

Neitsch, S.L., Arnold, J.G., Kiniry, J.R., and Williams, J.R. (2005b). *Soil and Water Assessment Tool Theoretical Documentation, Version 2005*. Temple, TX: Grassland, Soil, and Water Research Laboratory, Agriculture Research Service.

Neitsch, S.L., Arnold, J.G., Kiniry, J.R., and Williams, J.R. (2001). *Soil and Water Assessment Tool–Version 2000–User's Manual*. Temple, TX: Grassland, Soil, and Water Research Laboratory, Agriculture Research Service.

Nie, W., Yuan, Y., Kepner, W., Nash, M.S., Jackson, M., and Erickson, C. (2011). Assessing impacts of Landuse and Landcover changes on hydrology for the upper San Pedro watershed. *Journal of Hydrology*, 407: 105–114.

Parajka, J., Merz, R., and Blöschl, G. (2005). A comparison of regionalisation methods for catchment model parameters. *Hydrology Earth System Science*, 9: 157–171.

Praskievicz, S., and Chang, H. (2009). A review of hydrological modelling of basin-scale climate change and urban development impacts. *Progress in Physical Geography*, 33(5): 650–671.

Razavi, T., and Coulibaly, P. (2013). Streamflow prediction in ungauged basins: Review of regionalization methods. *Journal of Hydrological Engineering*, 18(8): 958–975.

Saltelli, A., Scott, E.M., Chan, K., and Marian, S. (2000). *Sensitivity Analysis*. Chichester, UK: John Wiley & Sons.

Samuel, J., Coulibaly, P., and Metcalfe, R. (2011). Estimation of continuous streamflow in Ontario ungauged basins: Comparison of regionalization methods. *Journal of Hydrological Engineering*, 16: 447–459.

Santhi, C., Srinivasan, R., Arnold, J.G., and Williams, J.R. (2006). A modeling approach to evaluate the impacts of water quality management plans implemented in a watershed in Texas. *Environmental Modelling Software*, 21(8): 1141–1157.

Sellami, H., La Jeunesse, I., Benabdallah, S., Baghdadi, N., and Vanclooster, M. (2014). Uncertainty analysis in model parameters regionalization: A case study involving the SWAT model in Mediterranean catchments (Southern France). *Hydrology Earth System Science*, 18: 2393–2413.

Stehr, A., Debels, P., Arumi, J.L., Romero, F., and Alcayaga, H. (2009). Combining the soil and water assessment tool (SWAT) and MODIS imagery to estimate monthly flows in a datascarce Chilean Andean basin. *Hydrological Science Journal*, 54(6): 1053–1067.

Van Griensven, A., Meixner, T., Grunwald, S., Bishop, T., Diluzio, M., and Srinivasan, R. (2006). A global sensitivity analysis tool for the parameters of multi-variable catchment models. *Journal of Hydrology*, 324: 10–23.

Van Liew, M.W., and Garbrecht, J. (2003). Hydrologic simulation of the little Washita river experimental watershed using SWAT. *Journal of the American Water Resources Association*, 39(2): 413–426.

Wang, W., Hu, S., Li, Y. and Cao, S. (2013). How to select a reference basin in the ungauged regions. *Journal of Hydrological Engineering*, 18(8): 941–947.

Wu, K., and Johnston, C.A. (2007). Hydrologic response to climatic variability in a Great Lakes Watershed: A case study with the SWAT model. *Journal of Hydrology*, 337: 187–199.

Wu, K., and Xu, Y.J. (2006). Evaluation of the applicability of the swat model for coastal watersheds in south eastern Louisiana. *Journal of the American Water Resources Association*, 42(5): 1247–1260.

Wu, Y., and Liu, S. (2012). Automating calibration, sensitivity and uncertainty analysis of complex models using the R package Flexible Modeling Environment (FME): SWAT as an example. *Environmental Modelling & Software*, 31: 99–109.

Yan, B., Fang, N.F., Zhang, P.C., and Shi, Z.H. (2013). Impacts of land use change on watershed streamflow and sediment yield: An assessment using hydrologic modelling and partial least squares regression. *Journal of Hydrology*, 484: 26–37.

Zhang, X., Srinivasan, R., and Hao, F. (2007). Predicting hydrologic response to climate change in the Luohe river basin using the SWAT model. *ASABE*, 50(3): 901–910.

Zhang, Y., Chiew, F.H.S., Zhang, L., and Li, H. (2009). Use of remotely sensed actual evapotranspiration to improve rainfall-runoff modeling in Southeast Australia. *Journal of Hydrometeorology*, 10(4): 969–980.

# 15 Impact of Land Use and Land Cover Changes on Nutrient Concentration in and around Kabar Tal Wetland, Begusarai (Bihar), India

*Rajesh Kumar Ranjan and Priyanka Kumari*

## CONTENTS

| | |
|---|---|
| 15.1 Introduction | 244 |
| 15.2 Study Area | 244 |
| 15.3 Methodology | 245 |
| 15.4 Results | 246 |
|     15.4.1 Land Use/Land Cover Distribution | 246 |
|     15.4.2 Major Nutrients Concentration | 246 |
| 15.5 Discussion | 248 |
| 15.6 Conclusion | 249 |
| Acknowledgment | 249 |
| References | 249 |

**ABSTRACT** Rapid unplanned development of industries and intensive agriculture are causing overexploitation of environmental resources including land, water, fossil fuels, and so on, which is resulting in changes of the land use/land cover of the earth's surface. Intensive agricultural practices using an excessive amount of fertilizers and of pesticides to increase the production of crops have led to the eutrophication of water bodies. In view of the significant importance of wetlands, an attempt has been made to analyze the impact of land use/land cover changes on the dynamics among nutrients in the Kabar Tal wetland located in the Begusarai district of Bihar. The impact assessment has been carried out by analyzing land use/land cover changes through multispectral, multitemporal (1990, 2010, and 2015) data in the Kabar Tal wetland and the changes in nutrient concentration during these periods. The study shows that the land use/land cover changes in the catchment have significantly affected the Kabar Tal wetland area. Drastic increases in agricultural land (10.95%) and fallow land (3.73%) have been observed during these periods. Open water cover decreased significantly by 7.22% from 1990–2015. In addition, increasing trends of major nutrients were reported during the 25-year period. These increases in nutrient concentrations correlate positively with the increase in agricultural area and negatively with water quantity. Overall, the wetland suffered serious loss during the 25-year period due to changes in land use/land cover, especially from unregulated agricultural activities. The unplanned intensive agricultural activities have led to the inflow of huge of fertilizers (nutrients), which have significantly increased nutrient concentrations while significantly decreasing the open water area, which indicates the shrinking of the wetland.

## 15.1 INTRODUCTION

Wetlands, which are among the most productive ecosystems in the world comparable to rain forests and coral reefs, are under increasing threat due to both identifiable point sources such as municipal and industrial wastewater and nonpoint sources including urban and agricultural runoff. Degradation causes changes in water quality, quantity, and flow rates in the wetlands. The major pollutants associated with urbanization are sediment, nutrients, oxygen-demanding waste, heavy metals, hydrocarbons, bacteria, and viruses (USEPA, 1994). Excessive amounts of fertilizers and animal waste reaching wetlands through runoff from agricultural activities can cause eutrophication in wetlands.

Land cover reflects the geomorphologic features, the biophysical state of the earth's surface and immediate subsurface, including soil materials, vegetation, and water (e.g., Prakasam, 2010). For the purpose of fulfilling the demand of growing populations and for subsequent development, humans always cause changes in the character of land that impact the wetland ecosystem. The characterization of land use in and around a wetland is essential for evaluating it in terms of water quality and quantity. Data from maps and aerial photographs can be used to estimate the potential for inputs of nutrients and sediments into wetlands (USEPA, 2002). The structure and function of stream ecosystems and wetlands are completely linked to the status and condition of their surrounding catchments (Cuffney et al., 2000; Berka et al., 2001). Changes in land use and land cover (LULC) in the catchments of wetland influence the water quality and quantity in the catchment itself and subsequently influence the quality and quantity of a wetland's water. Except for LULC changes, the inflow of untreated domestic sewage, dumping of solid wastes, and uncontrolled encroachments of catchments for agriculture and other purposes has threatened the sustenance of wetlands. LULC change (particularly rapid urbanization and increasing unplanned agricultural activities) and the impact of anthropogenic nutrient enrichment (eutrophication) on aquatic ecosystems including wetlands have been identified as one of the dominant causes for degradation of wetlands and have been reported globally (Carpenter et al., 1998; Mitsch and Gosselink, 2000; Alam et al., 2011; Ramachandra et al., 2013). Therefore, LULC change by humans in the past few decades is considered as one of the major factors responsible for wetland degradation. Many studies show that watershed dominated by agriculture and/or human settlement have significantly higher nutrient levels (Cuffney et al., 2000; Wang, 2001). Like urbanization, agricultural activities cause both direct problems of habitat conversion and indirect effects of water and soil pollution, impacting the biogeochemical cycling of nutrients (Vitousek et al., 1997).

Therefore, this study aims to identify the impact of LULC changes on nutrient concentration over the 25-year period of 1990 to 2015 in the Kabar Tal wetland in the Begusarai district of Bihar.

## 15.2 STUDY AREA

The Kabar Tal wetland (Figure 15.1), also known as the Kabar Tal Bird Sanctuary, is located at 86°05' E to 86°09' E, 25°30' N to 25°32' N in the Begusarai district of Bihar and is one of the many wetlands still existing in North Bihar, India. During dry season, the lake covers an area of 2600 hectares (ha) whereas during the monsoon season, it covers an area of 7,400 ha because it is connected to nearby water bodies and receives water from nearby agricultural areas (Ambastha et al., 2007b). Wetland water is used for irrigation and domestic purposes. Kabar Tal's climate is mostly a type with seasonal fluctuation and is shrinking at an alarming rate due to encroachment of the wetland for agricultural practices, which also leads to choking the inlet of water and alters to the wetland area into intensive agricultural lands (Ghosh et al., 2004). At present, Kabar Tal is also used as a drainage system for nearby villages (Ambastha et al., 2007b). Once the lake was a haven for migratory birds, providing one of the largest breeding grounds for them. The water quality of the wetland has been affected by waste coming from agricultural farms and human settlements, which pollute the wetland. The western part of the Kabar Tal has been reported to have massive sedimentation, which has almost obliterated the surface waters, including the channel flow of the middle and lower Gandak River. The eastern part of it has experienced a lower rate of sedimentation compared to the western part (Ghosh et al., 2004).

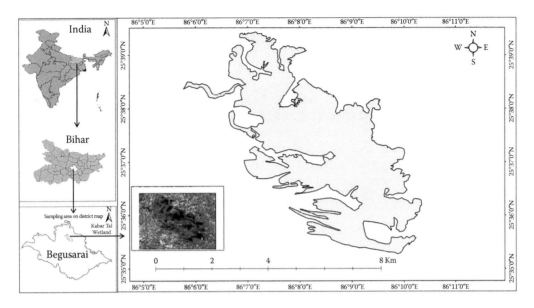

**FIGURE 15.1** Map of the study area (Kabar Tal wetland).

## 15.3 METHODOLOGY

Multispectral and multitemporal LANDSAT satellite images of Kabar Tal were acquired for three years, 1990, 2010, and 2015 (Table 15.1). All satellite images having the resolution of 30 meters were brought to Universal Transverse Mercator (UTM) projection in zone 44 N. ArcGIS 10.2.2 and ERDAS IMAGINE 2014 software were used for the layer stacking classification of images. Unsupervised classification was employed to perform the classification. Unsupervised classification bases its results (groupings of pixels with common characteristics) on the software analysis of an image without having sample classes provided by the user. The software uses techniques to determine which pixels are related and groups them into classes. The area of the classes was calculated in km² by using this formula: Area (km²) = Histogram × 30 × 30/1,000,000). Overall unsupervised classification accuracy is 82%–85% for 1990, 2010, and 2015. The Kappa coefficient is 0.62, 0.73, and 0.65 for 1990, 2010, and 2015, respectively. For the study of changes in nutrient concentration in Kabar Tal, data were taken from various published reports (Kumari, 2016). Because no water quality data were available for year 2010, 2012 water quality data were used for the analysis (Table 15.1).

On the basis of prior knowledge and information available on the study area, a classification scheme was developed. It gives a broad classification of the LULC of the study area as (1) open water, (2) fallow land, and (3) agricultural area. Open water is the area submerged with water particularly during the monsoon season, and fallow land is the area of the Kabar Tal having no agricultural activities and being without water.

**TABLE 15.1**
**Data Source**

| S. No. | Data Type | Data Production | Scale | Source |
|---|---|---|---|---|
| 1 | Landsat 5™ | May 1990 | 30 m | |
| 2 | Landsat 7 (ETM) | May 2010 | 30 m | |
| 3 | Landsat 8 (ETM) | May 2015 | 30 m | |
| 4 | Nutrient data | 1990, 2012, and 2015 | NA | Ramakrishna et al. (2002), Anand (2012), Gupta (2016) |

## 15.4 RESULTS

### 15.4.1 Land Use/Land Cover Distribution

Analysis of the area for the years 1990, 2010, and 2015 identified changes in the LULC of the study area (Table 15.2, Figure 15.2). Of the study area in 1990, open water covered about 12.02% of the area, 48.18% of the agricultural area, and about 39.80% of the fallow land. In 2010, about 5.7% of the study area was covered by open water, 54.22% covered by agricultural area, and fallow land covered about 39.98%. In 2015, about 36.07% of the study area was covered by fallow land, 59.13% was covered by agricultural area, and 4.80% was covered by open water. The magnitude of changes of land used pattern is reported in Table 15.3. The drastic increases in agricultural land (10.95%) and fallow land (3.73%) were observed during these periods, and the open water cover decreased significantly by 7.22% from 1990–2015.

### 15.4.2 Major Nutrients Concentration

The increasing trends of major cations (except calcium) and anion nutrients were reported for the 25-year period (Figure 15.3 and Tables 15.3 and 15.4). The phosphate concentration in 1990, 2012, and 2015 was 0.00071, 0.05935, and 0.029 mg/L, respectively. However, the concentration of nitrate reported was 0.00078, 0.00167, and 0.000844 mg/L, respectively. The concentration of sulfate drastically increased from 1.5 to 3.5 to 47 mg/L from 1990, 2010, and 2015, respectively. The increase in the concentration of

**TABLE 15.2**
**Areas of Land Use/Land Cover Classes**

| Class Name | Area (km²) 1990 | Area (%) 1990 | Area (km²) 2010 | Area (%) 2010 | Area (km²) 2015 | Area (%) 2015 |
|---|---|---|---|---|---|---|
| Open water | 9.6579 | 12.02 | 4.6179 | 5.70 | 3.8628 | 4.80 |
| Agricultural area | 38.7135 | 48.18 | 43.6023 | 54.22 | 47.5002 | 59.13 |
| Fallow land | 31.9725 | 39.80 | 32.1237 | 39.98 | 28.9809 | 36.07 |
| Total | 80.3439 | 100.00 | 80.3439 | 100.00 | 80.3439 | 100.00 |

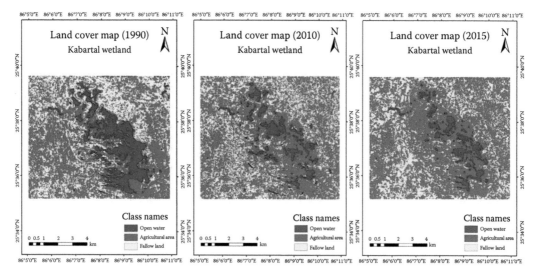

**FIGURE 15.2** Land use cover map of Kabar Tal wetland 1990, 2010, and 2015.

## TABLE 15.3
### Land Use/Land Cover Change: Trend, Rate, and Magnitude

| Class Name | Change 1990–2010 (Area in km²) | Change 1990–2010 (Area in %) | Change 2010–2015 (Area in km²) | Change 2010–2015 (Area in %) | Change 1990–2015 (Area in km²) | Change 1990–2015 (Area in %) |
|---|---|---|---|---|---|---|
| Open water | −5.04 | −6.32% | −0.75 | −0.9% | −5.79 | −7.22% |
| Agricultural area | +4.88 | +6.04% | +3.89 | +4.91% | +8.78 | +10.95% |
| Fallow land | +0.15 | +0.18% | −3.14 | −3.91% | +2.99 | +3.73% |

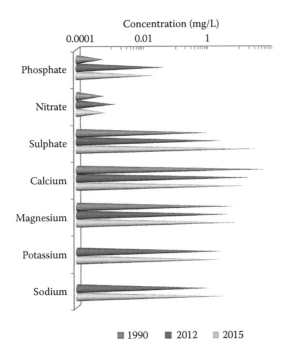

**FIGURE 15.3** Graph showing nutrients concentration (log scale) in and around Kabar Tal in 1990, 2012, and 2015.

## TABLE 15.4
### Change Detection in the Nutrient Concentration with Reference to Reports Published in 2002, 2012, and 2015 for the Kabar Tal

| Parameter | 1990 | 2012 | 2015 |
|---|---|---|---|
| PH | 8.4 | 7.3 | 7.9 |
| EC (µS/cm) | 103.5 | 270.0 | 257 |
| Phosphate (mg/L) | 0.00071 | 0.05935 | 0.029 |
| Nitrate (mg/L) | 0.00078 | 0.00167 | 0.000844 |
| Sulphate (mg/L) | 1.5 | 3.5 | 47.0 |
| Calcium (mg/L) | 93.5 | 34.4 | 22.7 |
| Magnesium (mg/L) | 8.7 | 7.0 | 12.9 |
| Potassium (mg/L) | – | 3.9 | 4 |
| Sodium (mg/L) | – | 1.653 | 4.75 |

magnesium was also reported as 7.0 to 12.9 mg/L from 2012 to 2015; however, a decrease of 93.5, 34.4, and 22.7 mg/L in the concentration of calcium was reported from 1990 to 2012 to 2015, respectively. In addition, the sodium concentration also increased from 1.65 to 4 mg/L from 2012 to 2015, respectively. No significant variation was observed in potassium concentration during these periods.

## 15.5 DISCUSSION

The natural wetland ecosystems used for agricultural purposes have lost much of their original character, leading to reduced biodiversity and performance of functions other than crop production (Hassan et al., 2005). The increase in the concentration of major nutrients (phosphorus and nitrogen) over the 25-year period (1990–2015) in Kabar Tal reported in the present study shows the eutrophic nature of the wetland (Wetzel, 2000; Carlson and Simpson, 1996). The nutrient dynamic in a wetland is influenced by various processes, such as transport, transformation, storage, and the release and removal of nutrients. The inflow regime of the Kabar Tal wetland depends on and is influenced by the Buri Gandak River and other water bodies such as the Bikrampur Chaurs and Nagri Jheel, particularly in the late monsoon months (Ambastha et al., 2007a). The elevated phosphorus content of these inflow water bodies enhance its influx in Kabartal (Shardendu et al., 2012). The nitrogen in the water is generally present in inorganic forms including nitrate, nitrite, and ammonia and organic forms including amino acids, urea, and so on. The most important source of $NO_3^-$ in wetland waters is biological oxidation of nitrogenous organic matter, which includes domestic sewage, agricultural runoff, and effluents from industries (Saxena, 1998; Bhatt et al., 2015). The eutrophication process in Kabartal is leading to the decline in the quality of wetland water at an alarming rate. In addition, the depth of the wetland is also declining rapidly due to an infestation of aquatic weeds such as Phragmatis and Hydrilla (Ghosh et al., 2004). The increase in nutrients resulted not only from agricultural runoff but also effluents released by the local inhabitants. Due to extensive overgrazing, unsustainable agricultural practices, and overexploitation of biomass for fuel, the need for fodder and timber has stripped the land of its natural vegetation cover, resulting in huge erosion and sedimentation that decreases water quantity and water quality (Ghosh et al., 2004). These changes in quality and quantity of water in the wetland not only impact the natural weathering processes, nutrient concentrations, and availability but also the biogeochemistry of the wetland. Sulphate can be naturally occurring or the result of municipal or industrial discharges. When naturally occurring, they are often the result of the breakdown of leaves that fall into a stream, of water passing through rock or soil containing gypsum and other common minerals, or of atmospheric deposition. Runoff from fertilized agricultural lands also contributes sulphate to water bodies. The increase in agricultural practices/area and anthropogenic activities and the decrease in water quantity in wetland have led to increased sulphate concentration in the recent year than in previous years (Table 15.5). The basic sources of calcium ion ($Ca^{2+}$) in

**TABLE 15.5**
**Correlation between Land Use and Nutrient Concentration**

|  | Phosphate | Nitrate | Sulphate | Open Water | Agricultural Area | Fallow Land |
|---|---|---|---|---|---|---|
| Phosphate | 1 |  |  |  |  |  |
| Nitrate | 0.905[a] | 1 |  |  |  |  |
| Sulphate | 0.019 | −0.408 | 1 |  |  |  |
| Open water | −0.788[a] | −0.451 | −0.631[a] | 1 |  |  |
| Agricultural area | 0.538[a] | 0.129 | 0.853[a] | −0.943[a] | 1 |  |
| Fallow land | 0.063 | 0.481 | −0.997[a] | 0.566[a] | −0.807[a] | 1 |

[a] Significance level of 0.05.

surface water are carbonate rocks (limestone, dolomite) that are dissolved by carbonic acid contained in water. When the availability of carbon dioxide is low, however, the reaction begins to proceed in a reverse direction, which leads to the precipitation of $CaCO_3$. Another source of $Ca^{2+}$ in natural water is gypsum, which is common in sedimentary rocks. The water availability in the wetland decreased drastically from 1990–2015, leading to fewer weather reactions and finally, the decrease in the calcium concentration was observed in Kabartal wetland water (Figure 15.3).

The current study also indicates that there is a strong correlation between changes in the LULC and nutrient concentration in the lake (Table 15.5). A strong correlation exists between the agricultural area and phosphate, nitrate, and sulphate concentrations. This is mainly due to the increase in agricultural land where farmers use chemical fertilizers (nitrogenous and phosphate) to enhance productivity, and the runoff of nutrients accumulate in the wetland. Fallow land shows a negative correlation with sulphate (−0.997) and the agricultural area (−0.807). From 1990–2015, the area of open water decreased, resulting in the shrinkage of one of the largest freshwater lakes. The study shows that intensive agricultural practices (from 1990 to till date) lead to decrease in open water area.

## 15.6 CONCLUSION

LULC change has become a major tool in current strategies to manage natural resources such as the wetland ecosystem. The current study shows that LULC changes in catchments have significantly affected the Kabar Tal wetland. It has suffered serious loss during 25-year period studied due to changes, especially unregulated agricultural activities, in LULC. These unplanned and intense agricultural activities have led to a huge inflow of fertilizers (nutrients) that significantly increase nutrient concentrations while significantly decreasing the open water area, indicating the shrinking of the wetland. As a consequence, the total area of the wetland has changed because it provides more of the area's water to agricultural areas and fallow land. These changes not only decrease the water depth (quantity) but also decrease the water quality and cause eutrophication. Because the wetland biogeochemistry involves the exchange or flux of materials between living and nonliving components, the increase in nutrient concentration (eutrophication) will impact the overall nutrient dynamics and biogeochemistry of the wetland.

## ACKNOWLEDGMENT

The authors would like to thank The Centre for Environmental Sciences, Central University of South Bihar, Patna. Dr. R. K. Ranjan thanks the Principal Chief Conservator of Forests, Government of Bihar and Divisional Forest Officer, Begusarai, Bihar, for granting permission to conduct sampling. The study was financed by SERB, Dept. of Science and Technology, Government of India (Startup Grant for Young Scientist Project No. SR/FTP/ES-1/2014).

## REFERENCES

Alam, A., Rashid, S. M., Bhat, M. S., and A. H. Sheikh, 2011. Impact of land use/land covers dynamics on Himalayan wetland ecosystem. *Journal of Experimental Sciences* 2: 60–64.

Ambastha, K., Hussain, S. A., and R. Badola, 2007(a). Resource dependence and attitudes of local people toward conservation of Kabartal wetland: A case study from the Indo-Gangetic plains. *Wetlands Ecology and Management* 15: 287–302.

Ambastha, K., Hussain, S. A., and R. Badola, 2007(b). Social and economic considerations in conserving wetlands of Indo-genetic plains: A case study of Kabartal wetland, India. *Environmentalist* 27: 261–273.

Anand, N., 2012. M.Sc. dissertation, Geochemical assessment of water and sediments from Kabartal wetland, Begusarai, Bihar, Central University of South Bihar, Patna

Berka, C., Schreier, H., and K. Hall, 2001. Linking water quality with agricultural intensification in a rural watershed. *Water, Air, and Soil Pollution* 127: 389–401.

Bhatt, N. A., Rainaand, R., and A. Wanganeo, 2015. Ecological investigation of zooplankton abundance in the Bhoj wetland, Bhopal of central India: Impact of environmental variables. *International Journal of Fisheries and Aquaculture* 7:81–93.

Carlson, R.E. and J. Simpson, 1996. A coordinator's guide to volunteer lake monitoring methods. North American Lake Management Society, p. 96.

Carpenter, S. R, Caraco, N. F., Correll, D. L., Howarth, R.W., Sharpley, A.N., and V. H. Smith, 1998. Nonpoint pollution of surface waters with phosphorus and nitrogen. *Ecological Applications* 8:559–568.

Cuffney, T.F., Meador, M.R., Porter, S.D., and M.E. Gurtz, 2000. Responses of physical, chemical and biological indicators of water to a gradient of agricultural land use in the Yakima River, Washington. *Environmental Monitoring and Assessment* 64: 259–270.

Ghosh, A. K., Bose, N., Singh, K. R. P., and R. K. Sinha, 2004. Study of spatial temporal changes in the wetlands of North Bihar through remote sensing. *13th International Soil Conservation Organization Conference*, Brisbane, Australia.

Gupta, D. 2016. Hydrogeochemistry of Kabar Tal, Begusarai, Bihar (India). M.Sc. dissertation, Central University of South Bihar, Patna.

Hassan, R., Steve, M., Carpenter, R., Chopra, K., and D. Capistrano, 2005. *Millennium Ecosystem Assessment. Ecosystems and Human Well-Being.* Washington DC: Island Press.

Kumari, P., 2016, Impact of land use and land cover changes on surface water quality using remote sensing and GIS: A case study of Kanwar Lake, Begusarai, Bihar. M.Sc. dissertation, Central University of South Bihar, Patna.

Mitsch, W.J., and J.G. Gosselink, 2000. *Wetlands*, John Wiley & Sons, New York, 2000. 3rd ed, pp. 920.

Prakasam, C., 2010. Land use and land cover change detection through remote sensing approach: A case study of Kodaikanal taluk, Tamil Nadu. *International Journal of Geomatics and Geosciences*, 1: 150–158.

Ramachandra, T.V., Meera, D.S., and B. Alakananda, 2013. Influence of catchment land cover dynamics on the physical, chemical and biological integrity of wetlands. *Environment and We -International Journal of Science and Technology*, 8: 37–54.

Ramakrishna, Muley, E.V., Siddiqui, S. Z., and A.K. Pandey, 2002. Limnology, Wetland Ecosystem Series 4, Fauna of Kabar Lake. Zoological Survey of India, 15–21.

Saxena, S., 1998. Settling studies on pulp and paper mill wastewater. *Indian Journal of Environmental Health*, 20: 273–280.

Shardendu, S., Sayantan, D., Sharma, D., and S. Irfan, 2012. Luxury uptake and removal of phosphorus from water column by representative aquatic plants and its implication for wetland management. *ISRN Soil Science*, 1–8.

USEPA, 1994. National Health and Environmental Effects Research lab. Western Ecology Division 2005 SW Carvallis.

USEPA, 2002. Methods for evaluating wetland condition: Land-use characterization for nutrient and sediment risk assessment. Office of Water, U.S. Environmental Protection Agency, Washington, DC. EPA-822-R-02-025.

Vitousek, P. M., Mooney, H. A., Lubchenco, J., and J. M. Melillo, 1997. Human domination of Earth's ecosystems. *Science* 277: 494–499.

Wang, X. 2001. Integrating water quality management and land use planning in a watershed context. *Journal of Environmental Management* 61: 25–36.

Wetzel, R.G., 2000. Freshwater ecology: Changes, requirements, and future demands. *Limnology* 1, 3–9.

# 16 Evaluation of Spectral Mapping Methods of Mineral Aggregates and Rocks along the Thrust Zones of Uttarakhand Using Hyperion Data

*Soumendu Shekhar Roy and Chander Kumar Singh*

## CONTENTS

16.1 Introduction ........................................................................................................................252
    16.1.1 Geological Setting .................................................................................................253
16.2 Methodology ......................................................................................................................254
    16.2.1 Preprocessing Hyperion Data ................................................................................254
    16.2.2 Bad Band Designation ...........................................................................................255
    16.2.3 Radiometric Calibration ........................................................................................255
    16.2.4 Data Dimension Reduction ....................................................................................256
        16.2.4.1 Minimum Noise Fraction Transformation ...............................................256
        16.2.4.2 Endmember or Pure Pixel Extraction ......................................................256
        16.2.4.3 n-Dimensional Visualizer .........................................................................257
16.3 Spectral Analysis ................................................................................................................259
    16.3.1 Spectral Library .....................................................................................................259
    16.3.2 Endmember Spectral Library .................................................................................259
    16.3.3 Spectral Analyst .....................................................................................................259
        16.3.3.1 Binary Encoding ......................................................................................259
        16.3.3.2 Spectral Angle Mapping ..........................................................................260
16.4 Spectral Mapping Method's Weights and Scores ..............................................................261
16.5 Weight Scores of Different Mapping Methods and the Results ........................................262
    16.5.1 Field Spectra Data ..................................................................................................270
    16.5.2 Evaluation of the Spectral Mapping Method .........................................................272
16.6 Geological Interpretations of the Results ..........................................................................272
16.7 Conclusion .........................................................................................................................272
References ..................................................................................................................................273

**ABSTRACT** By virtue of its narrow bandwidth and higher spectral resolution, hyperspectral data (Hyperion, AVIRIS) can provide an accurate spectral mapping process for surface characterization of geological lithounits. A detailed spectral classification of lithounits as a function of its mineral phase assemblage (rock assemblage) can be further used for geological interpretation of any tectonic event in any area of interest. The lesser Himalayas (Kumaun) have a series of thrusts and faults among the metasedimentary lithounits of regional metamorphic grade. Identification of these lithounits and their respective mineral assemblage to analyze the grade of the rock undergoing metamorphism, lithomapping, mineral phase identification, and so on. Use of a spectroscopic technique (aerial survey) depends on an accurate spectral mapping method. Spectral angle mapper (SAM), binary encoding (BE), and spectral feature fitting (SFF) are the prevalent methods for spectral classification of endmembers or spectral signatures. But to define the weight and evaluation of different spectral mapping methods, validating the data with field spectroscopy to understand its geological implication is needed. The choice of spectral mapping method will proportionally affect the accuracy of arial imagery mapping. The metamorphic spectral classes thus obtained will provide an insight into the geothermal history of the rock assemblages under study.

## 16.1  INTRODUCTION

Aerial imagery and remote sensing satellite data have great importance for geological field study. They constitute a major role in reconnaissance field survey. Commonly used multispectral imagery can give us a brief idea regarding a field area's location, land use and land cover (LULC) distribution and even the probable location of a thrust or fault by following physiographic trends and ridges. But the broader aspects of geoscience remain untouched, and the main reason can be traced to the lack of spectral resolution for which the geological information in terms of spectra remains unclear for the researchers. This can be overcome by the utilization of hyperspectral data. The term *hyper* denotes "many" or "abundance" due to the presence of a high number of image bands (220–240) the spectral information does not get suppressed. The bandwidth (~1 nm) of the Hyperion data is another factor in the availability of such a huge amount of information. The bandwidth is so low that the spectral data are not normalized over the same pixel.

Utilizing higher spectral resolution data for better surface characterization in understanding the nature of geotectonic episodes of any field area involves a numerous stages and substages of analysis and interpretation. The lesser Himalaya (better known as Kumaun Himalaya) was selected because various literature reports of structural evolution and petrographic studies have revealed the presence of major and minor faults over the area Main Frontal Thrust or Himalayan Frontal Thrust (MFT/HFT), Main Boundary Thrust (MBT), and Main Central Thrust (MCT) including the major thrusts of the Himalayas along with several minor thrust zones such as Ramgarh Thrust, Almora Thrust, Vaikrita Thrust, and Munsiari Thrust. The lesser Himalaya extends between the MBT and MCT zones. Valdiya (1980), Rupke (1974), and Karunakaran and Rao (1979) have classified different lithounit groups of this highly stressed zone. The Almora and Ramgarh groups constitute the lesser Himalaya Group of metasedimentary rocks (Srivastava and Mitra, 1996). Ramgarh Thrust is between the Outer Himalaya and Ramgarh Group of Gneissic and schistosed metasedimentary units. The South Almora Thrust (SAT) separates the subunits of the lesser Himalayas. The Almora Group, also known as the Almora Klippe, of quartzite, schist, and gneisses is the hanging wall block that is thrust along the Damta shale–phyllite formation (Valdiya, 1980; Ghose et al. (1974); Srivastava and Mitra, 1996). This thrust known as the North Almora Thrust (NAT) along which mylonitization can be observed indicates the transition zone from brittle to ductile zones (Srivastava and Mitra, 1996).

By using hyperspectral data, it is possible to identify metamorphic units' texture such as gneiss, phyllite, and possibly mylonite (Trouw, 2009), and the nature of the thrust acting on the given area can be derived. Different experiments on the spectroscopic nature of rocks have been conducted in Jet Propulsion Laboratory (JPL–NASA), United States Geological Survey (USGS), and others, and they can provide the spectral library of rocks that had been geologically emplaced on similar conditions by thrusting at thermal range (Balridge et al., 2009) and VNIR-SWIR range in USGS mineral and rock spectral library (Kokaly et al., 2017). By matching the spectral endmembers of the Hyperion scene with those of standard spectra using proper scaling and spectral mapping methods, the nature of the thrust over an area can be deduced. However, it is to be noted that the laboratory condition as well the rock condition of two different scenarios undergoing the same process may or may not be identical in every point because a large number of variables such as inhomogeneity of mineral phases, presence of microstructures, and so on could be involved. Furthermore, the different spectral mapping methods involved in classifying these lithounits can affect the result, depending on several factors such as the threshold selected for a class, the weight factors assigned for a variable to be classified along the atmospheric condition during which the sensor recorded data for the scene. The current research is a comparative study of those spectral mapping methods that are utilized for the spectral classification of surface.

The Hyperion data from the Earth observer (EO1) mission, which started November 2001, can provide hyperspectral data to March 2017 when the mission was terminated. Hyperion data along with Advanced Land Imager (ALI) imagery archive data can be accessed from USGS EarthExplorer portal without any charge. The data available for Hyperion scene constitute the following:

- L1R—which at sensor radiance is corrected but is not geometrically corrected.
- L1T—which is terrain corrected at sensor radiance in GeoTIFF format and with geometric correction.
- L1GST—this datum is at sensor radiance corrected with geometric correction.

For this study Hyperion level L1GST data was selected with WGS datum of Almora region, west of the town of Almora on December 19, 2012, which for spectral analysis needs further radiometric calibration to obtain the rectified reflectance values and then the atmospheric correction using the Fast Line-of-sight Atmospheric Analysis of Spectral Hypercubes (FLAASH) module or Quick Unsupervised Anisotropic Clustering (QUAC) module to classify or characterize the surface features.

### 16.1.1 Geological Setting

The Lesser Himalaya or Kumaon Himalaya shows several klippe that tectonically overlie metasedimentary rocks of Precambrian era (Srivastava and Mitra, 1994). The sedimentary units consist of Precambrian Damta (shale greywacke, siltstone) and Deoban (limestone) Groups and Precambrian–Cambrian Jaunsar and Mussoorie Group (Valdiya, 1980; Srivastava and Mitra, 1994). These klippe have the lithology of amphibolite-grade metapelite, quartzite, and gneiss belonging to the Precambrian Almora Group. The Ramgarh Group, which occurs under the Almora unit separated by the Ramgarh Thrust, belongs in the greenschist grade (Srivastava and Mitra, 1996).

The NAT marks the northern limit of this klippe, and the SAT marks the southern extent consisting of a sequence of muscovite–biotite–quartz schist and augen gneiss (Valdiya, 1980). The mineral assemblage of this klippe shows an amphibolitic grade of metamorphism (Ghose et al., 1974).

The Hyperion imagery obtained for the current study covers the Jaunsar lithounit in the south and passes through the Ramgarh and Almora units in the north. The thrust zones of the Ramgarh Group and both NAT and SAT pass over the Hyperion data.

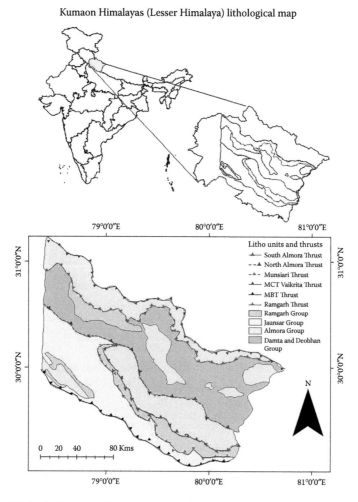

**FIGURE 16.1** Lithological map and thrust zones of Lesser Himalaya-Kumaun. (From Valdiya, K.S., *Geology of Kumaun lesser Himalaya*, Wadia Institute of Himalayan Geology, Uttarakhand, India, 1980.)

The structural succession in the field area according the geological map studied by Valdiya (1980) can be constructed as (Figure 16.1):

Almora Group
Almora Thrust
Ramgarh Group
Ramgarh Thrust
Jaunsar and Mussoorie Groups
Damta and Deoban Groups

## 16.2 METHODOLOGY

### 16.2.1 Preprocessing Hyperion Data

Hyperspectral data consist of a spectral range from (400 nanometre to 2500 nanometre) in its 242 bands. This huge amount of band range may pose a much higher level of complexity for spectral classification since the variances that are not correlated are to be considered. This can be achieved by data dimension reduction by which the unwanted information can be subtracted and analysis of the requisite amount to provide a simpler and correct result. The preprocessing of the data has been carried forward on ENVI 5.3, which is provided with FLAASH module, a MODTRAN atmospheric tool (that requires a license). Only after processing the data will the reflectance values for proper surface characterization be obtained.

### 16.2.2 BAD BAND DESIGNATION

Of the 242 bands of Hyperion data, the Hyperion visible-near-infrared (VNIR) sensor has 70 bands, and the shortwave-infrared (SWIR) sensor has 172 bands. The bands from 1 to 7, 58 to 76, and 225 to 242 have zero data values (Barry, 2001). Except for these bands, the atmospheric water absorption windows, which create strong noise values in the data, must be taken into consideration. The bands listed from 121 to 126, 167 to 180, and 222 to 224 need to be removed in the preprocessing method for better spectral analysis (Datt et al., 2003). Thus, bad band removal provides a spectral subsetting of 175 spectral channels of the 242 bands.

### 16.2.3 RADIOMETRIC CALIBRATION

For FLAASH atmospheric correction or any other atmospheric correction like QUAC, ATmospheric CORrection (ATCOR), and so on, every methodological module requires the pixel values to be properly and uniformly calibrated as per their algorithm's need. The Level 1 radiance (L1R) (corrected), Level 1 terrain (L1T) (corrected), and Level 1 radiometric corrections and systematic geometric corrections (L1Gst) are provided in a BSQ (band interleaved by sequence)) format, which needs to be converted into BIL (band interleaved by line) format. Radiometric calibration converts the image into a radiance image of BIL format. The radiance offset and gain values provided in the metadata of the imagery is used for the said process provides a uniformly calibrated image with radiance gains (Apan, 2003). In the absence of this information for any L1 data, the values of 0.025 radiance gains for VNIR bands and 0.0125 radiance gains for SWIR bands need to be applied. This can be achieved by dividing the VNIR bands and SWIR bands by factors of 40 and 80, respectively, utilizing band math.

For radiance calibration after applying the radiance gains and offset, all data need to be scaled in $\mu W/(cm^2 * sr * nm)$ from the original pixel values of $W/(cm^2 * sr * nm)$, which is obtained by dividing the radiance image by a factor of 0.10, otherwise known as the FLAASH factor.

After the radiance is obtained, it is converted to reflectance values for spectral analysis. The reflectance image without any further correction, can be obtained by dividing the radiance image by the irradiance values for each and every channel provided in the metadata because the reflectance value is the ratio of these two parameters. But by doing so, the atmospheric disturbances (atmospheric absorptions) will not be taken into account and can lead to an erroneous result. The best option is to apply an atmospheric correction module rather than going for over the top atmospheric correction. It can be achieved by calibrating the imagery to apparent surface reflectance, which yields the most accurate results when using spectral indices. This is especially important for hyperspectral sensors such as AVIRIS and EO-1 Hyperion. Calibrating imagery to surface reflectance also ensures consistency when comparing indices over time and from different sensors.

FLAASH is a model-based radiative transfer program developed by Spectral Sciences, Inc. It uses MODTRAN4 radiation transfer code to correct images for atmospheric water vapor, oxygen, carbon dioxide, methane, ozone absorption, and molecular and aerosol scattering.

This concludes the preprocessing steps, but it should be ensured whether the reflectance data are normalized or not. This can be obtained by running simple statistical analysis. If the results are not normalized and there are negative values, use the formula in Band math:

$$(b1 \text{ le } 0)*0 + (b1 \text{ ge } 10000)*1 + (b1 \text{ gt } 0 \text{ and } b1 \text{ lt } 10000)*float(b1)/10000 \quad (16.1)$$

where b1 is a channel or band, which will be applied for every channel to scale the reflectance values within the range of 0 to 1 where 1 is the 100% reflectance as for white body reflectance reference

## TABLE 16.1
### Data Used in This Study and Their Sources

| Sr. No. | Data | Source |
|---|---|---|
| 1 | Hyperspectral data | Earth Explorer of USGS portal |
| 2 | Field reflectance data | ASD spectroradiometer, DTRL, DRDO |
| 3 | Spectral library of reference rock | NASA, Jet Propulsion Laboratory https://speclib.jpl.nasa.gov/ |
| 4 | Lithological map and thrust zones | Literature survey (Valdiya, 1980; Srivastava and Mitra, 1996) |

and 0 being the black body reference or complete absorption. This equation has to be applied to every band obtained for reflectance data (Table 16.1).

### 16.2.4 Data Dimension Reduction

#### 16.2.4.1 Minimum Noise Fraction Transformation

The obtained reflectance data can further have reduced to lesser number of bands or spectral channels containing the required information by subtracting the bands with higher signal to noise ratio and the correlated bands. This is obtained by minimum noise fraction (MNF). For hyperspectral data, ENVI uses the MNF transform algorithm to determine the inherent dimensionality of image data, to segregate noise in the data, and to reduce the computational requirements for subsequent processing (Harris Geospatial Documents archives ENVI).

The MNF transformed and implemented in ENVI is a linear transformation that consists of the following separate principal component analysis rotations (Green et al., 1988).

The first rotation uses the principal components of the noise covariance matrix to decorrelate and rescale the noise, resulting in transformed data in which the noise has unit variance but no correlation band to band.

The second rotation uses the principal components derived from the original image data after the noise has been decorrelated with unit variance and zero correlation (band to band) by the first rotation and rescaled by the noise standard deviation. The final eigenvalues and the associated images determine the inherent dimensionality. Thus, the huge data dimension is divided into two parts, one part having a higher eigenvalue and coherent image and the other having noise with near unit eigenvalues and no correlation. Thus, MNF reduces the noise from the reflectance data as the second part results from the twofold rotation.

#### 16.2.4.2 Endmember or Pure Pixel Extraction

The pixel purity index (PPI) is used to find the most spectrally pure (extreme) pixels in multispectral and hyperspectral images. These typically correspond to mixing endmembers. The PPI is computed by repeatedly projecting n-dimensional scatter plots on a random unit vector where n denotes the number of image spectral channel (MNF transformed). ENVI software records the extreme pixels in each projection (those pixels that fall onto the ends of the unit vector), and it notes the total number of times each pixel is marked as extreme. A PPI in which each pixel value corresponds to the number of times that pixel was recorded as extreme is created (Harris Geospatial Documents archives ENVI).

To obtain the PPI, a threshold value has to be set beyond which the corresponding pixels will be considered as pure spectra or spectral endmember. In this study a value of 2.5 was selected. The value of 2.5 flags all pixels greater than two DN values from the extreme pixels (both high and low) as extreme. This threshold selects the pixels on the ends of the projected vector. The threshold should be approximately two to three times the noise level in the data. Mention of the favorable number iterations and iteration per block is also required to obtain the PPI values. Specifying a block size means that data points are plotted every $n$th iteration instead of each iteration. Normally thousands of iterations are required for obtaining the PPI; in this case, it was set for about 10,000 times (Figure 16.2).

# Evaluation of Spectral Mapping Methods of Mineral Aggregates and Rocks

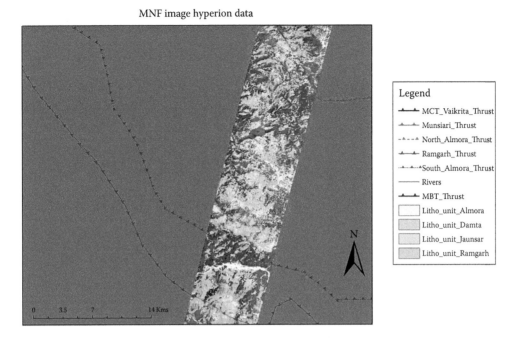

**FIGURE 16.2** The minimum noise fraction image obtained from the Hyperion scene.

### 16.2.4.3 n-Dimensional Visualizer

To visualize the pure pixels over the Hyperion reflectance scene, the PPI values are projected over the MNF band space and observed over the spectral bands. The *n-D Visualizer* helps us to locate, identify, and cluster the purest pixels and the most extreme spectral responses (endmembers) in a data set in n-dimensional space. The n-D Visualizer was designed to help visualize the shape of a data cloud that results from plotting image data in spectral space (with MNF bands as plot axes) (Figure 16.3).

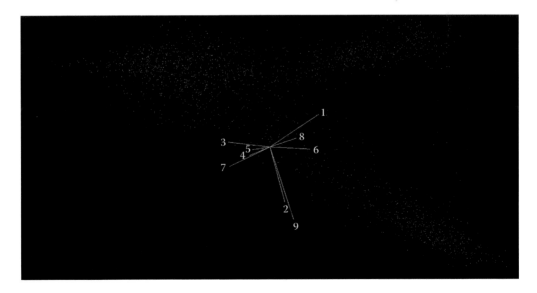

**FIGURE 16.3** n-dimensional space for endmember extraction (pure pixel) at 10,000 iterations, 2.5 threshold for PPI over the Hyperion image with the axes of MNF bands, which are the dimensions of this subspace (ENVI n-dimension Visualizer).

The clusters of pure pixels obtained from PPI aid in characterizing the group of pixels that belongs to the same surface feature. By plotting all the pure pixels on the n-Dimension where n is the number of spectral bands and then rotating each band keeping one axis (band) fixed, the pixels that do not change their distance with respect to the axis, center, or any other pixels can be identified, and then that pixel or group of pixels can be termed as same spectral group or class or same surface feature by virtue of reflectance property.

For Hyperion data, it is an impossible or very time-consuming process to locate each and every group of similar surface reflectance response based on an eye estimate, so clustering has to be done based on variables such minimum distance (squared error) to club, cluster, or merged similar surface features.

Thus, characterized surface groups were obtained, which were members that needed to be identified spectrally using spectral analyst (Table 16.2).

The endmembers obtained through data dimension reduction methods were converted to the region of interest similar to the training data set in supervised classification.

### TABLE 16.2
### End Members Extracted by n-Dimensional Visualizer on PPI Band

| End Member | Pixel Location | |
|---|---|---|
| Class No. Assigned | X Value | Y Value |
| N_D Class 6 | 1209 | 2353 |
| N_D Class 8 | 1460 | 1506 |
| N_D Class 14 | 1154 | 2355 |
| N_D Class 16 | 1291 | 2591 |
| N_D Class 17 | 1200 | 2337 |
| N_D Class 18 | 1284 | 2605 |
| N_D Class 19 | 1206 | 2388 |
| N_D Class 20 | 1149 | 2307 |
| N_D Class 22 | 1220 | 2382 |
| N_D Class 23 | 1366 | 2228 |
| N_D Class 24 | 1283 | 2630 |
| N_D Class 25 | 1198 | 2391 |
| N_D Class 27 | 1224 | 2399 |
| N_D Class 28 | 1142 | 2335 |
| N_D Class 29 | 1592 | 873 |
| N_D Class 30 | 1470 | 1459 |
| N_D Class 32 | 1208 | 2355 |
| N_D Class 33 | 1221 | 2378 |
| N_D Class 35 | 1413 | 2125 |
| N_D Class 36 | 1729 | 306 |
| N_D Class 37 | 1243 | 2396 |
| N_D Class 38 | 1244 | 2392 |
| N_D Class 39 | 1200 | 2352 |
| N_D Class 40 | 1244 | 2393 |
| N_D Class 42 | 1290 | 2592 |
| N_D Class 44 | 1236 | 2387 |

## 16.3 SPECTRAL ANALYSIS

### 16.3.1 Spectral Library

To match the spectral response obtained from the reflectance data, a standard reference is needed to determine the type of rock surface obtained from the scene. This can be achieved by taking the spectral response of different metamorphic, sedimentary, or igneous rocks studied under laboratory conditions and preparing a library to match the response that will result from identifying the endmembers. However, regarding the condition in which the standard samples were studied, the mineralogical composition of those rocks also can vary to a certain extent. The best possible way to nullify such effects is to prepare your own spectral library of samples (rocks, vegetation, man-made constructions) in the field with a spectroradiometer with a preferable range from at least 426.82 nanometers to 2365.20 nanometers (Hyperion RANGE-VNIR to SWIR). This, however, increases the cost of study as well labor, but the result from such spectral library yields a higher percentage of accurate results. The rock samples under the spectral studies need to be analyzed further for mineralogical data as well.

In this study, the Johns Hopkins University (JHU) Spectral Library was used. Except for man-made materials, all spectra in the Johns Hopkins Library were measured under the direction of John W. (Jack) Salisbury. Most measurements were made by Dana M. D'Aria, either at the JHU in Baltimore, Maryland, or at the USGS in Reston, VA.

ENVI spectral library resources have access to the JHU Spectral Library. The metamorphic rock spectral library (meta.crs) was used in this study and was taken as the standard reference. The spectral library under consideration needed to be resampled in terms of the image spectra obtained (Grove et al., 1992). The lstandard spectral samples of the JHU Spectral Library are provided in the following discussion.

### 16.3.2 Endmember Spectral Library

To match with the spectral library of the standard materials, a spectral library of the endmembers obtained from pure pixels was created and were resampled. The pure pixels thus obtained were converted into the region of interest, and the spectral signatures from respective pixels were collected as endmembers. This spectral library was used for spectral matching and assigning the closest possible metamorphic class name by spectral analyst tool.

### 16.3.3 Spectral Analyst

The *spectral analyst* method is an identification process to compare the pixel's signature with the standard reference. The Spectral Analyst helps to identify materials based on their spectral characteristics. The Spectral Analyst uses ENVI techniques such as binary encoding (BE, spectral angle mapper (SAM), and spectral feature fitting (SFF) to rank the match of an unknown spectrum to the materials in a spectral library.

The output of the Spectral Analyst is a ranked or weighted score for each of the materials in the input spectral library. The highest score indicates the closest match and higher confidence in the spectral similarity. Similar materials should have relatively high scores, but unrelated materials should have low scores.

#### 16.3.3.1 Binary Encoding

The BE classification technique encodes the data and endmember spectra into zeros and ones based on whether a band falls below or above the spectrum mean, respectively. An exclusive (OR) function compares each encoded reference spectrum with the encoded data spectra and produces a classification image (Mazer, 1988). All pixels are classified as the endmember with the greatest number of bands that match unless a minimum match threshold is specified, in which case some pixels could be unclassified if they do not meet the criteria.

Considering a single element (spatial resolution) of a pixel of $L$th dimension ($L$ bands) as

$$\vec{X_{ij}} = [X_{ij}(1), X_{ij}(2), \ldots X_q(1) \ldots, X_q(L)] \quad (16.2)$$

and its spectral mean as

$$\mu_{ij} = \left[1\frac{1}{L}\sum_{l=1}^{L} X_{ij}(1)\right] \quad (16.3)$$

an $L$ bit binary code vector $Y_{ij}^a$ is constructed

$$\vec{Y_{ij}^a} = H\{X_{ij} - \mu_{ij}\}$$

where $H(u)$ is a unit step operator defined by (16.4)

$$H(u) = \{1, U \geq 0 \quad (16.4)$$

$$0, U < 0,$$

This vector thus constructed is a binary representation of spectral amplitude, but the information is within the local slope measured from the spectral curve at each wavelength:

$$Y_{ij}^b(1) = \{1, [X_{ij}(1+1) - X_{ij}(1-1)] \geq 0$$
$$0, [X_{ij}(1+1) - X_{ij}(1-1)] < 0 \quad (16.5)$$

where $l = 1, 2, 3\ldots\ldots L$ spectral channels.

An additional $U_{ij}^b(l)$ binary code is constructed from this, and the two vector codes together form the 2L binary code, which represents the spectrum of the location $i, j$ pixel.

The similarity measure used to determine spectral signature matches was the Hamming distance (Viterbi and Omura, 1979), which is computed from:

$$D_h(Y_{ij}, Y_{mn}) = \sum_{l=1}^{2L} Y_{ij}(1)(\text{XOR})Y_{mn}(1) \quad (16.6)$$

But for every spectral data occurring naturally in the real world, the chance or probability to match over this 2L extent, that is, 100% match, will be the rarest of rare occasions, so an acceptance distance $d_a$ or threshold encoding distance was provided such that $Y_{mn}$ of $m, n$ spatial location will be spectrally identical with $Y_{ij}$ only if:

$$Y_{ij} = Y_{mn} \text{ if } D_h < d_a \quad (16.7)$$

In ENVI, the threshold is set from a range of 0 to 1 where zero is the null tolerance level and 1 represents 100% matching threshold.

### 16.3.3.2 Spectral Angle Mapping

SAM is a spectral mapping method that provides rapid mapping of the spectral similarity of image spectra to reference spectra (Boardman, 1993). The reference spectra can be either laboratory or field spectra or extracted from the image. The pre-assumption required for this method is that the

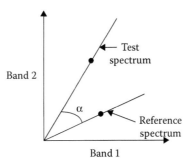

**FIGURE 16.4** Schematic diagram of the methodology of the SAM method.

data should be reduced to "apparent reflectance" with all dark current and path radiance biases removed (Kruse, 1993). The algorithm calculates the "angle" between the two spectra and thus determines the spectral similarity between the two. The method treats both target spectra and reference spectra as vectors in a space with dimensionality equal to the number of bands.

The simplest way to describe this methodology is to visualize the spectral response on a two-dimensional band. The lines connecting each spectrum point and the origin contain all possible positions for that material corresponding to the range of possible illuminations. Poorly illuminated pixels will fall closer to the origin (the dark point) than pixels with the same spectral signature but greater illumination. However, the angle between the vectors is the same regardless of their length (Kruse, 1993). The SAM algorithm (Boardman, 1993) generalizes this geometric interpretation to $n$th dimensional space.

The algorithm consists of the arc-cosine of the dot product of the target spectrum ($t$) and reference spectrum ($r$) (Figure 16.4).

$$a = \cos^{-1}\left(\frac{(\vec{t}\cdot\vec{r})}{\|\vec{t}\|\|\vec{r}\|}\right) \quad (16.8)$$

For $n$ number of bands, the equation becomes

$$a = \cos^{-1}\left(\frac{\sum_{i=1}^{nb} tr}{\left(\sum_{i=1}^{nb} t^2\right)^{1/2}\left(\sum_{i=1}^{nb} r^2\right)^{1/2}}\right) \quad (16.9)$$

## 16.4 SPECTRAL MAPPING METHOD'S WEIGHTS AND SCORES

In the spectral analyst, normally the three different spectral matching methods give equal weight whose sum should be of unit value. However, the necessary weight according to the need of the user can also be implemented if any spectral feature of a known location is to be maximized or enhanced by any of the methods. The best method is to provide the default 1.000 value to the SFF, but in such case, the most suitable spectral library includes the field spectra. If such conditions are not available, it is further modified to a multirange spectral feature that is a piecewise spectral matching of target and reference spectra. If the absorption features or spectral facies are known they can be applied to the multirange spectral feature fitting (SFF) for eg. Al-OH vibration zone or Fe-OH vibration zones for aluminosilicates (quartzites) and Amphiboles respectively in metamorphic rocks (Longhi 2001).

A field spectrometer reading finally will provide the cross-validation and thus an evaluation of the spectral matching algorithm.

## 16.5 WEIGHT SCORES OF DIFFERENT MAPPING METHODS AND THE RESULTS

The endmembers extracted by the data dimension reduction method were matched by the three methods SAM, BE, and SFF. But a difference in weight may result in a variation in the spectral mapping result. To select the best method applicable to any area of geological interest if the mapped lithounit is matched with field spectra data.

*Case 1: Equal weight*

In this case, the weight factor of 0.333 is applied to each of the spectral matching methods to identify each endmember extracted by PPI. The spectral library of metamorphic rock (Meta.crs) of the JHU Spectral Library resource was selected and resampled from the micrometer to nanometer scale (Grove et al., 1992) (Table 16.3). Here the endmember obtained is designated with the highest score or is the most likely to match the spectra.

### TABLE 16.3
### Result of the Spectral Mapping Methods for Corresponding Spectral Endmember Classes for Equal Weight to the Three Spectral Mapping Methods (SAM, BE, and SFF)

| End Members | Spectral Method | SAM | BE | SFF (Single Range) |
|---|---|---|---|---|
| | Weight | 0.33 | 0.33 | 0.33 |
| CLASS 4 | Meta_crs_spec_library | CHLORITIC GNEISS | | |
| CLASS 6 | Meta_crs_spec_library | CHLORITIC GNEISS | | |
| CLASS 14 | Meta_crs_spec_library | CHLORITIC GNEISS | | |
| CLASS 16 | Meta_crs_spec_library | CHLORITIC GNEISS | | |
| CLASS 17 | Meta_crs_spec_library | CHLORITIC GNEISS | | |
| CLASS 18 | Meta_crs_spec_library | CHLORITIC GNEISS | | |
| CLASS 19 | Meta_crs_spec_library | CHLORITIC GNEISS | | |
| CLASS 20 | Meta_crs_spec_library | CHLORITIC GNEISS | | |
| CLASS 22 | Meta_crs_spec_library | CHLORITIC GNEISS | | |
| CLASS 23 | Meta_crs_spec_library | SERPENTINE MARBLE | | |
| CLASS 24 | Meta_crs_spec_library | SERPENTINE MARBLE | | |
| CLASS 25 | Meta_crs_spec_library | CHLORITIC GNEISS | | |
| CLASS 27 | Meta_crs_spec_library | CHLORITIC GNEISS | | |
| CLASS 28 | Meta_crs_spec_library | CHLORITIC GNEISS | | |
| CLASS 29 | Meta_crs_spec_library | CHLORITIC GNEISS | | |
| CLASS 30 | Meta_crs_spec_library | CHLORITIC GNEISS | | |
| CLASS 32 | Meta_crs_spec_library | CHLORITIC GNEISS | | |
| CLASS 33 | Meta_crs_spec_library | CHLORITIC GNEISS | | |
| CLASS 35 | Meta_crs_spec_library | CHLORITIC GNEISS | | |
| CLASS 36 | Meta_crs_spec_library | CHLORITIC GNEISS | | |
| CLASS 37 | Meta_crs_spec_library | CHLORITIC GNEISS | | |
| CLASS 38 | Meta_crs_spec_library | CHLORITIC GNEISS | | |
| CLASS 39 | Meta_crs_spec_library | PURPLE QUARTZITE | | |
| CLASS 40 | Meta_crs_spec_library | CHLORITIC GNEISS | | |
| CLASS 42 | Meta_crs_spec_library | CHLORITIC GNEISS | | |
| CLASS 44 | Meta_crs_spec_library | CHLORITIC GNEISS | | |

The spectra of these endmembers obtained from the imagery are resampled as a spectral library creating the reference spectra, and spectral angle classification is executed with a tolerance of 0.1 radian for each class (Figures 16.5 and 16.6).

Endmember spectral matched with highest probability from the spectral library and thus were designated as the corresponding metamorphic rock.

Serpentinized marble (metamorphosed limestone) is major class obtained along with chloritic gneiss, which seems to repeat at a regular interval with the thrust zone (Table 16.4).

*Case 2*

The BE method is given the 1.00 weight factor (100%) over SAM and SFF. Here the endmember obtained from the pure pixels was provided as the spectral subunit with a matching threshold of 0.98 (98%), thus providing the Hamming distance for classification (Table 16.5).

The BE spectral classifier when provided with weight factor higher than SAM shows the dominance of quartzite in the Hyperion scene compared to the previous scenario. BE results show the presence of quartzite and gneiss as the predominant spectral feature in the scenario (Figure 16.7 and Table 16.6).

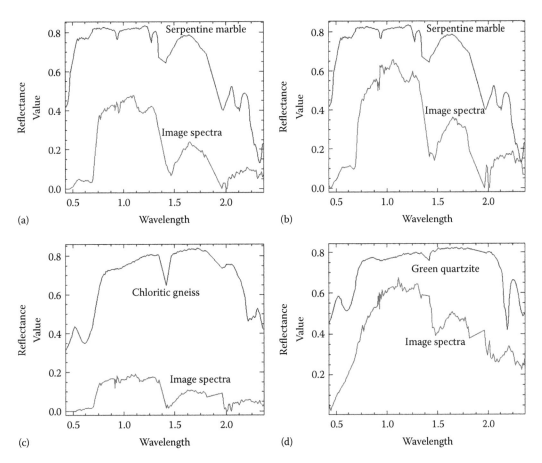

**FIGURE 16.5** Spectral curve matching of endmembers and the most suitable rock spectra from the spectral library according to the equal weightage defined to the spectral mapping methods. (a) and (b) Image end members matching with serpentine marble, (c) image end members matching with chloritic gneiss, and (d) image end members matching with green quartzite.

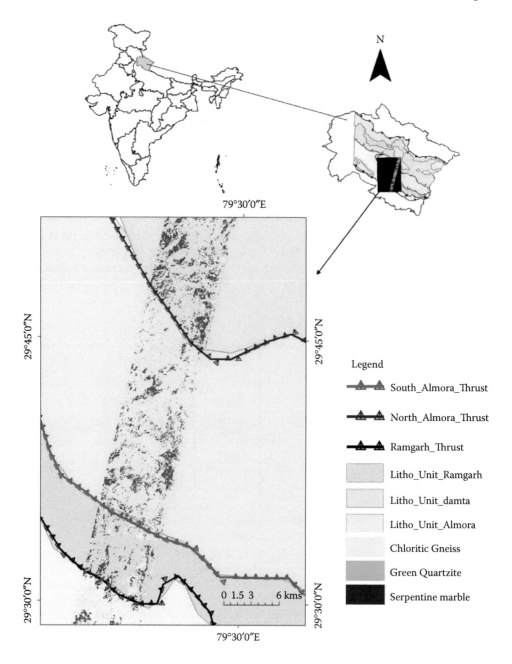

**FIGURE 16.6** Classified image of Hyperion data according to the equal weightage provided to SAM and BE classifier spectral mapping methods.

## TABLE 16.4
Percentage of Endmember Classes in the Study Area (Equal Proportion in Weight for SAM, BE, SFF, and Single Range)

| Classes | Pixel Count | Percentage |
| --- | --- | --- |
| Unclassified | [3404743] | 96.79 |
| CHLORITIC_GNEISS | [159] | 0.01 |
| CHLORITIC_GNEISS | [191] | 0.01 |
| CHLORITIC_GNEISS | [169] | 0.01 |
| CHLORITIC_GNEISS | [14626] | 0.42 |
| CHLORITIC_GNEISS | [154] | 0.00 |
| CHLORITIC_GNEISS | [161] | 0.01 |
| CHLORITIC_GNEISS | [172] | 0.01 |
| SERPENTINE_MARB | [42331] | 1.20 |
| SERPENTINE_MARB | [39043] | 1.11 |
| CHLORITIC_GNEISS | [52] | 0.00 |
| CHLORITIC_GNEISS | [4] | 0.00 |
| CHLORITIC_GNEISS | [1817] | 0.05 |
| CHLORITIC_GNEISS | [3445] | 0.10 |
| CHLORITIC_GNEISS | [164] | 0.01 |
| CHLORITIC_GNEISS | [52] | 0.00 |
| CHLORITIC_GNEISS | [14] | 0.00 |
| CHLORITIC_GNEISS | [2503] | 0.07 |
| CHLORITIC_GNEISS | [7104] | 0.20 |
| CHLORITIC_GNEISS | [7] | 0.00 |
| CHLORITIC_GNEISS | [8] | 0.00 |
| QUARTZITE_PURPL | [6] | 0.00 |
| CHLORITIC_GNEISS | [12] | 0.00 |
| CHLORITIC_GNEISS | [160] | 0.01 |
| CHLORITIC_GNEISS | [102] | 0.00 |
| CHLORITIC_GNEISS | [529] | 0.02 |
| CHLORITIC_GNEISS | [1] | 0.00 |

## TABLE 16.5
### Result of the Spectral Mapping Methods with 100% Weight on BE

| End Members | Spectral Method | SAM | BE | SFF |
|---|---|---|---|---|
| | Weight | 0 | 1 | 0 |
| Class 4 | Meta_crs_spec_library | | CHLORITIC GNEISS | |
| Class 6 | Meta_crs_spec_library | | PURPLE QUARTZITE | |
| Class 14 | Meta_crs_spec_library | | GREEN QUARTZITE | |
| CLASS 16 | Meta_crs_spec_library | | CHLORITIC GNEISS | |
| CLASS 17 | Meta_crs_spec_library | | CHLORITIC GNEISS | |
| CLASS 18 | Meta_crs_spec_library | | GREEN QUARTZITE | |
| CLASS 19 | Meta_crs_spec_library | | GRAY QUARTZITE | |
| CLASS 20 | Meta_crs_spec_library | | GRAY QUARTZITE | |
| CLASS 22 | Meta_crs_spec_library | | GRAY QUARTZITE | |
| CLASS 23 | Meta_crs_spec_library | | PURPLE QUARTZITE | |
| CLASS 24 | Meta_crs_spec_library | | GREEN QUARTZITE | |
| CLASS 25 | Meta_crs_spec_library | | CHLORITIC GNEISS | |
| CLASS 27 | Meta_crs_spec_library | | GREEN QUARTZITE | |
| CLASS 28 | Meta_crs_spec_library | | GREEN QUARTZITE | |
| CLASS 29 | Meta_crs_spec_library | | GREEN QUARTZITE | |
| CLASS 30 | Meta_crs_spec_library | | GREEN QUARTZITE | |
| CLASS 32 | Meta_crs_spec_library | | GREEN QUARTZITE | |
| CLASS 33 | Meta_crs_spec_library | | GREEN QUARTZITE | |
| CLASS 35 | Meta_crs_spec_library | | GREEN QUARTZITE | |
| CLASS 36 | Meta_crs_spec_library | | GREEN QUARTZITE | |
| CLASS 37 | Meta_crs_spec_library | | GRAY QUARTZITE | |
| CLASS 38 | Meta_crs_spec_library | | CHLORITIC GNEISS | |
| CLASS 39 | Meta_crs_spec_library | | PURPLE QUARTZITE | |
| CLASS 40 | Meta_crs_spec_library | | CHLORITIC GNEISS | |
| CLASS 42 | Meta_crs_spec_library | | CHLORITIC GNEISS | |
| CLASS 44 | Meta_crs_spec_library | | GRAY QUARTZITE | |

# Evaluation of Spectral Mapping Methods of Mineral Aggregates and Rocks 267

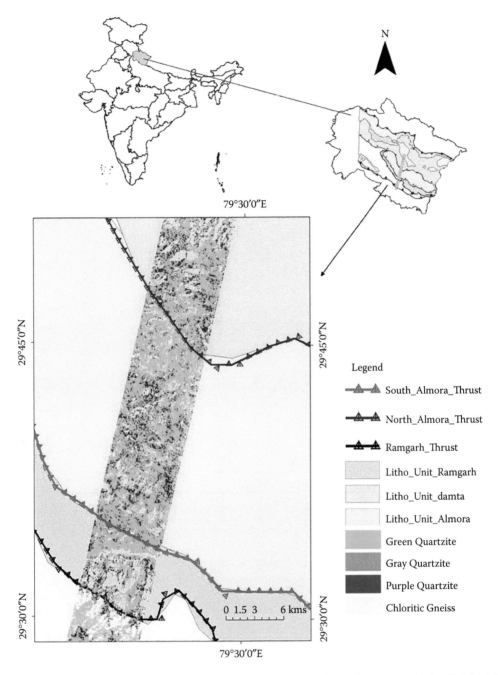

**FIGURE 16.7** Classified image of Hyperion data according to the 100% weightage provided to BE classifier spectral mapping method.

## TABLE 16.6
### Percentage of Spectral Endmember Classes in BE Classification

| Classes | Pixel Count | Percentage |
|---|---|---|
| Unclassified | [5590062] | 92.25 |
| GREEN_QUARTZITE | [2039] | 0.03 |
| CHLORITIC_GNEISS | [283] | 0.01 |
| CHLORITIC_GNEISS | [355] | 0.01 |
| GREEN_QUARTZITE | [18026] | 0.30 |
| GRAY_QUARTZITE_ | [686] | 0.01 |
| GRAY_QUARTZITE_ | [57] | 0.00 |
| GRAY_QUARTZITE_ | [209] | 0.00 |
| PURPLE_QUARTZIT | [121136] | 2.00 |
| GREEN_QUARTZITE | [85968] | 1.42 |
| CHLORITIC_GNEISS | [57] | 0.00 |
| GREEN_QUARTZITE | [3] | 0.00 |
| GREEN_QUARTZITE | [1071] | 0.02 |
| GREEN_QUARTZITE | [11993] | 0.20 |
| GREEN_QUARTZITE | [58563] | 0.97 |
| GREEN_QUARTZITE | [822] | 0.01 |
| GREEN_QUARTZITE | [2829] | 0.05 |
| GREEN_QUARTZITE | [844] | 0.01 |
| GREEN_QUARTZITE | [3075] | 0.05 |
| GRAY_QUARTZITE_ | [237] | 0.00 |
| CHLORITIC_GNEISS | [27] | 0.00 |
| PURPLE_QUARTZIT | [22] | 0.00 |
| CHLORITIC_GNEISS | [17] | 0.00 |
| CHLORITIC_GNEISS | [5660] | 0.09 |
| GRAY_QUARTZITE_ | [7] | 0.00 |
| CHLORITIC_GNEISS | [277] | 0.01 |

*Case 3*

Because the SFF tries to fit the spectral curve onto the spectral projection of the endmember, a single-range SFF can lead to an erroneous interpretation; however, if the absorption patterns of the target material are known, multiple SFF can be used. Absence of that information leads to a different scenario in which SFF (single range) has been excluded from the mapping method and the result should be observed by considering the other two spectral mappings with a varied proportion of importance.

A weight factor of 40% SAM and 60% of BE gave us the following result (Figure 16.8 and Table 16.7). In the spectral classification of 60% weight factor on BE and 40% on spectral angle, surface spectral classification with repetitive occurrence of gneissic and phyllitic zones and uniform distribution of quartzitic units was observed.

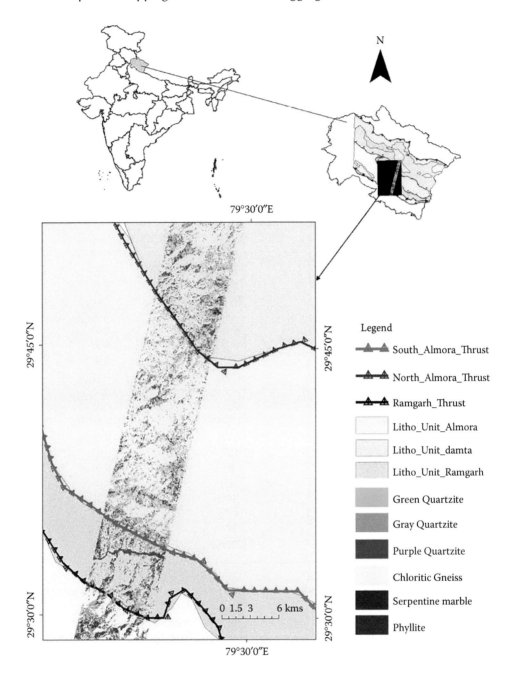

**FIGURE 16.8** Classified image of Hyperion data according to a weightage factor of 0.6 (60%) of binary encoding and 0.4 (40%) of SAM classifier.

### TABLE 16.7
### Percentage of Endmember Considering SAM and BE Methods

| Classes | Pixel Count | Percentage |
|---|---|---|
| Unclassified | [2296882] | 94.53 |
| GRAY_QUARTZITE | [154] | 0.01 |
| GRAY_QUARTZITE | [161] | 0.01 |
| GRAY_QUARTZITE | [172] | 0.01 |
| PURPLE_QUARTZIT | [41031] | 1.69 |
| GREEN_QUARTZITE | [33842] | 1.39 |
| GREEN_QUARTZITE | [4] | 0.00 |
| GREEN_QUARTZITE | [1817] | 0.08 |
| GREEN_QUARTZITE | [3415] | 0.14 |
| GREEN_QUARTZITE | [164] | 0.01 |
| GREEN_QUARTZITE | [52] | 0.00 |
| GREEN_QUARTZITE | [14] | 0.00 |
| GREEN_QUARTZITE | [2483] | 0.10 |
| GREEN_QUARTZITE | [7022] | 0.29 |
| GRAY_QUARTZITE | [7] | 0.00 |
| PURPLE_QUARTZIT | [6] | 0.00 |
| GRAY_QUARTZITE | [102] | 0.00 |
| PURPLE_QUARTZIT | [1] | 0.00 |
| GREEN_QUARTZITE | [159] | 0.01 |
| GREEN_QUARTZITE | [14510] | 0.60 |
| CHLORITIC_GNEISS | [528] | 0.02 |
| CHLORITIC_GNEISS | [190] | 0.01 |
| CHLORITIC_GNEISS | [169] | 0.01 |
| CHLORITIC_GNEISS | [52] | 0.00 |
| CHLORITIC_GNEISS | [8] | 0.00 |
| CHLORITIC_GNEISS | [12] | 0.00 |
| CHLORITIC_GNEISS | [160] | 0.01 |
| MARBLE_META | [4776] | 0.20 |
| MARBLE_META | [1] | 0.00 |
| PHYLLITE_META | [22016] | 0.91 |

The thematic maps of the endmember classes obtained can be observed to vary with any change of weight or exclusion of any of the spectral mapping methods in the spectral analysis. Evaluation of the efficiency of the spectral mapping method required a match with spectral classes obtained from the field reflectance data.

#### 16.5.1 FIELD SPECTRA DATA

Spectra of the rocks of Almora quartzite regions were collected by a joint expedition of TERI School of Advanced Studies and Defence Terrain Research Laboratory (DTRL) of Defence Research Development Organisation (DRDO) of the Ministry of Indian Defence in the November 2016.

The field spectra were mostly of Almora quartzite schist and Damta quartzite–shale phyllite and were not covered by the Hyperion scene but were used to observe how much the lithological unit spectrally matched the field spectra.

When the field spectra were selected as the reference spectral library for SAM classification for the rock surface between NAT and SAT, it was observed that spectral classes showed the presence of quartzites and quartzite–shale intercalations in the study area (Figure 16.9 and Table 16.8).

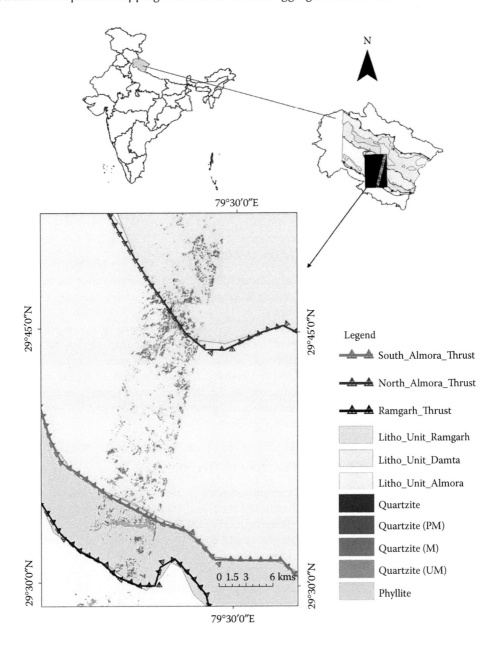

**FIGURE 16.9** Classified image of Hyperion data according to field spectral data of phyllite and quartzite data collected using ASD-Spectroradiometer.

**TABLE 16.8**
**The Class Distribution Result of the Field Spectral Mapping for the Study Area between North and South Almora Thrusts**

| Classes | Pixel Count | Percentage |
| --- | --- | --- |
| Unclassified | [5985761] | 98.78 |
| QUARTZITE_15C80 | [444] | 0.01 |
| QUARTZITE_PM_15 | [13] | 0.00 |
| QUARTZITE_M_15C | [549] | 0.01 |
| QUARTZITE_U_15C | [42244] | 0.70 |
| PHYLLITE_SHALE | [30467] | 0.50 |

### 16.5.2 Evaluation of the Spectral Mapping Method

The field spectral data show a much more accurate distribution of quartzite over the thrust zones with localized quartzite–phyllite zones south and north of Almora Thrusts, indicating stress-rich mineralization of foliated metamorphic rocks.

It was observed that the results correspond highly with the second and third scenarios in spectral analyst method, that is, consideration of SAM and BE. The reason may correspond to the fact as these two methods seek to find the degree of correlation in terms of spectral angle (radian) and similar measurement of Hamming distance, respectively. SFF as considered over the whole range as a single unit did not yield a result because of the fact that absorption patterns and ranges were not taken into account. Prior field knowledge regarding the target medium and absorption feature ranges for multiple-range SFF can provide better results.

## 16.6 GEOLOGICAL INTERPRETATIONS OF THE RESULTS

It is apparent that the lithounit under the Hyperion scene is rich in quartzite (BE classification), and the area has a repetition of gneissic bands (chloritic) and a fair amount quartzite phyllite (field spectra). This fact indicates that this lithozone belongs to the foliated metamorphic classes of ductile nature. The metamorphic texture like gneiss and phyllite shows that these units come under stable ductile assemblages of aseismic zones (Wise et al., 1984; Winter, 2010). However, they occur in repetitive zones, which suggests that the existence of seismic zone assemblages of non-foliated, fault structure-related cataclastic (brittle deformation) and mylonitised units (Passchier and Trouw, 2005) might be present within this zone, comply with supporting theory that thrust belts pass through this region. The Hyperion scene after spectral mapping has zoned out this repetitive band of gneiss and phyllites, but due to the lack of availability of a spectral library of cataclastic rocks or mylonitised units, it is not possible to spectrally delineate the thrust zones with higher accuracy.

The variation in quartzite lithounit (green quartzites) as per the BE classifier suggests that the lithounit of Almora quartzite have underwent a change, which may or may not be directly link with the thrusting but there is a probability. The only way to confirm is to study the rock petrographic section and mineralogical analysis.

## 16.7 CONCLUSION

The different spectral methods used to evaluate the lithological classes of the lesser Himalaya primarily of the Almora and Ramgarh section reveal the presence of stable gneissic and phyllitic textures in a quarzitic terrain. Mineral mapping suggests the terrain is mineralogically homogenous.

The lithounits as per literatures suggested that the Almora group of rocks is of quartzite biotite muscovite schist grade, whereas those of Ramgarh was of greenschist grade. The abundance of chloritic gneiss and quartzite suggest a similar result. Because the Spectral Library of John Hopkins University was used to identify the metamorphic spectral classes, which may or may not differ with the rock conditions in the field, and field spectra of the said zone were absent, the postclassification cannot be verified. The spectral mapping methods used in this study only after correlating with field geological map can be assessed for accuracy assessment; however, according to pertinent literature (Srivastava and Mitra, 1996; Valdiya, 1980), the geological map complies with spectral classes.

The weight factor for proper delineation of lithounits suggests that BE and SAM have a higher accuracy in identifying quartzite compared to SFF in spectral analyst tool because SFF was applied over a single range, which excludes the different absorption features of the rock surface over multiple ranges (Longhi et al., 2001).

There is also a probability that the reason the spectral mapping of endmembers differs in SFF compared to SAM and BE is that the terrain had a heterogeneous rock surface feature, metasedimentary units, quartzite shale intercalations, and phyllites as observed in field spectra data between the NAT and the SAT. So, it is necessary to cross check a highly heterogeneous surface scene with different mineral spectra libraries along with rock assemblage spectra. For an endmember, the score obtained from spectral matching methods against reference spectras of two different type of rock (sedimentary and metamorphic) spectra library will confirm the lithounit type. For heterogeneity in surface characterization, the spectral signatures obtained from the Hyperion scene needs to be treated as intimate, intrinsic spectral mixtures instead of a single spectral class. The spectral mapping for highly heterogeneous and nonuniform terrain needs a linear spectral separation into its component mineral phases.

The different gneissic and quartzitic zones delineated by the different spectral mapping methods with different weight factors reveal that there is the existence of strain zones as the repetition of gneissic and phyllite suggests, which could correlate with the thrust zones (Ramgarh, SAT, and NAT). Because the lithology over which it exists is not homogenous, that is, there is quartzite rich in muscovite and biotite schistosed units along with phyllite .Delineating such units undergoing thrust requires spectrally identifying the different mineralogical phases in similar type of lithological unit undergoing deformation.

Thus identification of the metamorphic rock assemblage through hyperspectral data of the lesser Himalaya (Kumaon) reveals a region with highly strained lithounit of gneissic quartzites and quartzite, the shale and phyllite intercalation and their spatial pattern suggest the existence of thrusting. To quantify the nature, orientation of thrusts and geothermal history of these metamorphic assemblages needs to be disintegrated into their mineral phases. Reflectance data form hyperspectral data can be used in the linear unmixing model into mineral phases using a mineral spectral library. Thus, spectral mapping of remote areas of geological interest will provide a building block for understanding the geothermal history of the area.

## REFERENCES

Apan, A., Held, A., Phinn, S., and Markley, J. 2003. September. Formulation and assessment of narrow-band vegetation indices from EO-1 Hyperion imagery for discriminating sugarcane disease. *Proceedings of the Spatial Sciences Institute Biennial Conference (SSC 2003): Spatial Knowledge without Boundaries*, pp. 1–13.

Baldridge, A.M., Hook, S.J., Grove, C.I., and Rivera, G. 2009. The ASTER spectral library version 2.0. *Remote Sensing of Environment*, 113:711–715.

Barry, P. 2001. EO-1/ Hyperion Science Data User's Guide, Level 1_B. TRW Space, Defence and Information Systems, Redondo Beach, CA.

Boardman, J.W., 1993, October. Automating spectral unmixing of AVIRIS data using convex geometry concepts, Fourth Annual JPL Airborne Geoscience Workshop, 11.

Datt, B., McVicar, T.R., Van Niel, T.G., Jupp, D.L.B., and Pearlman, J.S. 2003. Pre-processing EO-1 Hyperion hyperspectral data to support the application of agricultural indices. *IEEE Transactions on Geoscience and Remote Sensing,* 41(6):1246–1259.

Green, A.A., Berman, M., Switzer, P., and Craig, M.D. 1988. A transformation for ordering multispectral data in terms of image quality with implications for noise removal. *IEEE Transactions on Geoscience and Remote Sensing,* 26(1):65–74.

Ghose, N., Chakrabarti, B., and Singh, R. 1974. Structural and metamorphic history of the Almora group, Kumaun Himalaya, Uttar Pradesh. *Geology,* 4:171–174.

Grove, C.I., Hook, S.J. and Paylor III, E.D. 1992. *Laboratory reflectance spectra of 160 minerals, 0.4 to 2.5 micrometers.* Pasadena, CA: Jet Propulsion Laboratory.

Karunakaran, C., and Rao, A.R. 1979. Geological Survey of India. *Miscellaneous Publication,* 41 Part IV:1–66.

Kokaly, R.F., Clark, R.N., Swayze, G.A., Livo, K.E., Hoefen, T.M., Pearson, N.C., Wise, R.A. et al. 2017. USGS Spectral Library Version 7. *U.S. Geological Survey Data Series,* 1035:61. doi:10.3133/ds1035.

Kruse, F.A., Lefkoff, A.B., Boardman, J.W., Heidebrecht, K.B., Shapiro, A.T., Barloon, P.J., and Goetz, A.F.H. 1993. The spectral image processing system (SIPS)—Interactive visualization and analysis of imaging spectrometer data. *Remote Sensing of Environment,* 44(2–3):145–163.

Longhi, I., Sgavetti, M., Chiari, R., and Mazzoli, C. 2001. Spectral analysis and classification of metamorphic rocks from laboratory reflectance spectra in the 0.4–2.5 μm interval: A tool for hyperspectral data interpretation. *International Journal of Remote Sensing,* 22(18):3763–3782.

Mazer, A.S., Martin, M., Lee, M., and Solomon, J.E. 1988. Image processing software for imaging spectrometry data analysis. *Remote Sensing of Environment,* 24(1):201–210.

Passchier, C.W., and Trouw, R.A.J. 2005. *Microtectonics.* 2nd ed. Berlin, Germany: Springer–Verlag.

Rupke, J. 1974. Stratigraphic and structural evolution of the Kumaon Lesser Himalaya. *Sedimentary Geology,* 11(2–4):8189–87265.

Srivastava, P., and Mitra, G. 1994. Thrust geometries and deep structure of the outer and lesser Himalaya, Kumaon and Garhwal (India): Implications for evolution of the Himalayan fold-and-thrust belt. *Tectonics,* 13(1):89–109.

Srivastava, P., and Mitra, G. 1996. Deformation mechanisms and inverted thermal profile in the North Almora Thrust mylonite zone, Kumaon Lesser Himalaya, India. *Journal of Structural Geology,* 18(1):27–39.

Trouw, R.A.J, Passchier, C.W, and Wiersma, D.J. 2009. *Atlas of Mylonites-and Related Microstructures.* New York: Springer Science & Business Media.

Valdiya, K.S. 1980. *Geology of Kumaun lesser Himalaya.* Wadia Institute of Himalayan Geology, Uttarakhand, India.

Viterbi, A.J., and Omura, J.K. 1979. *Principles of Digital Modulation and Coding.* Mineola: Courier Corporation.

Winter, J.D. 2010. *An Introduction to Igneous and Metamorphic Petrology.* New York: Prentice Hall.

Wise, D.U., Dunn, D.E., Engelder, J.T., Geiser, P.A., Hatcher, R.D., Kish, S.A., and Schamel, S. 1984. Fault-related rocks: Suggestions for terminology. *Geology,* 12(7):391–394.

# 17 Assessment of Flood-Emanated Impediments to Kaziranga National Park Grassland Ecosystem—A Binocular Vision with Remote Sensing and Geographic Information System

*Surajit Ghosh, Subrata Nandy, Debarati Chakraborty, and Raj Kumar*

## CONTENTS

| | | |
|---|---|---|
| 17.1 | Grassland Ecosystem | 276 |
| 17.2 | Kaziranga National Park | 277 |
| | 17.2.1 History and Geographic Association | 277 |
| | 17.2.2 Topography and Climate | 278 |
| | 17.2.3 Vegetation and Ecosystem in Kaziranga National Park | 280 |
| | 17.2.4 Components of Grassland Ecosystems of Kaziranga National Park | 280 |
| | 17.2.5 Conservation Areas | 280 |
| 17.3 | Brahmaputra River | 281 |
| 17.4 | Flooding and Kaziranga National Park | 281 |
| 17.5 | Role of Space Technology | 282 |
| | 17.5.1 Land Use and Land Cover Mapping | 283 |
| | 17.5.2 Flood Mapping through Active Microwave Sensor | 284 |
| | 17.5.3 River Water Level Monitoring | 286 |
| 17.6 | Remote Sensing and Geographic Information System Integration: Towards Near Real-Time Monitoring and Assessment | 287 |
| | 17.6.1 Mobile Geographic Information System | 287 |
| | 17.6.2 Android | 288 |
| | 17.6.3 Overall System Architecture Design and Deployment | 288 |
| 17.7 | Conclusion | 288 |
| References | | 289 |

**ABSTRACT** Kaziranga National Park (KNP), a UNESCO World Heritage Site, harbors one of the last unmodified natural grasslands of Northeast India. The park has proudly been the humble abode of two remarkable megaherbivores, the great Indian rhino and the elephant, and also provides a prime habitat for the Royal Bengal Tiger. It is a haven to various economically important graminoids. Each year, KNP's grassland ecosystem encounters a battle with flood affecting its plant and animal components. Remote sensing and geographic information systems can offer a binocular viewpoint for regular assessment of the park's flood-prone areas, which will contribute to long-term conservation practices and better natural resource management of its grassland ecosystem.

Natural disasters are the leading causes of natural and private property damage. They also lead to long-term adverse effects on biodiversity of the damaged area. Water-related hazards are responsible for about 90% of all natural hazards. Flood is a frequent event in the lower catchment of the Brahmaputra where KNP is situated. The park's land cover has ~74% of grasslands (excluding river and other water bodies), making it a unique habitat for great Indian one-horned rhino. Flood has adversely affected the park's alluvial grassland, which emerged as one of the leading causes of the inundation of the vast grasslands causing habitat degradation and fragmentation. Habitat loss is one of the significant threats to the biodiversity. As flood always disrupts the natural habitat, it unfailingly impacts the rhino feeding habitat, which consequently takes a considerable time for recovering its preflood biodiversity status. According to the red list category of the International Union for Conservation of Nature and Natural Resources, the rhino comes under the vulnerable B1ab (iii) category and, hence, the annual impact of the flood on its prime feeding habitat (grasslands) is a matter of grave concern.

In the research reported in this chapter, different remotely sensed data sets such as SARAL/AltiKa, Landsat 8, and Sentinel 1A were used to assess the flood situation of the park during August 2017. SARAL/AltiKa was used to obtain the water level of that Brahmaputra River, Landsat 8 OLI data were used for making the park's land use and land cover (LULC), and Sentinel 1A was employed to detect the study area's flood situation. The advantages of integrating real-time assessment data for better management of the park's grassland ecosystem are thoroughly discussed.

## 17.1 GRASSLAND ECOSYSTEM

Grasses are one of the most versatile life forms on the earth. The grassland ecosystem represents the world's major food and biodiversity resources. Grassland ecosystems occupy about 25% of earth's vegetation cover. They occur in every continent of the planet except Antarctica. They flourish on almost all soil types and thrive well in all climatic zones. Grasslands thrive in a vast temperature range from temperate to tropical with mean annual temperatures varying from near 0°C to around 26°C. Mean rainfall per annum in arid grasslands varies from about 250 mm/yr to even well over 1000 mm/year in mesic grasslands. A grassland is characterized by a diverse assemblage of plants dominated by large rolling fields of grasses (Poaceae) and sedges (Cyperaceae) and may include a nonwoody, nongraminoid, herbaceous vegetation cover of rushes (Juncaceae) and forbs/phorbs (Nippert et al., 2013). All grasslands support an array of animal community, including some of the most species-rich grazing food webs of the planet. Grasslands often exhibit appreciative belowground plant biomass (high root: shoot ratio). Consequently, they have proportionally high inputs of plant root litter in comparison to surface litter. This often in combination with the comparatively slow rate of decomposition due to periodic water limitation causes high organic matter and nutrient accumulations in the soil (Nippert et al., 2006). The ensuing high fertility of grassland soils is one of the reasons they have been so widely exploited for agricultural purposes.

The three prime factors controlling the origin, maintenance, and structure of natural grasslands are climate, fire, and grazing (Nippert et al., 2013). Majority of the grassland ecosystems have co-evolved with large body-sized grazers (Stebbins, 1981). Another important characteristic of grasslands is that it regularly experiences extremely high intra- and interannual variability in rainfall. In comparison to other biomes, grasslands are more responsive to variation in rainfall amounts (Nippert et al., 2006). The existence of grasslands in areas experiencing high rainfall, signifies either the clear felling of forests or the interplay of edaphic and fluvial factors (e.g., Terai grasslands, northern India). Maintenance of such mid-successional grasslands, specifically when these also act as a wildlife habitat, depends upon their careful planning and management practices.

Moreover, grasslands share an old and intricate relationship with humans (Nippert et al., 2013). All major cereals—maize (corn), rice (paddy), wheat, barley, and sorghum—are grass. From the very beginning of the Neolithic revolution, grasses have been a consort of humans in a true sense. They are needed in almost every aspect of life ranging from food to fodder and construction materials to decorative to medicinal purposes. Hence, grasslands were not only exploited by microbes, plants, and animals alone but also by humans from the very beginning of our civilization.

Indian grasslands have evolved at various altitudes and in different geographical zones under diverse climate regimes, and each harbors a rich array of flora and fauna (Singh et al., 1983). About 24% of the geographical areas of India are occupied by grasslands. They also aid high density of domestic livestock. Major types of Indian grasslands are Alpine moist meadows (Greater Himalaya), Alpine arid pastures or steppes (trans-Himalaya), hillside grasslands (mid-elevation ranges of Himalaya), chaurs (Himalayan foothills), wet-alluvial or Terai grasslands of Gangetic and Brahmaputra flood plains, phumdi (Manipur), Banni and Vidis (Gujarat), savannas (western and peninsular India), plateaus and valley grasslands (Satpuras and Maikal hills), dry grasslands (Andhra Pradesh and Tamil Nadu), and Shola grasslands of Western Ghats. India has nearly 95 National Parks and 500 wildlife sanctuaries. Only a handful of these protected areas have grasslands. Notable ones are Velavador National Park and Sailana Florican Sanctuary in Gujarat, Desert National Park in Rajasthan, Kaziranga National Park (KNP) and Manas Tiger Reserve in Assam, and Keibul Lamzao National Park in Manipur.

## 17.2 KAZIRANGA NATIONAL PARK

### 17.2.1 History and Geographic Association

KNP is one of the oldest Indian wildlife conservancy reserves (Figure 17.1). It was first notified in 1905. KNP was established as a reserved forest in 1908 with an area of 228.82 km$^2$. Later in 1974, it became a National Park under the Wildlife Protection Act (WPA) of 1972 with an area of 429.93 km$^2$. The present area of the park is now 859.4 km$^2$ (according to the Government Notification No. FOR/WL/722/48/45, dated 11-02-1974 effective 01-01-74) and is one of the most massive expanses of protected area in the sub-Himalayan zone (Vasu and Singh, 2016). The park is a critically important World Heritage Site. KNP extends from 26°30′–26°50′N latitude and 92°50′–93°41′E longitude and is positioned in the Kaziranga-Karbi Anglong landscape of Assam. It is situated within the Indo-Burma ecozone (Rodgers and Panwar, 1988). It is bounded on the north by the mighty Brahmaputra River and the Karbi Anglong hills in the south (Vasu and Singh, 2016).

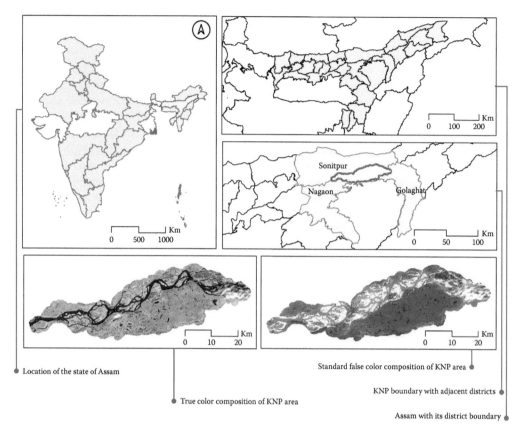

**FIGURE 17.1** Study area of Kaziranga National Park.

### 17.2.2 Topography and Climate

The entire KNP is situated on alluvial floodplain of the Brahmaputra and its tributaries with many meandering scars and scrolls. The ground of KNP consists of both new and old alluvium formed by annual replenishment of the Brahmaputra and is prone to inundation during floods (Figure 17.2). The soil of the park is formed of grey silt and fine-to-medium sands forming the recent composite floodplain. Alterations of the Brahmaputra River lead to striking examples of interactions of riverine

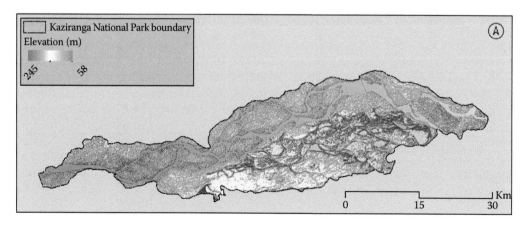

**FIGURE 17.2** Topography of Kaziranga National Park.

and fluvial processes, thus qualifying the park for World Heritage site designation under natural criteria (ii) of International Union for Conservation of Nature and Natural Resources (IUCN). Erosion and landmass formation is still ongoing along the northern boundary of KNP. Mild and dry winter occurs between November and February (5°C–25°C). Hot and humid monsoon persists in the park from June to October. July and August are the hottest months (maximum temperature 35°C).

KNP receives an average precipitation of 2452 mm (based on India's water portal data). Analysis of 100 years (1901–2000) of rainfall data using the Mann–Kendall (MK) test shows significant spatial and temporal (interannual) variability. Furthermore, rainfall data for the KNP region (Figure 17.3) shows a declining trend. The nonparametric MK test was applied in this study because it is recommended for trend analysis of hydrometeorological data (Li et al., 2009). The rate of decline of precipitation for the region is 0.89 mm yr$^{-1}$. Furthermore, around 67% of the precipitation is concentrated in the months of June to September (southwest monsoon). Overall Nagaon district receives above average rainfall (Figure 17.4), but the spatial and temporal (interannual) variability is significantly high. Due to the rise and fall of the riverine water table, KNP is submerged during monsoon (Vasu and Singh, 2016).

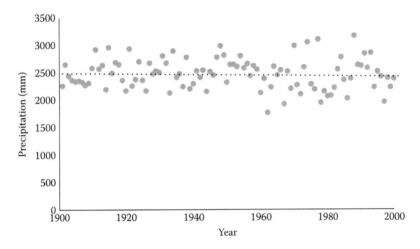

**FIGURE 17.3** Rainfall trend of the Kaziranga National Park region.

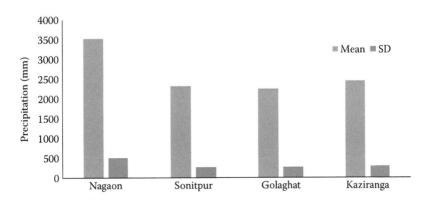

**FIGURE 17.4** Rainfall distribution of the Nagaon, Sonitpur, Golaghat, and Kaziranga districts.

### 17.2.3 VEGETATION AND ECOSYSTEM IN KAZIRANGA NATIONAL PARK

Kaziranga is situated in the Brahmaputra valley and is mostly composed of vast stretches of marshlands, tall grasses, dense forests patches, and numerous small water bodies crisscrossed by four main rivers—the Brahmaputra, Diphlu, Mora Diphlu, and Mora Dhansiri. It forms a part of the northeast Brahmaputra Valley Province. The land is quite level all over the park, and the landscape with its scattered trees and grasslands is a look-alike of the typical savannah. These picturesque swamplands are dotted with water hyacinth, water lilies, and lotus plants. The park is situated against the backdrop of the Mikir and Karbi Anglong hills.

The dominant biomes of KNP include Assam alluvial plains, semievergreen forests, tropical moist mixed deciduous forests, eastern Dillenia swamp forests, wetlands, and eastern wet alluvial grasslands (Champion and Seth, 1968). However, the main flora of the park is the dense, tall, and hygrophilous elephant grass. These are considered as wet savannahs that adapted to the very wet climate and evolved due to high seasonal fluctuations in rainfall (Vasu and Singh, 2016).The tall grass stands are intermixed with small swamplands generated by the receding floodwaters of the river. Usually swamp areas are beels, or low-lying areas submerged for some months during floods, abandoned river channels, and so on.

About 74% area, especially the western part (excluding river and other water bodies) of KNP, includes grassland, making the park a unique haven for great Indian one-horned rhino. The wet alluvial KNP grassland can be visibly sorted into tall and short grass patches and savannas. Tall "elephant" grasses flourish on the higher grounds and dominate more than 61% of the park area whereas short grasses on the lower grounds embracing the water bodies or beels occupy only 3% (Khatri and Barua, 2011). Their population has been retained by annual flooding and burning over the years. During the Mora Diffolu's backflow from the west, large areas in the western range are initially inundated, making an ideal environment for the development of short grass areas along the water bodies (Vasu and Singh, 2016).

### 17.2.4 COMPONENTS OF GRASSLAND ECOSYSTEMS OF KAZIRANGA NATIONAL PARK

To understand any ecosystem or predict the probable aftermath of disturbance it requires identification of the most important interactive chains within it. KNP grasslands may be considered as disturbance-dependent communities because they are frequently subjected to the regular events of annual floods and activities of grazers. The major herbivores of KNP grasslands are Indian rhino (*Rhinoceros unicornis*), Indian elephants (*Elephus maximus*), Asiatic wild buffalo (*Bubulas bubalis*), gaur (*Bos gaurus*), hog deer (*Axis porcinus*), sambar deer (*Cervus unicolor*), swamp deer or Barasingha (*Cervus duvauceli*), and so on. Rhinos being megagrazers exert high influence on their feeding habitat.

A recent study clearly demonstrates rhinos role in shaping and maintaining grassland ecosystem diversity (Cromsigt and Te Beest 2014). They are primarily grazers; hence, Indian rhinos of KNP are most dependent on the grassland ecosystem of the park. Thus, how this mega grazer is affected by regular flood events that submerge their prime habitat and how they cope with this is a matter of great interest.

KNP grasslands further extend support to fringe villages of the Mishing tribe. They rely on the park for meeting a variety of needs including firewood, agricultural implements, medicine, detergents, construction materials, fishing, forest food, implements, country drinks, fiber, spices, brooms, and thatch grasses (Kutum et al., 2011).

### 17.2.5 CONSERVATION AREAS

KNP is the humble abode of world's highest Indian rhino and Asiatic wild buffalo population. With about 1165 individuals, the elephant population of KNP is one of the most remarkable populations of the animal in Asia (Singh, 2017). KNP braces more than 35 mammalian species of which 15 are included in Schedule I of the WPA of 1972. It provides the optimal habitat for the Royal Bengal Tigers

(*Panthera tigris*) to attain their highest ecological density (currently 116; Singh, 2017). Although KNP exhibits a range of natural values and provides shelter for many threatened species and migratory birds, the showstopper of the park is the Indian rhino (*Rhinoceros uncornis*) with a population of 2401. KNP is the foothold of the one-horned Indian rhino (Singh, 2017). Being haven to two-third of world's great one-horned rhinos, KNP also fulfills natural criteria (iv) of the IUCN for qualifying as a World Heritage site. The grassland ecosystem of KNP is highly threatened due to uncontrolled agricultural expansion, conversion into woodlands, and degradation due to invasion by exotic aggressive weeds (Vasu and Singh, 2016). Poaching is also a matter of grave concern for the park authority.

## 17.3 BRAHMAPUTRA RIVER

The Brahmaputra River is the world's sixth largest in terms of water resources, carrying an estimated 629.05 km$^3$ of water annually. Of the total 2906 km length of the Brahmaputra, 918 km flows through India, including 640 km through the state of Assam alone. The Brahmaputra has 41 tributaries, 26 flowing in the north bank and 15 in the south bank. The Brahmaputra valley has the average width of 80–90 km whereas the river itself is 6–10 km wide. The average rainfall varies between 2480 and 6350 mm. Major natural causes of flood are geomorphology and geology of the region, physiographic and seismic condition of the valley, and excessive rainfall. Whereas the anthropogenic causes contributing to the situation are drainage congestion due to man-made embankments, human encroachment of riverine areas is also a problem (Floods in Assam, 2017).

The Brahmaputra basin has been widening continuously since the last century. The mighty Brahmaputra is infamous for being flood prone and causing erosion. Of the eight northeastern states of India, Assam is the most affected by flood and erosion. These two hazards severely affect the economic, political, social, and cultural milieu of Assam.

For flood mitigation in the Brahmaputra valley, multipurpose dam (capable of hydroelectric generation) construction has been recommended. However, the idea has been proven to be contradictory and not viable in itself in many previous similar projects. In case of the mighty Brahmaputra river, this is likely to prove even more disastrous considering its characteristics described previously. The flood banks of Brahmaputra River are characterized by its enormous volume of sediment load leading to continuous changes in channel morphology, stream bank erosion, rapid bed aggradation, and bank line recession and erosion (Sarma et al., 2007). The river has many braided channels along its course in the alluvial plains. Bank oscillation causes a shifting of outfalls of its tributaries that bring newer areas under water (Kotoky et al., 2005). The total maximum daily sediment load is a significant source of sediment and nutrient pollution adversely affecting the water quality, vegetation composition, and fish spawning habitat.

## 17.4 FLOODING AND KAZIRANGA NATIONAL PARK

Floods are an integral part of hydrological cycles and can be defined as extremely high flows or levels of rivers that inundate floodplains or terrains bordering major river channels. Floods are considered to be the most devastating and recurrent natural disaster, causing major damage to floodplains (Dhar and Nandargi, 2003). In India, floods are broadly classified into three categories, namely riverine, dam-break, and storm-surge. As per the 11th plan working group (Planning Commission, 2007), the total flood-affected area in India is 456,400 km$^2$, of which 5140 km$^2$ lies in Assam, accounting for 2.3% of its total geographical area.

Although the annual average rainfall of the upper catchment area of the Brahmaputra and its tributaries do not show much variability, flooding intensity in its basin often varies because of the intensity of rain in concentrated spells. In the event of a flood, the floodwater from the Brahmaputra River enters the park from the northern boundary and through the Diphlu and the Mori Dhansiri rivers. While the water from the submerged high lands recedes quickly, the low lying areas of the tiger reserve form basins, especially around the beels of the southern boundary and the western part of the park remains under water for a substantial period of time even after flood waters have receded in other places. There are approximately 60 rivers and small water channels and 200 water bodies in the core area of KNP. Flooding is an integral part of the ecological system of the tiger reserve. It not only helps in maintenance of vegetation status but also contributes to the process of silt deposition and soil formation, which replenish the dry beels and adds fertility to the soil. Flooding also maintains water quality by catalyzing the clearing of water hyacinths in stagnated waters. The beels and rivers of Kaziranga are the home of several freshwater fish species. Due to annual flooding and stagnation of water in beels for other periods of the year, Kaziranga is regarded as the breeding ground for tropical freshwater fish. Although not well studied, the function of Kaziranga as a nursery for freshwater fishes is highly significant considering its preference as a protein source for the people living in that part of the world. However, flooding in KNP is not without its negative impact on wildlife with regard to the shortage of fodder, malnutrition, and animal casualty.

The most significant stresses on KNP grasslands are the restriction of fields to protect the regularly expanding populace of the one-horned Indian rhinoceros, erosion that has caused the loss of an immense area in the last few decades, and the siltation resulting in the lower water retention capacity. Flood is a regular event in the lower catchment of the Brahmaputra where KNP is situated. The increasing level of multiwave floods in last decade is threatening the future of the park. Flood adversely affects the alluvial grassland of KNP. It causes habitat degradation as well as fragmentation. Some plants in KNP capable of partially tolerating submergence are *Acorus calamus, Brachiaria eruciformis, Saccharum narenga, Murdannia nudiflora,* and *Oryza rufipogon.* However, the plants' density, use in animal diets, and availability during floods still lack study.

Due to recurrent flood causing enhanced siltation, many beels in KNP have shrunk. Moreover, sand deposition in areas harboring small grass has contributed to degrading their use for grazers. The grasses of these regions while emerging through sandy deposits, become coarse, thick, and sometimes unpalatable to the foragers.

During flood periods, most of the animals including the rhinos must migrate from the park and take shelter on the nearby high grounds in Karbi Anglong Hills. The three major of the wildlife corridors still existing in the park connecting Kaziranga to the Karbi Anglong Hills (Menon et al., 2005) are (1) the Panbari corridor (in the east), (2) the Kanchanjuri corridor (in the west), and (3) the Kukurakata-Bagser (Burapahad) corridor (also in the west). But recent increases in the number of tea gardens, human habitations, and enhanced agrarian activities along the southern boundary of the park have made the food availability and mobility of wild animals through the wildlife corridors to the hills during high flood season increasingly difficult. Moreover, prolonged periods of floods causing submergence severely threaten the biodiversity status of grassland and beels. Prolonged flooding also leads to severe degradation of plant biomass, and the entire ecosystem takes time to return to its previous plant biodiversity status, making the herbivores and megaherbivores of KNP more vulnerable.

## 17.5 ROLE OF SPACE TECHNOLOGY

The earth is composed of around 70% water of which only approximately 3% is freshwater. Even from the available 3% freshwater around 99% is stored in glaciers, ice caps, and groundwater. Only the remaining 1% of the freshwater is available in lakes, ponds, rivers, and so on. This limited availability of freshwater is also due to the ever-increasing population and changing climate. Moreover, the availability of water varies spatially and temporally. Considering these facts, the conservation of the water bodies is a matter of grave concern. The traditional methods of assessing and monitoring water resources are time consuming, costly, and limited to accessible areas. However, the advent of satellite-based remote sensing to map the dynamics of surface water bodies in any part of the globe has made monitoring easier because its use provides synoptic and dynamic coverage of the earth (Thakur et al., 2017). The use of remote sensing and geographic information system (GIS) provides

### TABLE 17.1
### Key Studies in Kaziranga National Park Using Remote Sensing and GIS

| Sl. No. | Key Issues Addressed | References |
|---|---|---|
| 1 | Land cover change and habitat modeling | Kushwaha et al. (2000) |
| 2 | Habitat modeling | Kushwaha and Roy (2002) |
| 3 | Land cover change | Kushwaha (2008) |
| 4 | Rapid assessment of flood | Ghosh et al. (2016) |

near real-time monitoring of the extent of floods and of water levels. Overlaying the flood area on the land use and land cover (LULC) provides information about flood-affected areas. The delineation of flood-prone areas is one of the most important steps in flood management. Therefore, space technology can contribute substantially to flood disaster management. In addition, many of these datasets are freely accessible. Thus, the main aim of this study was to use a freely available remote sensing data set to assess the 2017 flood situation of KNP. Key studies in KNP using remote sensing and GIS are listed in Table 17.1.

### 17.5.1 LAND USE AND LAND COVER MAPPING

Because of its temporal and spatial coverage, geospatial technology proves to be an important asset for monitoring over time. Begun in July 1972, Landsat currently is longest running program collecting satellite imagery of the earth. The Landsat 8 satellite has acquired images of the earth continuously since February 11, 2013. It has two sensors, the Operational Land Imager (OLI) and Thermal Infrared Sensor (TIRS).

The current study used Landsat 8 satellite data for LULC analysis. Its free satellite data were downloaded from USGS EarthExplorer website (http://earthexplorer.usgs.gov/). OLI data were used to map the LULC of KNP. The band description of Landsat 8 OLI sensor is given in Table 17.2. To generate LULC classes, six bands (band 2 to band 7) of Landsat 8 data (geo coded with Universal Transverse Mercator [UTM] projection, spheroid and datum WGS 1984,

### TABLE 17.2
### Landsat 8 Band Description

| | | | | Band Combination for | | | | |
|---|---|---|---|---|---|---|---|---|
| Sl. No. | Spectral Band | Wavelength (µm) | Spatial Resolution (m) | False Color 5, 4, 3 | Agriculture 6, 5, 2 | Water 5, 6, 4 | Urban 7, 6, 4 | True color 4, 3, 2 |
| 1 | Band 1—coastal/ aerosol | 0.433–0.453 | 30 | | | | | |
| 2 | Band 2—blue | 0.450–0.515 | 30 | | | | | |
| 3 | Band 3—green | 0.525–0.600 | 30 | | | | | |
| 4 | Band 4—red | 0.630–0.680 | 30 | | | | | |
| 5 | Band 5—near Infrared | 0.845–0.885 | 30 | | | | | |
| 6 | Band 6—shortwave infrared | 1.560–1.660 | 30 | | | | | |
| 7 | Band 7—shortwave infrared | 2.100–2.300 | 30 | | | | | |
| 8 | Band 8—panchromatic | 0.500–0.680 | 15 | | | | | |
| 9 | Band 9—cirrus | 1.360–1.390 | 30 | | | | | |
| 10 | Band 10—long-wave infrared | 10.30–11.30 | 100 | | | | | |
| 11 | Band 10—long-wave infrared | 10.30–11.30 | 100 | | | | | |

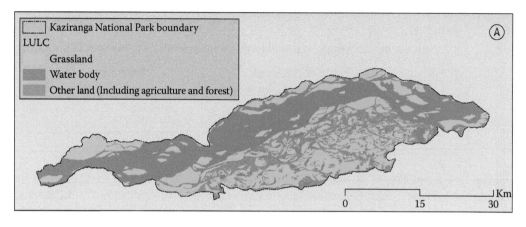

**FIGURE 17.5** Land use land cover map.

**TABLE 17.3**
**Land Use Land Cover Area in Percentage**

| Sl. No. | LULC Class | Area (%) |
|---|---|---|
| 1 | Grassland | 39.89 |
| 2 | Water body | 47.01 |
| 3 | Other land (including agriculture and forest) | 13.10 |

Zone 46 North) of the 2016 season were used. The supervised classification approach was used to make the LULC map. Three classes, grassland, water body, and other land (including agriculture and forest) were generated (Figure 17.5 and Table 17.3). A total number of 45 sample classes (15 for each class) were selected. Accurate classification depends on many factors such as time of satellite data and preprocessing. Collecting an ample number of homogenous training sites for each class is very important to obtain perfect classification. Assessment of classification results is important for the classification procedure. Classification accuracy was assessed through Google Earth imagery. Overall accuracy of approximately 87% was achieved.

### 17.5.2 Flood Mapping through Active Microwave Sensor

Remote sensing helps to monitor flood events in a cost-effective manner. Satellite images of different dates provide information about changes in flood patterns. The application of microwave remote sensing to map water is primarily used in flood inundation mapping because during the flood period, availability of cloud-free optical data is limited. The microwaves can easily penetrate the clouds, and the flooded region below the clouds can easily be identified (Thakur et al., 2017). The brightness of features in a radar image depends on the strength of the backscattered signal. The amount of energy that is backscattered depends on the illuminating signal (wavelength, polarization, viewing geometry, etc.) and the characteristics of the illuminated surface (roughness, shape, orientation, dielectric constant, etc.). The water surface behaves as being partially diffused and a partially specular reflector. Reflection produced specularly is similar at different wavelengths, but that of absorption and backscatter produce distinctive spectral signatures. The roughness of water surface also affects its reflectance properties. In radar images, the appearance of rough surfaces is bright and of smooth surfaces is dark. Smooth water surface acts as specular reflectors of radar waves and yields very low returns to the antenna, resulting in a dark appearance in the image whereas rough water surfaces return radar signals of varying strengths. Because water acts as a smooth surface, the dark pixels in the image represents water. Flooded areas can easily be identified in such a data set. For further information, one can see Thakur et al., 2017. In the current study, the Sentinel 1A Interferometric Wide Swath data set was used for detecting flood in the park. Sentinel 1A is a polar orbiting satellite that works day and night in the C- band range. Therefore, Sentinel data set is most suitable for assessing flood conditions, specifically in cloud cover condition. Data sets of February 25, 2017, and August 12, 2017, were used to assess the flood situation. These two data sets were chosen so that one set represents dry season and the other flooding season. The flood situation in Assam became very critical during the second week of August 2017 when it affected several districts of Assam. The expanse of Brahmaputra River increased by three to five times in some parts during this period (February to August 2017) (Figure 17.6). Approximately 70% of grassland was inundated on August 12, 2017 (Figures 17.5 through 17.7).

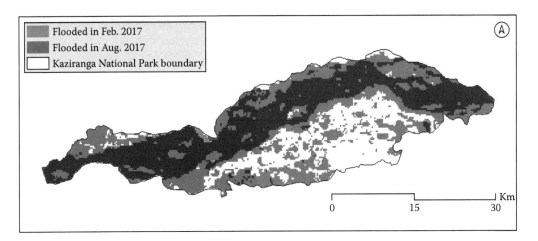

**FIGURE 17.6** Flood map of Brahmaputra River.

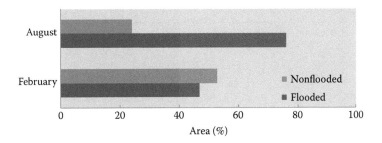

**FIGURE 17.7** Flood area statistics of grassland.

### 17.5.3 RIVER WATER LEVEL MONITORING

The satellite radar altimetry has the unique credibility and reliability to study oceanic sciences and continental hydrology. Remote sensing, especially altimetry from space, appears to provide complementary assistance for monitoring water level for the last three decades. Satellite with Argos and AltiKa (SARAL/AltiKa), launched on February 25, 2013, a joint venture of Indian Space Research Organisation (ISRO) and the Centre National d'Etudes Spatiales (CNES), is one of the pioneer missions in the history of satellite radar altimetry. It was the first monofrequency (Ka-band, 35.75 GHz) mission with the highest sampling rate (40 Hz) and lowest pulse width (2 ns) to monitor narrow rivers. The applications of radar altimetry in inland hydrology have been significantly increased in recent years in India after launching the SARAL mission.

Instantaneous water levels can be calculated by radar altimetry range measurement with centimeter level accuracy. The satellite altimetry takes measurements at its nadir with a footprint and averages everything in that footprint. The conventional pulse-limited altimeter measures the range between the satellite and the observed surface by using the time delay technique (Fu and Cazenave, 2001). The water level ($H$) above the reference ellipsoid is calculated using the altimetry-measured range ($R$), satellite altitude ($H_{sat}$), and correction factors (atmospheric and instrumental correction factors as well as Polar tide and solid earth tide effects) as follows:

$$H = H_{sat} - (R + C_{iono} + C_{dry} + C_{wet} + C_{st} + C_{pt}) \qquad (17.1)$$

where $C_{iono}, C_{dry}, C_{wet}, C_{st}$, and $C_{pt}$ denote the corrections applied because of the delay in pulse propagation through the ionosphere, humid and dry atmospheres ("dry" and "wet"); Polar tide and solid earth tide effects considered in the calculation of water level, respectively, were also considered in computing water level. $H_{sat}$ is the precise satellite altitude obtained from radio-positioning systems. In case of river water-level estimation, retracking algorithms are used to improve the estimation of surface water level using space-based altimeters (Ghosh et al., 2015). The altimetry data can be validated with the gauge data. The common reference is established by shifting the altimetry levels by the difference between the gauge data and altimetry-derived water-level time series means (Michailovsky et al., 2012). AltiKa, an onboard altimetry system, is utilized for studying inland water. SARAL/AltiKa was used to retrieve water levels of the Brahmaputra River in KNP. SARAL/AltiKa data were processed using the ICE-1 retracking method (Figure 17.8). The water level variation

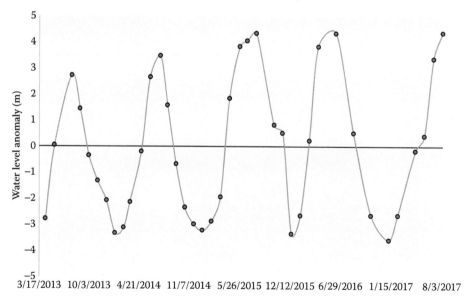

**FIGURE 17.8** Water level anomalies derived from SARAL/AltiKa.

gives an indication of the status of inundation of the flooded area (Ghosh et al., 2016). The water level increased over 6 m from February to August 2017.

## 17.6 REMOTE SENSING AND GEOGRAPHIC INFORMATION SYSTEM INTEGRATION: TOWARDS NEAR REAL-TIME MONITORING AND ASSESSMENT

### 17.6.1 MOBILE GEOGRAPHIC INFORMATION SYSTEM

Information technology has touched every aspect of our society. Spatial data and the information related to it fulfill a major requirement of both natural resource and disaster management. The usefulness of availing near real-time information on a portable device is undeniable, especially during rescue operations. In comparison to conventional modes used for field data collection, viz paper-based surveys and reports, the use of mobile-based technology can be considered a breakthrough. Conventional paper-based processes are time consuming, laborious, and erroneous. Thus, the necessity of procuring and visualizing spatial as well as nonspatial data on mobile devices has been acknowledged in research and industry.

A GIS-based mobile application on the Android platform, *River App*, was developed to show information regarding the water level and its extent in KNP (Figure 17.9). Information is updated on a monthly basis, and the figures provided reflect the current river level compared with historical time series data. The system is built on Android architecture in accordance with Google guidelines for application development. The app uses Google maps API and chart library to display the data based on the location chosen. For design, the XML has been used because it is lightweight and fast. It also displays the location on the Google map. The Android app is compatible up to Android OS 8.0, that is, Android Oreo.

**FIGURE 17.9** Interface of *River App* android application.

### 17.6.2 Android

Android is an extremely famous mobile OS launched by Google. The Linux kernel serves as its bedstone, and the OS is preliminarily meant for touchscreen mobile devices such as tablets and smartphones. The OS user interface imitates real-world touch gestures such as tapping and pinching and swiping for manipulating on-screen icons and has a virtual keyboard. There are several applications for conttrolling cars, televesions, and even wristwatch through Android. Other variants of Android are also available on digicams, game consoles, PCs, and other electronics. The Android applications ("apps") reaching more than 2.7 million as of February 2017 can be obtained from the Google Play store. The app has the largest installed base of any OS and boasts an active user number of 2 billion as of May 2017. The application has a desktop widget and can be placed directly on the Android's home application. The widget receives its content from web servers using a RESTful API based on HTTP requests and JSON data format. Documentation of the API has been published on the Google Project web page and can be freely used by all third-party developers.

Recently more and more mobile map apps with specific functions are being released on by the App Store. Most of them come with a standardized style due to the default style of Google Maps. With the flood of diverse open-source mapping platforms including Leaflet Application Programming Interface, OpenStreetMap API, and Mapbox API, there are immense possibilities of designing literally customized web maps with various inter activities.

Our study is focused on developing a prototype mobile map app for displaying geotagged images in contrast to that of Google Maps API and other available open-source mapping API such as Leaflet API. The work flow for generating a user-defined mobile map, which we believe will aid to learn Mobile Cartography, has been described. Thus, the current study encompasses the basic practicality of display points of interest with allied information as well as advanced methods for representing geospatial data through mobile mapping techniques including jQuery mobile, CARTO API, MarkerCluster plugin, Mapbox studio, and web mapping.

These available tools can be used for making the entire design technique convenient and will motivate cartographers to represent geographic information much more efficiently. As a result, this work explores more approaches for cartographers creating customized mobile maps with open source mapping platforms.

### 17.6.3 Overall System Architecture Design and Deployment

The architectural design of the system embraces two major entities, namely mobile application and server application (Tejassvi, 2014). The first application is capable of sending and receiving data from the sever whereas the second needs to be installed and integrated with a database. Once the server encounters the data, they can be used to update the database. The mobile application needs an authentication for connecting to the server. Connectivity to the database should be managed only at the server end.

Deployment of the application requires configuring some resources, most preferably an Android Smartphone with GPRS, Android OS version 8.0 and previous versions, in built GPS. For accommodating information recovery from the mobile application and vice versa, the server and database should work 24 hours, 7 days a week.

## 17.7 CONCLUSION

Protected areas are the setting stones of biodiversity conservation programs regardless of their origins. These areas are places of immense natural, social, and cultural value. They are the remaining places where intricate evolutionary processes are still dynamic with minimum anthropogenic perturbations. Conservation of protected areas is our essential duty. Assessment of probable risk and its probable extent comes prior to taking protective measures. KNP is located in a flood-prone zone of the lower catchment region of river Brahmaputra. The risks to the various ecosystems of KNP by flooding require regular assessment and comparison with available historical flood data. The grassland ecosystem of KNP is home to world's largest population of Indian rhinos. Several other grazers, birds, arthropods, annelids, mollusks, and so forth, many of which are still unexplored, also depend on the park, which also harbors several economically and culturally important graminoids. Thus, the implementation of proper measures for assessing the flood hazards of this ecosystem will ensure a step toward the proper resource management of this unique ecosystem. The grassland ecosystem of the park is intertwined with several inland water bodies that play a crucial role in maintaining the equilibrium of the park's grassland ecosystem. The role of KNP's inland water bodies are especially critical during floods; hence, appropriate habitat risk modeling of its grassland ecosystem will help to better understand their role and consequently will help to develop appropriate policies for their management.

However, proper disbursement of existing policies will facilitate an improved involvement of local stakeholders. The primary stakeholders of KNP are local tribes, for example, the Mishings, tourists, and researchers. All stakeholders should be informed about the ecological and economic benefits of the grassland ecosystem and of rhinos through colorful poster campaigning as well as interactive sessions. We hope that these informational modes will help to engage the attention of stakeholders regarding the actual problems. Moreover, the long-existing traditional knowledge of local communities should also be explored for employing localized resolutions on local topography. A repository of local submergence tolerant plant species, their appropriate location mapping, and the proper management of these resources can positively contribute to maintaining the biodiversity status of the KNP's grassland ecosystem.

Mobile GIS is an integrated framework for accessing geospatial data and services through devices such as smartphones and tablets via wireless networks. Approximately 67% of the park's grassland was inundated on August 12, 2017. With the experience regarding the 2016 and 2017 floods, it is recommended that dedicated mobile apps should be developed to disseminate near real-time flooding information in KNP. The usage of near real-time monitoring data will provide better assessment and appropriate monitoring of the flood situation of the park's grassland ecosystem. This therefore will assist in meeting the long-term goal of maintaining the specific biodiversity of the KNP grassland ecosystem.

## REFERENCES

Champion, H.G. and Seth, S.K. 1968. *A revised survey of forest types of India*. Government of India Press, New Delhi. XXVII + 404 pp.

Cromsigt, J.P.G.M. and Te Beest, M. 2014. Restoration of a megaherbivore: Landscape-level impacts of white rhinoceros in Kruger National Park, South Africa. *Journal of Ecology* 102: 566–575.

Dhar, O.N. and Nandargi, S. 2003. Hydrometeorological aspects of floods in India. In *Flood Problem and Management in South Asia*. Springer, Dordrecht, the Netherlands, pp. 1–33.

Floods in Assam—Causes, Effects and Solutions. 2017. My India, https://www.mapsofindia.com/my-india/government/why-india-cant-afford-to-ignore-assam-flood-situation. Last accessed October 21, 2017.

Fu, L.L. and Cazenave, A. 2001. *Satellite Altimetry and Earth Sciences: A Handbook of Techniques and Applications* (Vol. 69). Academic Press London, UK.

Ghosh, S., Kumar Thakur, P., Garg, V., Nandy, S., Aggarwal, S., Saha, S.K., Sharma, R. and Bhattacharyya, S. 2015. SARAL/AltiKa waveform analysis to monitor inland water levels: A case study of Maithon reservoir, Jharkhand, India. *Marine Geodesy* 38(sup1): 597–613.

Ghosh, S., Nandy, S. and Senthil Kumar, A. 2016. Rapid assessment of recent flood episode in Kaziranga National Park, Assam using remotely sensed satellite data. *Current Science* 111: 1450.

Khatri, P.K. and Barua, K.N. 2011. Structural composition and productivity assessment of the grassland community of Kaziranga National Park, Assam. *Indian Forester* 137(3): 290–295.

Kotoky, P., Bezbaruah, D., Baruah, J. and Sarma, J.N. 2005. Nature of bank erosion along the Brahmaputra river channel, Assam, India. *Current Science* 88(4): 634–640.

Kushwaha, S.P.S. 2008. *Mapping of Kaziranga Conservation Area, Assam*. Indian Institute of Remote Sensing, Dehra Dun, India. IIRS/FED/Kaziranga/36/ 8026/2008.

Kushwaha, S.P.S. and Roy, P.S. 2002. Geospatial technology for wildlife habitat evaluation. *Tropical Ecology* 43(1): 137–150.

Kushwaha, S.P.S., Roy, P.S., Azeem, A., Boruah, P. and Lahan, P. 2000. Land area change and rhino habitat suitability analysis in Kaziranga National Park, Assam. *Tigerpaper* 27(2): 9–17.

Kutum, A., Sarmah, R. and Hazarika, D. 2011. An ethnobotanical study of Mishing tribe living in fringe villages of Kaziranga National Park of Assam, India. *Indian Journal of Fundamental and Applied Life Sciences* 1(4): 45–61.

Li, Z., Liu, W.Z., Zhang, X.C. and Zheng, F. 2009. Impacts of land use change and climate variability on hydrology in an agricultural catchment on the Loess Plateau of China. *Journal of Hydrology* 377: 35–42.

Menon, V., Tiwari, S., Easa, P. and Sukumar, R. 2005. Right of passage, Elephants Corridors of India. Final Technical Report. Wildlife Trust of India.

Michailovsky, C.I., McEnnis, S., Berry, P.A.M., Smith, R., Bauer-Gottwein, P. 2012. River monitoring from satellite radar altimetry in the Zambezi River basin. *Hydrology and Earth System Sciences* 16(7): 2181–2192.

Nippert, J.B, Knapp, A.K and Briggs, J.M. 2006. Intra-annual rainfall variability and grassland productivity: Can the past predict the future? *Plant Ecology* 184(1): 65–74.

Nippert, J.B., Ocheltree, T.W., Orozco, G.L., Ratajczak, Z., Ling, B. and Skibbe, A.M. 2013. Evidence of physiological decoupling from grassland ecosystem drivers by an encroaching woody shrub. *PLoS One* 8(12): 81630.

Planning Commission, 2008. Eleventh Five Year Plan, 2007–2012. Government of India.

Rodgers, W.A. and Panwar, H.S. 1988. Planning a wildlife protected area network for India: An exercise in applied biogeography. *Tropical Ecosystems: Ecology and Management*. Wiley Eastern Limited, New Delhi, 93–107.

Sarma, J.N., Borah, D. and Goswami, U. 2007. Change of river channel and bank erosion of the Burhi Dihing river (Assam), assessed using remote sensing data and GIS. *Journal of the Indian Society of Remote Sensing* 35(1): 93–100.

Singh, J.S., Lauenroth, W.K. and Milchunas, D.G. 1983. Geography of grassland ecosystems. *Progress in Physical Geography* 7(1): 46–80.

Singh, S. 2017. Protection measures considerably increase population of one-horned rhino. Press Information Bureau Government of India Special Service and Features.

Stebbins, G.L. 1981. Co-evolution of grasses and herbivores. *Annals of the Missouri Botanical Gardens* 68: 75–86.

Tejassvi, T. 2014. An android application to support flash flood disaster response management in India. Master's dissertation. University of Twente Faculty of Geo-Information and Earth Observation (ITC).

Thakur, P.K., Nikam, B.R., Garg, V., Aggarwal, S.P., Chouksey, A., Dhote, P. and Ghosh, S. 2017. Hydrological parameters estimation using remote sensing and GIS for Indian region–A review. *Proceedings of the National Academy of Sciences, India Section A: Physical Sciences*. (In press) 87(4): 641–659.

Vasu, N.K. and Singh, G. 2016. Grasslands of Kaziranga National Park: Problems and approaches for management. *Envis Bulletin*, 17: 104–113. http://wiienvis.nic.in/WriteReadData/Publication/19_Grassland%20Habitat_2016.pdf. Last accessed October 21, 2017.

# Supplementary Information

## A.1  TERRESTRIAL SOLAR RADIATION

It is a solar radiation received at the top of the atmosphere. As an average value, it was measured as 1367 Wm$^{-2}$. Iqbal (1983) stated that at a certain time, the terrestrial solar radiation is a function of the solar zenith angle at a certain latitude and the distance between sun and earth.

In AHVRR hydrological analysis system (AHAS) algorithms and theory (Parodi, 2002), the following sequence of equations was used to retrieve the terrestrial shortwave radiation from a solar zenith angle and the date of the satellite's passing.

$$S\downarrow exo = SC \times E_o \times \cos(SZA) \tag{A.1}$$

$$E_o = 1.00011 + 0.034221 \cdot \cos(d_a) + 0.00128 \cdot \sin(d_a) \tag{A.2}$$
$$+ 0.000719 \cdot \cos(2.d_a) + 0.000077 \cdot \sin(2.d_a)$$

$$d_a = 2\pi(d_n - 1)/365 \tag{A.3}$$

where:
  $S\downarrow exo$ is the instantaneous terrestrial solar radiation (Wm$^{-2}$)
  $E_o$ is the eccentricity correction factor
  $d_a$ is the day angle (radians)
  $d_n$ is the number of Julian days
  $SC$ is the solar constant 1367 (Wm$^{-2}$)
  SZA is the solar zenith angle (radians)

## A.2  SHORTWAVE RADIATION ESTIMATION

Terrestrial shortwave radiation $S\downarrow exo$ was used to find the incoming shortwave radiation. The transmissivity of the atmosphere used was 0.75 as the default value in the absence of solarimeter as suggested by Parodi (2002).

$$S\downarrow = 0.75 \times S\downarrow exo \tag{A.4}$$

## A.3 ALBEDO ESTIMATION FROM LANDSAT7-ETM+ IMAGE

*Surface albedo* is defined as the ratio of the reflected radiation to the incident shortwave radiation. Surface albedo images from the LANDSAT7-ETM+ Image of the study area were calculated for the actual atmospheric conditions computed in SEBAL through the following steps:

### Various Intermediate Parameterization Used in the SEBAL Algorithm and Other Ground Station Data Are Shown in the Following Table.

**Required Ground Station Data**

| Parameter | December 23, 2010 | February 9, 2011 | March 29, 2011 | April 14, 2011 |
|---|---|---|---|---|
| Wind speed u (m/s) | 0.06 | 1.15 | 0.83 | 0.42 |
| Wind measure height z (m) | 5 | 5 | 5 | 5 |
| Blending height zb (m) | 100 | 100 | 100 | 100 |
| Height momentum z* (m) | 5 | 5 | 5 | 5 |

**Incoming Instantaneous Radiation (Watt $m^{-2}$)**

| Parameter | December 23, 2010 | February 9, 2011 | March 29, 2011 | April 14, 2011 |
|---|---|---|---|---|
| Shortwave | 585 | 1034 | 871 | 923 |
| Long wave | 258 | 274 | 324 | 304 |
| Average daily radiation [watt $m^{-2}$] | | | | |
| Shortwave incoming | 350 | 350 | 450 | 450 |
| Long wave net | −138 | −119 | −123 | −133 |

**Other Required Data**

| Parameter | December 23, 2010 | February 9, 2011 | March 29, 2011 | April 14, 2011 |
|---|---|---|---|---|
| rho*cp (J/K/m) | 1154 | 1154 | 1154 | 1154 |
| $T_o$ wet (°C) | 16.80 | 16.80 | 23.00 | 22.70 |
| $T_o$ dry (°C) | 23.5 | 20.30 | 28.00 | 30.50 |
| **Dry Pixels' Initial Conditions** | | | | |
| $H_{dry}$ (watt $m^{-2}$) | 225 | 671 | 526 | 501 |
| $rah_{dry}$ (s $m^{-1}$) | 26.21 | 89.74 | 69.47 | 78.66 |
| $\Delta T_{dry}$ (°C) | 24.6 | 52.2 | 31.7 | 77.60 |
| **Coefficient ($T_o-T_a$) versus $T_o$ Initial (After Iterations)** | | | | |
| $rah_{dry}$ (s $m^{-1}$) | 17.07 | 14.82 | 17.78 | 16.17 |
| DT(dry) (°C) | 4.8 | 8.6 | 8.1 | 7.0 |
| $B_{(slope)}$ | 3.67 | 2.46 | 2.03 | 0.91 |
| $A_{(offset)}$ | −61.73 | −41.36 | −46.63 | −41.27 |

# Supplementary Information

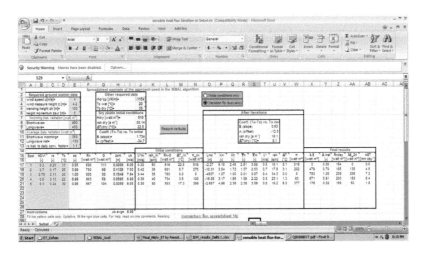

Calculations_for_05_H_Automatic_GMD_model.xls

## A.4 SPECTRAL RADIANCE FOR EACH BAND ($L_\lambda$)

For LANDSAT7-ETM+ with header file data on gains and biases, a simpler equation for $L_\lambda$ is given:

$$L_\lambda = (\text{Gain} \times DN) + \text{Bias} \tag{A.5}$$

where Gain and Bias refer to the values given in the header file of LANDSAT7-ETM+ images.

## A.5 THE REFLECTIVITY FOR EACH BAND ($\rho_\lambda$)

For LANDSAT7-ETM+, a simpler equation for $\rho_\lambda$ is given as (Table A.1) (Irish, 2007):

$$\rho_\lambda = \pi \times \frac{d_r}{\text{ESUN}_\lambda \cos(\theta)} \tag{A.6}$$

where:
  $\text{ESUN}_\lambda$ is the mean solar exoatmospheric irradiance for each band (W/m²/μm)
  cos ($\theta$) is the cosine of the solar incidence angle (from nadir), 90°-$\beta$ (sun elevation angle ($\beta$) from header file of LANDSAT7-ETM+)
  $d_r$ is (dimensionless), the inverse squared relative earth-sun distance, $d_r$ is computed using the following equation by FAO-56 (Allen et al., 1998).

$$d_r = 1 + 0.033 \cos\left(\text{DOY}\,\frac{2\pi}{365}\right) \tag{A.7}$$

where DOY is the day of year.

**TABLE A.1**
**Landsat7-ETM+ Solar Spectral Irradiances (ESUN$_\lambda$)**

| Band | W/m²/μm |
|---|---|
| 1 | 1969.00 |
| 2 | 1840.00 |
| 3 | 1551.00 |
| 4 | 1044.00 |
| 5 | 225.700 |
| 7 | 082.070 |
| 8 | 1368.00 |

## A.6  THE ALBEDO AT THE TOP OF THE ATMOSPHERE ($\alpha_{toa}$)

For LANDSAT7-ETM+, a simpler equation for $\alpha_{toa}$ is given as per Igusky, 2008:

$$\alpha_{total} = \sum \omega_\lambda \cdot \rho_\lambda \tag{A.8}$$

where $\omega_\lambda$ is a weighting coefficient for each band of LANDSAT7-ETM+, computed as follows:

$$\omega_\lambda = \frac{ESUN_\lambda}{\sum ESUN_\lambda} \tag{A.9}$$

$$\alpha_{total} = \omega_1 \cdot \rho_1 + \omega_2 \cdot \rho_2 + \omega_3 \cdot \rho_3 + \omega_4 \cdot \rho_4 + \omega_5 \cdot \rho_5 + \omega_7 \cdot \rho_7$$
$$\tag{A.10}$$
$$= 0.293\rho_1 + 0.274\rho_2 + 0.231\rho_3 + 0.156\rho_4 + 0.034\rho_5 + 0.012\rho_7$$

## A.7  SURFACE ALBEDO (A)

Surface albedo (Figure A.1) is computed by correcting the $\alpha_{toal}$ for atmospheric transmissivity as with following equation (Igusky, 2008).

$$\alpha = \frac{\alpha_{total} - \alpha_{path\_radiance}}{\tau_{sw}} \tag{A.11}$$

where values for $\alpha_{path\,radiance}$ range between 0.025 and 0.04, but for SEBAL, a value of 0.03 is recommend.

$$\tau_{sw} = 0.75 + 2 \times 10^{-5} \times Z \text{ (elevation (Z) for Delhi 293 m)}$$

$$\tau_{sw} = 0.568 \tag{A.12}$$

$$\alpha = \frac{\alpha_{total} - 0.03}{0.568} \tag{A.13}$$

Supplementary Information

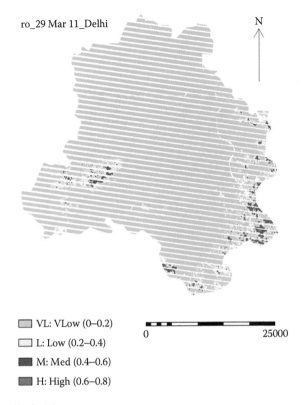

**FIGURE A.1** Surface albedo ($\alpha$).

## A.8 OUTGOING LONG-WAVE RADIATION (L ↑)

The outgoing long-wave radiation is the thermal radiation flux emitted from the earth's surface to the atmosphere (W/m²). It is computed by using three commonly used vegetation indices—normalized difference vegetation index (NDVI), soil-adjusted vegetation index (SAVI), and leaf area index (LAI)—proposed by Yao et al. (2008) in SEBAL through the following steps:

1. Any one of these indices can be used to predict various characteristics of vegetation, according to preferences of the user (Figures A.2 –A.4).

$$\text{NDVI} = \frac{\text{NIR} - R}{\text{NIR} + R} \quad (A.14)$$

NIR and $R$, reflectance of near infrared, red band, respectively

$$\text{SAVI} = \frac{(1+L)(\text{NIR} - R)}{\text{NIR} + R + L} \quad (A.15)$$

$L$ is taken as 0.5

$$\text{LAI} = -\frac{\ln\left(\frac{0.69 - \text{SAVI}}{0.59}\right)}{0.91} \quad (A.16)$$

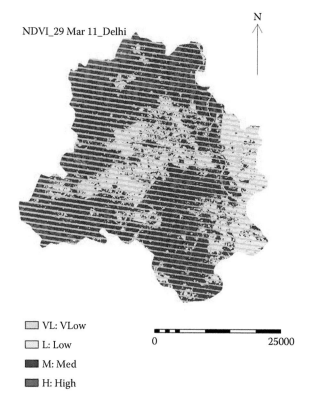

**FIGURE A.2** Normalized difference vegetation index (NDVI).

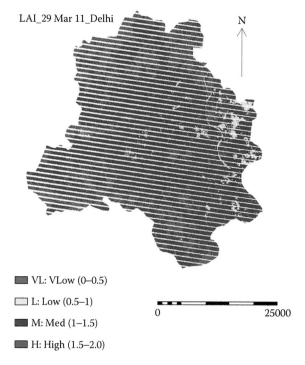

**FIGURE A.3** Leaf area index (LAI).

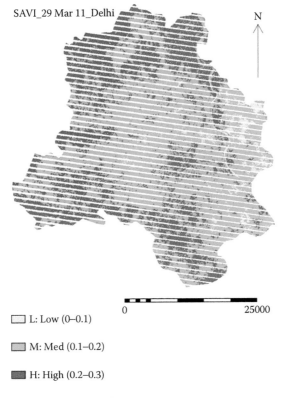

**FIGURE A.4** Soil-adjusted vegetation index (SAVI).

2. Surface emissivity ($\varepsilon$) is the ratio of the thermal energy radiated by the surface to the thermal energy radiated by a blackbody at the same temperature. An emissivity representing surface behavior for thermal emission in the broad thermal spectrum (6–14 μm), expressed as $\varepsilon_o$, and is used later to calculate total long wave radiation emission from the surface (Figure A.5).

The surface emissivity is computed by using the following empirical equations (Lim et al. 2012), where NDVI > 0

$$\varepsilon_0 = 0.95 + 0.01 \text{LAI};$$

$$\text{for LAI} < 3 \quad (A.17)$$

3. The thermal radiance from the surface is calculated by using following equation as:

$$Ts = \frac{K_2}{\ln\left[\left(\dfrac{\varepsilon_{NB} K_1}{R_c}\right) + 1\right]} \quad (A.18)$$

$$\varepsilon_{NB} = 0.97 + 0.0033 \text{LAI},$$

$$\text{for LAI} < 3 \quad (A.19)$$

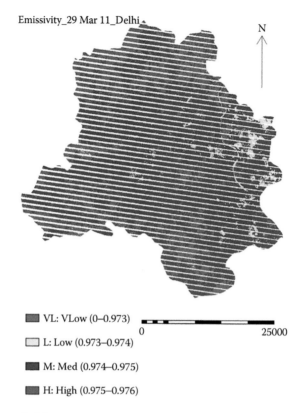

**FIGURE A.5** Surface emissivity.

and $R_C$ is calculated by Wukelic et al. (1989) as

$$R_c = \frac{L_6 - R_p}{\tau_{NB}} - (1 - \varepsilon_{NB})R_{sky} \tag{A.20}$$

where:
$L_6$ is the spectral radiance (Wm$^{-2}$ sr$^{-1}$ μm$^{-1}$) of band 6
$R_P$ is the path radiance in the 10–12.5 μm band (Wm$^{-2}$ sr$^{-1}$ μm$^{-1}$)
$R_{sky}$ is the narrow band downward thermal radiation from a clear sky (Wm$^{-2}$ sr$^{-1}$ μm$^{-1}$)
$\varepsilon_{NB}$ is the narrow band emissivity corresponding to the satellite thermal sensor wave length band, that is, narrow band transmissivity of air (10.4–12.5 μm range)
$\tau_{NB}$ $K_1 = 666.09$ and $K_2 = 1282.71$
$T_S$ is the surface temperature (Kelvin) (Figure A.6).

For clear sky atmospheric $\tau_{NB} = 0$ condition, $R_P = 0$, $R_{sky} = 0$
From the equation $O$ to $U$, given as

$$L\uparrow = \sigma.\varepsilon_0.(T_s)^4 \tag{A.21}$$

Supplementary Information

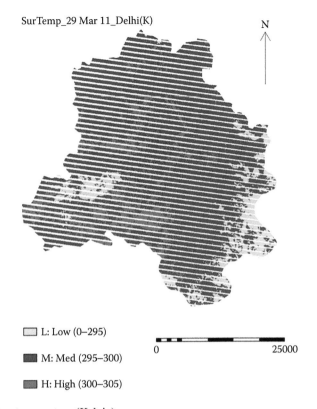

**FIGURE A.6** Surface temperature (Kelvin).

## A.9 INCOMING LONG-WAVE RADIATION ($L\downarrow$)

The amount of shortwave solar radiation reaching the land surfaces $L\downarrow$ depends on the atmospheric absorption and scattering of the shortwave radiation.

$$L\downarrow = \sigma.\varepsilon_a.(T_a)^4 \quad\quad\quad (A.22)$$

where:
  $\sigma$ is the Stefen Botlzmans constant ($5.67 \times 10^{-8}$ W.m$^{-2}$.K$^{-4}$)
  $T_a$ is the near-surface air temperature (in Kelvin)
  $\varepsilon_a$ is the effective atmospheric (apprant) emissivity and given (Allen et al., 2000) as

$$\varepsilon_a = 0.85 \times (-\ln \tau_{sw})^{0.09} \quad\quad\quad (A.23)$$

Substituting Equation A.23 into Equation A.22 and using $T_{cold}$ from the "cold" pixel for $T_a$ yields the following equation

$$L\downarrow = 2.67 \times 10^{-8} \times 0.85 (-\ln \tau_{sw})^{0.09} \times (T_{cold})^4 \quad\quad\quad (A.24)$$

## A.10  SELECTING THE HOT AND COLD PIXELS

Hot and cold pixels were selected to calibrate the model. The cold pixel was chosen from an agricultural field of full cover crop with the highest NDVI. The hot pixel was chosen in a dry and bare agricultural field with a NDVI near 0. The hot and cold pixels were chosen with the aid of a spreadsheet with selected points in the center of fields.

## REFERENCES

Allen, R.G., Pereira, L.S., Raes, D. and Smith, M. 1998. Crop evapotranspiration-guidelines for computing crop water requirements. FAO Irrigation and Drainage Paper 56. Rome, Italy: Food and Agriculture Organization of the United Nations.

Igusky, K. 2008. Quantifying albedo and surface temperature over different land covers: Implications for carbon offsets. Masters project submitted in partial fulfillment of the requirements for the Master of Environmental Management degree, Nicholas School of the Environment and Earth Sciences, Duke University.

Iqbal, M. 1983. *An Introduction to Solar Radiation*. Academic Press, New York, p. 223.

Irish, R. Landsat 7. *Science Data Users Handbook*. National Aeronautics and Space Administration, Report, February 17, 2007.

Lim, H. S., Mat Jafri, M. Z., Abdullah, K. and AlSultan, S. 2012. Application of a simple mono window land surface temperature algorithm from LANDSAT ETM+ Over Al Qassim, Saudi Arabia. *Sains Malaysiana*, **41**, 841–846.

Parodi, N. G., 2002, AHVRR, hydrological analysis system algorithms and theory–Version 1.3:10-1. Retrieved from https://www.itc.nl/research/products/_docs/AHAS_algorithms_and_theory_v3.pdf.

Wukelic, G. E., Gibbons, D. E., Martucci, L. M. and Foote, H. P. 1989. Radiometric calibration of Landsat Thematic Mapper thermal band. *Remote Sensing of Environment* 28: 339–347.

Yao, Y., Qin, Q., Zhu, L. and Yang, N. 2008. Relating surface albedo and vegetation index with surface dryness using LANDSAT ETM+ Imagery. In Proceedings: The Institute of Electrical and Electronics Engineers Geoscience and Remote Sensing Society (IGARSS), 2008.

# Index

Note: Page numbers followed by f and t refer to figures and tables respectively.

## A

Ablation zone, 31–32, 31f, 44
ABMs, *see* Agent-based models (ABMs)
Accumulation area ratio (AAR), 44–45
Advanced land imager, 253
Advanced spaceborne thermal emission and reflection radiometer (ASTER) DEM, 42, 52, 137, 227
Advanced Training and Users Manual, 178–179
Advanced very high resolution radiometer (AVHRR), 52
Aerosol optical depth (AOD), 151–152, 156, 168–169, 169f, 170t
Agent-based models (ABMs), 75, 77t, 79
Agricultural land, 7, 15t, 17t, 18, 18t, 19f, 19t, 20, 21t, 22t, 23, 51, 60, 82t, 83, 111, 116t, 119, 174, 227, 243–244, 246, 248–249
Agriculture, 3, 8, 13, 17f, 19f, 20, 23, 25, 68, 88, 103, 152–153, 193f, 199f, 208, 243–244, 283t, 284, 284f, 284t
*Agriculture*, 55–56, 58–61, 58f, 59f, 60f, 62f
Almora Group, 252–254, 254f, 273
Almora Klippe, 252
ALOS PRISM, 43
AltiKa, 276, 286, 286f
Andreassen, L. M., 36t
Android, 287–288
ANN, *see* Artificial neural networks (ANN)
AOD, *see* Aerosol optical depth (AOD)
ArcGIS, 2, 14, 125, 135–137, 195, 245
ARIMA, *see* Autoregressive integrated moving average (ARIMA) modeling
Artificial neural networks (ANN), 36t, 75, 88–89, 91f, 93, 94f, 97, 101–102, 101f, 102t, 104t, 105
ASTER (advanced spaceborne thermal emission and reflection radiometer) DEM, 42, 52, 137, 227
Atmospheric correction, 4, 90, 253, 255
Autoregressive integrated moving average (ARIMA) modeling, 151–152, 156, 167–169, 168f, 168t
AVIRIS, 252, 255

## B

Bada Shigri Glacier, 31f
Bahr, D. B., 40t, 44, 126t
Banaras Hindu University (BHU), 89, 103
Bare soil, 12, 15, 15t, 17t, 18t, 19t, 20, 21t, 22t, 23–25
Basin asymmetry, 132, 135t, 137, 142, 144t, 148
Basin slope, 132, 135t, 136–137, 145–146, 146f
Bastiaanssen, W. G. M., 177
Bayr, J. J., 36t
Beas Basin, 28, 34–35
Berthier, E., 46
Bhambri, R., 37t
Bhardwaj, A., 37t
BHU, *see* Banaras Hindu University (BHU)

Binary encoding, 252, 259–260, 269f
Biodiversity, 12–13, 52, 78t, 80, 248, 276, 282, 288–289
Bishop, M. P., 36t
Bruckl, E., 39, 126
Built-up area, 2, 3f, 4, 7, 11, 15t, 17t, 18–20, 19t, 21t, 22t, 23–25, 58–60, 58f, 59f, 65–67, 67f, 83, 110, 116, 117f, 119, 193f, 227
  dense, 1, 8, 17, 55, 60–61, 60f, 62f, 63–64, 64f, 65f
  residential, 51
  sparse, 55, 60f, 61, 62f, 63–64, 64f, 65f

## C

Catchments of Western Himalayas, 123–129, 128t
  abstract, 123
  area change and surface area, 127t
  data and methods, 125–126
  glacier areas, 125, 127f
  ice reserved in Stok and Matoo village catchments, 127, 128f
  introduction, 123–124
  results and discussion, 127–128
  satellite data, 125t
  scaling parameters, 126t
  study region, 124–125, 124f, 124t
  thickness and volume estimation, 125–126
  total glaciated area in Stok and Matoo village catchments, 127
Cellular automata model, 75, 79
CERES-rice model in Madhya Pradesh, 207–220
  abstract, 207–208
  calibrated genetic coefficients, 210t
  composite of break phases during RMM phases, 213f
  data sets, 210t
  data used, 209
  detrend analysis of simulated yield, 217–218
  difference between observed and simulated yield, 215f
  excess and deficit rainfall years, 212t
  genetic coefficients for different sensitivity experiments, 216t
  introduction, 208
  Madden–Julian oscillation indices, 208, 212–213
  material and method, 209–210
  model-simulated and MODIS evapotranspiration, 218f
  observed anomaly and detrend anomaly of rice yield, 217f
  observed rainfall and mean temperature, 212f
  observed rainfall and temperature variability, 210–212
  rainfall anomalies and Madden–Julian oscillation indices, 212–213
  rainfall, soil moisture, and evapotranspiration, 211f
  RMSE and SD of MODIS and simulated evapotranspiration, 219t
  RMSE of simulated yield for all genetic coefficients, 216t

CERES-rice model in Madhya Pradesh (*Continued*)
  RMSE of simulated yield for both genetic coefficients, 215t
  RMSE of SM from remote sensing and model simulations during drought, 219f
  sensitivity analysis, 207–208, 210, 215–217, 224–226, 228, 230
  simulated and remotely sensed evapotranspiration, 218–219
  simulated and remotely sensed soil moisture, 219
  soil physical characteristics, 209t
  study region, 209, 209f
  variability in crop yield, 214–215
  year-to-year variations in observed rice yield, 214f
Chaohai, L., 126
Chhota Shigri Glacier, 28–29, 29f, 31f, 46
Christaller, W., 80
Clarke, K. C., 79
Classification accuracy, 88–89, 95–97, 99–104, 104t, 245, 284
Classification success index (CSI), 88, 95, 99t, 100t, 101–102, 102t
Cogley, J. G., 40
Co-registration of optically sensed images and correlation (COSI-Corr), 41
CORONA, 43
COTS software, 43
Crevasses, 27, 32, 32f
CSI, *see* Classification success index (CSI)

## D

Debris-covered glaciers, 34, 36–38, 37f, 39f
Decision Support System for Agrometeorology Transfer (DSSAT), 207–208, 210, 210t, 214–216, 220
Defence Research Development Organisation (DRDO), 256t, 270
Defence Terrain Research Laboratory (DTRL), 256t, 270
Deglaciation pattern, 35f
Della Seta, M., 135t
Dense vegetation, 15t, 17f, 17t, 18–20, 18f, 18t, 19f, 19t, 21t, 81, 82f, 82t, 83, 83t, 91f, 94f, 97f, 98f, 98t, 99t, 100t, 101f, 102f, 102t, 103, 103t, 115, 116f, 116t, 117t, 118t, 119
Desert National Park, 277
Dharanirajan, K., 135t
Drainage basin asymmetry, 132, 136–137, 142, 144t
Drainage density, 132, 135–136, 138–139, 140f, 147–148, 226
Drainage morphometry, *see* Morphometry and tectonic activity in Dehgolan Basin
DRDO, *see* Defence Research Development Organisation (DRDO)
DSSAT, *see* Decision Support System for Agrometeorology Transfer (DSSAT)
DTRL, *see* Defence Terrain Research Laboratory (DTRL)

## E

El Hamdouni, R., 135t, 142
Elongation ratio, 132, 135t, 136–137, 142–144, 144t, 145f, 148
Endmembers, 252–253, 256–260, 257f, 262–263, 262t, 263f, 265t, 268, 268t, 270t, 273
ENVI, *see* Environment for Visualizing Images (ENVI)
Environment for Visualizing Images (ENVI), 43, 93, 255–256, 257f, 259–260
EPCO, *see* Plant evaporation compensation factor (EPCO)
Equilibrium line altitude (ELA), 44–46
ESA, *see* European Space Agency (ESA)
European Space Agency (ESA), 208, 210–211, 210t
Evapotranspiration in Killinochi, 197–198, 200t
Evapotranspiration in Madhya Pradesh, 208, 210t, 211f, 218–219, 218f, 219t
Evapotranspiration in New Delhi, 173–187, 186f
  abstract, 173–174
  comparison with conventional method, 185–186
  daily, 179–181
  evaporative fraction, 179, 180f, 181f, 182f, 183f
  introduction, 174–175
  LAI and NDVI, 184f
  lysimeter, 175, 183t, 185t
  methodology, 175–183
  net radiation, 178f
  remote sensing, 176–177, 176t
  seasonal variation, 187f
  sensible heat flux, 178–179, 180f
  soil heat flux, 177–178, 179f
  study area, 175, 176f
  surface energy balance algorithm for land model, 177
  surface radiation balance equation, 177
Evapotranspiration in Western Himalaya, 224–225, 227, 235f, 236f, 237f
  leaf area index (LAI), 184–185
  potential evapotranspiration (PET), 227
  satellite-based, 239
EVT rate, 196t, 198f
Exposed rock, 11–12, 15t, 17–18, 17t, 18t, 19t, 20, 21t, 22t, 23–25

## F

False color composite (FCC), 15, 33–34, 90, 91f
Farinotti, D., 38, 41
Farmer Welfare and Agriculture Development Department of Madhya Pradesh, 210
Fast line-of-sight atmospheric analysis of spectral hypercubes (FLAASH), 253, 255
Fault line, 134f, 142, 146–148, 147f
FCC, *see* False color composite (FCC)
FLAASH, *see* Fast line-of-sight atmospheric analysis of spectral hypercubes (FLAASH)
F-measure, 88, 95
Fog over North India, 151–170
  abstract, 151
  aerosol optical depth (AOD), 151–152, 156, 168–169, 169f, 170t
  autoregressive integrated moving average (ARIMA) modeling, 151–152, 156, 167–169, 168f, 168t
  average daily persistence, 160f
  average monthly visibility, 159f
  change in frequency by decade, 165t
  climatology, 154, 158t
  correlation between the time series and number, 162f
  dataset and methodology, 154–156
  decadal frequency, 157f
  decadal time series and trend analysis over IGP, 164
  decadewise trend analysis, 164f–165f

# Index

duration and variability, 158–160
geospatial analysis, 155–156
Indo–Gangetic Plain (IGP), 151–157, 153f, 154f, 158t, 159f, 160–162, 162f, 163f, 164–165, 166f–167f, 167, 169
intensity and persistence, 160–161
introduction, 152
long-term climatology, 157–158
Mann–Kendall statistical test, 155, 161t, 162f
monthly mean frequency, 157f
percentage frequency of duration, 159f
relationship between fog occurrence and aerosol optical depth, 168–169
spatial variability, 164–167, 166f–167f, 169, 227
statistics, 158t
study area, 152–153, 153f
time series and trend analysis, 155, 162f–163f
trend analysis, 161–164
trend analysis of winter, 163f
Forest, 15t, 82t, 193f, 196t, 198t, 199f, 204, 203f, 284, 284f, 284t
reserved, 277; *see also* Random forest-based classification (RF)
Forest, 55–56, 60–61, 60f, 62f, 65–66, 66f, 67f, 68, 68f
FRAGSTATS, 115, 117, 118f, 119
Frey, H., 40, 126

## G

Geodetic method, 45–46
Geographic information system (GIS), 2, 12–13, 20, 23–25, 28, 74, 80, 88, 109–110, 112, 119, 124, 132, 135, 148, 152, 276, 282, 283t, 287
mobile, 287, 289
Geospatial analysis, 155–156
Ghosh, S., 283t
Glacier area, 33, 36t, 38t, 40, 40t, 45–46, 125, 127
Glacier boundary, 31, 33f, 34f, 35f
Glacier flow mechanics method, 41–42
Glaciers, *see specific name of glacier*
Glaciers, spatio-temporal monitoring, 27–47
abstract, 27–28
accumulation area ratio (AAR), 44–45
ASTER DEM, 137, 227
automated classification, 36–38
digital elevation model (DEM), 14f, 28, 37f, 39f, 41–46, 125, 143f, 227–228
future, 47
glacier flow mechanics method, 41–42
identifying geomorphological features from space, 28–33
introduction, 28
manual delineation of area/length, 33–34
mapping and inventory, 33–35
Nye theory, 41–42
remote sensing, 41–42
thickness area relation method, 39–40
thickness estimation, 38–42
3D-DEM, 142, 143f, 146–148
volume, 40, 40t
volumetric estimation, 42
Glaciofluvial activity, 28–29
Google
Earth, 15, 125, 284
Maps, 287–288
Play, 288
Project, 288
Grassland ecosystem in Kaziranga National Park, 275–289
abstract, 276
Android, 287–288
Brahmaputra River, 281, 285f
components of ecosystem, 280
conservation areas, 280–281
flooding, 281–282
flood mapping through active microwave sensor, 284–285, 285f
grassland ecosystem, 276–277
history and geographic association, 277
Landsat 8 band description, 283t
land use and land cover mapping, 283–284, 284f, 284t
mobile geographic information system, 287
overall system architectural design and deployment, 288
rainfall, 279f
remote sensing and GIS, 283t, 287–288
*River App*, 287, 287f
river water level monitoring, 286–287
space technology, 282–287
study area, 278f
topography and climate, 278–279, 278f
vegetation and ecosystem, 280
water level anomalies, 286f
Great Himalayan Range, 39f
Grey information theory, 225
Groundwater, 191–196, 197f, 228, 282
delay, 229t
parameters, 230–231, 231f, 233
recession factor, 230
recharge, 111, 193, 196, 198, 201, 203–204, 203f, 204, 229t, 230–233
"revap," 229t, 230
Groundwater flow model (MODFLOW), 191, 195, 197f
Gupta, D., 245t
Gupta, R. P., 36t

## H

Hack, J. T., 132, 135t, 136
Hack's Index, 132, 135t, 136–137, 142
Hall, D. K., 33, 36t
Hendon, H. H., 210, 212
Himachal Pradesh, 28, 29f, 34, 46, 153, 165, 226–227, 226f
Himalayan Frontal Thrust (HFT), 252
Horton, R. E., 132, 135t, 136–138, 147
Hue component, 39f
Huss, M., 38
Hydrological modeling in Killinochi, 191–204
abstract, 191–192
aquifer, 194, 194f, 196f
boundary conditions, 197–201
data collection and processing, 195
evapotranspiration rate estimated based on data from WRB, 200t
extinction depth estimates based on land use classes, 198t
ground water simulation model, 195
head contours in meters, 202f
horizontal hydraulic conductivity distribution, 197f

Hydrological modeling in Killinochi (*Continued*)
   introduction, 192–194
   land use classes in basins, 196t, 199f
   land use land cover, 198–199
   methodology, 195–201
   rainfall and EVT, 198f
   rainfall time series, 197f, 202f
   recharge rate, 203f
   saturated hydraulic conductivity values of soil groups, 201t
   soil types, 199, 201f
   spatial and temporal discretization of model domain, 195–197
   study area, 193, 193f
Hydrological modeling in Western Himalaya, 223–239
   abstract, 224
   available water capacity in soil, 233
   average monthly ET simulated from SWAT and MODIS, 237f
   box-and-whisker plot, 236f
   calibration and validation, 228, 229t, 234–236
   comparison of actual ET between SWAT simulation and MODIS data, 235f
   data preparation, 227–228
   maximum canopy storage, 234
   groundwater parameter, 230
   groundwater recharge to deep aquifer, 233
   groundwater "revap" coefficient, 230
   hydrologic response units parameter, 233–234
   hydrologic simulations (water balance), 236–237
   introduction, 224–226
   limitations, 238
   material and methods, 227–229
   mean monthly value, 238f
   model setup, 228
   plant uptake compensation factor, 233
   saturated hydraulic conductivity, 233
   sensitivity analysis, 230
   sensitivity of water balance components, 231f, 232f, 234f
   Sirsa River Basin, 226f
   soil and water assessment tool (SWAT), 224–228, 229t, 231–232, 234–236, 235f, 236f, 237f, 238–239
   soil depth, 232-33
   soil evaporation compensation coefficient, 233–234
   soil parameter, 232
   study site, 226–227
   threshold depth of water in shallow aquifer, 230
   threshold water level in shallow aquifer, 230
Hyperion data used in spectral mapping methods of mineral aggregates in Uttarakhand, 251–273, 267f
   abstract, 252–254
   bad band designation, 255
   binary encoding, 252, 259–260, 269f
   class distribution result, 272t
   data dimension reduction, 256–258
   endmember or pure pixel extraction, 256, 258t, 263f, 264f, 265t, 268t, 270t
   endmember spectral library, 259
   field spectra data, 270, 271f
   geological interpretation, 273
   geological setting, 253
   introduction, 252–253
   lithological map and thrust zones, 254f
   minimum noise fraction transformation, 256, 257f
   n-dimensional visualizer, 257–258, 257f, 258t
   preprocessing data, 254–255, 256t
   radiometric calibration, 255–256
   spectral analysis, 259–261
   spectral analyst, 259–261
   spectral angle mapping, 260–261
   spectral library, 259
   spectral mapping, 266t, 272t
   weights and scores, 261–262
   weight scores of different mapping methods and results, 262–272, 262t, 263f, 264f, 269f
Hyperplane, 93
Hyperspectral data, 252–256, 256t, 273
Hypsometric curve, 132, 135t, 136, 139–141, 141f
Hypsometric integral, 132, 135t, 136, 141f

# I

IARI, *see* Indian Agricultural Research Institute (IARI)
IASH, 38
IDRISI Selva v. 17.0, 81, 84
IGP, *see* Indo–Gangetic Plain (IGP)
Image classification, 8, 15, 37t, 89–90, 92–93
Image processing of satellite data, 90–91
IMD, *see* India Meteorological Department (IMD)
India Meteorological Department (IMD), 207, 209–212, 210t
Indian Agricultural Research Institute (IARI), 173–175, 176f, 187
   Water Technology Centre, 181, 185–186
India urbanization, *see* Urbanization in India
Indo–Gangetic Plain (IGP), 151–157, 153f, 154f, 158t, 159f, 160–162, 162f, 163f, 164–165, 166f–167f, 167, 169
Intensity Hue Saturation image, 39f
International Association of Hydrological Sciences: "Predictions in Ungauged Basins," 225
International Union for Conservation of Nature and Natural Resources (IUCN), 276, 279, 281
International Water Management Institute (IWMI), 191–192, 195, 204
IUCN, *see* International Union for Conservation of Nature and Natural Resources (IUCN)
IWMI, *see* International Water Management Institute (IWMI)

# J

Jaccard coefficient, 95
Jawaharlal Nehru Agricultural University (JNKVV), 209–210
Jefferies Matusita Distance Method, 88, 92
Ji, W., 12
JNKVV, *see* Jawaharlal Nehru Agricultural University (JNKVV)
John Hopkins University (JHU) Spectral Library, 259, 262, 273
Johnston, C. A., 225

# K

Kamp, U., 37t
Kamusoko, C., 111
Kannen, N., 225, 230, 233
Karunakaran, C., 252

# Index

Kaushal, A., 36t
Kaziranga National Park grassland ecosystem, 275–289
  abstract, 276
  Android, 288
  Brahmaputra River, 281, 285f
  components of ecosystem, 280
  conservation areas, 280–281
  flooding, 281–282
  flood mapping through active microwave sensor, 284–285, 285f
  grassland ecosystem, 276–277
  history and geographic association, 277
  Landsat 8 band description, 283t
  land use and land cover mapping, 283–284, 284f, 284t
  mobile geographic information system, 287
  overall system architectural design and deployment, 288
  rainfall, 279f
  remote sensing and GIS, 283t, 287–288
  *River App*, 287f
  river water level monitoring, 286–287
  space technology, 282–287
  study area, 278f
  topography and climate, 278–279, 278f
  vegetation and ecosystem, 280
  water level anomalies, 286f
Keibul Lamzao National Park, 277
Keller, E. A., 135t, 137
Kendall, M. G., 161–162, 163t; *see also* Mann–Kendall statistical test
Khamis-Mushyet: digital elevation model map, 14f
Kulkarni, A. V., 126
Kundu, S., 80
Kushwaha, S. P. S., 283t

## L

LAI, *see* Leaf area index (LAI)
Land change modeling, 73–80, 74f, 76t–78t, 84
Land conversion, 52, 77t
Land features modeling, 87–105
  artificial neural networks (ANN), 93–94, 94f, 101–102, 101f
  classification accuracy, 88–89, 95–97, 99–104, 104t, 245, 284
  classification algorithms, 104t
  classification success index (CSI), 88, 95, 99t, 100t, 101–102, 102t, 103t, 239
  F-measure, 88, 95
  image classification, 8, 15, 37t, 89–90, 92–93
  image processing of satellite data, 90–91
  introduction, 88–89
  Jaccard coefficient, 95
  Jefferies Matusita Distance Method, 92
  marginal rates, 88, 94–95, 233
  methodology, 90–96, 91f
  postprocessing summary of classification accuracy, 103–104
  random forest-based classification (RF), 88–89, 91f, 94, 97, 101–103, 102f, 104t, 105, 112
  results and discussion, 96–104
  selected measures, 91f, 94–95, 99t, 100t, 101, 102t, 103t
  separability analysis, 88, 91–92, 91f, 97, 97f, 98t
  study area and materials, 89–90, 90f
  support vector machines (SVMs), 88–89, 91f, 92–93, 97, 98f, 99t, 100–105, 100t, 104t, 112
  training and testing pixels used in classification, 91t, 97f
  transformed divergence method, 88–89, 92
  Z-test, 89, 95–96, 104–105, 104t
Landsat 5, 1–2, 3f, 4–5, 55t, 125t, 245t
Landsat 7, 55t, 81, 176, 178
  enhanced thematic mapper (ETM), 115, 175–177, 185–186, 245
Landsat 8, 1–2, 3f, 4–5, 16f, 55t, 125t, 132, 134f, 135, 245t, 276, 283, 283t
Landsat MSS, 33–34
Landscape patterns in Khamis-Mushyet using geoinformation technology, 11–26
  abstract, 11–12
  classification description, 15t
  digital elevation model map of Khamis-Mushyet, 14f
  gains and losses and net change, 21–22
  introduction, 12–13
  land use and land cover, 17f, 17t, 18f, 18t, 19f, 19t, 20–24
  methodology, 14–16, 16f
  nonlinear, 89
  result, 17–20
  study area and data used, 13–14, 14f
  urban expansion, 24–25, 25f
Land surface temperature (LST), 1–2, 3f, 4, 7–8, 20, 23–24, 51–52, 54
  Chennai, 59–60, 59f, 60f
  Delhi, 61–63, 62f
  estimation, 5–6, 56–58
  India, 59–68, 67f
  Kolkata, 63–64, 64f
  Kuttanad, 7f
  mono-window algorithm, 5–6
  Mumbai, 64–66, 65f, 68f
Land use and land cover (LU/LC), 6f, 198–199, 227, 252
  classes in basins, 196t, 199f
  extinction depth estimates based on land use classes, 198t
  India, 51, 55–56, 74, 243–249
  Kaziranga National Park, 276, 283
  Khamis-Mushyet, 11–12, 17, 17f, 17t, 18f, 18t, 19f, 19t, 20–24, 21t
  Kuttanad, 6
  Lucknow District, 81–83, 82f, 82t, 83t
  mapping, 55–56, 283–284, 284f, 284t
  normalized difference water index, 6–7
  predictive modeling, 82f, 83t
  Sri Lanka, 195
  urbanization, 55–56; *see also* Slope, Land use, Exclusion, Urban extent, Transportation, Hillshade (SLEUTH) model
Land use and land cover in Kabar Tal wetland, Begusarai (Bihar), 243–249, 246f
  abstract, 243
  areas, 246t
  data sources, 245
  distribution, 246
  introduction, 244
  methodology, 245
  nutrients, 246–248, 247f, 247t, 248t
  study area, 244, 245f
  trend, rate, and magnitude, 247t

Lateral moraines, 29, 30f, 32–33
Latin Hypercube sampling, 225
Leaf area index (LAI), 183–185, 183t, 184f, 225, 228, 295, 296f, 297
Leica Geosystems
    ERDAS IMAGINE 9.1, 2, 43
    ESRI ArcGIS 10.2.1, 2
LH-OAT sample, 228, 230
Linear imaging self-scanning (LISS-IV), 88–90, 90f, 91f, 105
Lougeay, R., 36t
LST, *see* Land surface temperature (LST)
Lucknow case study, 73–74, 80–84, 81f, 82f, 82t, 83t, 109, 115–119, 115f, 116f, 117t

## M

Madden–Julian oscillation indices, 208, 212–213
Madhya Pradesh, *see* CERES-rice model in Madhya Pradesh
Main Boundary Thrust (MBT), 252, 254f, 257f
Main Central Thrust (MCT), 252, 254f, 257f
Main Frontal Thrust (MFT), 252
Manas Tiger Reserve, 27
*Mangrove*, 55, 64–67, 66f, 67f, 68f
Mann–Kendall statistical test, 155, 161t, 162f
MAPE, *see* Mean absolute percentage error (MAPE)
Markov chain model, 75
Marshland, 55–56, 58–59, 58f, 59f, 60f, 68, 280
Mass balance, 28, 33, 38, 43–46
Maximum likelihood classifier (MLC), 4, 15, 16f, 17, 55
MBT, *see* Main Boundary Thrust (MBT)
MCT, *see* Main Central Thrust (MCT)
Mean absolute percentage error (MAPE), 167, 168t
MFT, *see* Main Frontal Thrust (MFT)
Mitra, G., 256t
MLC, *see* Maximum likelihood classifier (MLC)
MLP, *see* Multi-layer perceptron (MLP)
Mobile geographic information system, 287
Moderate-resolution imaging spectroradiometer (MODIS), 52, 151, 156, 209, 210t, 218f, 218t, 236, 239
    ET, 218, 224–226, 228, 234–236, 235f, 236f, 237f, 238
MODFLOW, *see* Groundwater flow model (MODFLOW)
MODIS, *see* Moderate-resolution imaging spectroradiometer (MODIS)
MODTRAN, 255
Monin Obhukov Similarity (MOS) theory, 179
Mono-window algorithm, 5–6, 56, 81
Morainic loop, 30f
Moriasi, D. N., 228, 235
Morphometry and tectonic activity in Dehgolan Basin, 131–148, 135t
    abstract, 132
    basin slope, 132, 135t, 136–137, 145–146, 146f
    drainage basin asymmetry, 132, 136–137, 142, 144t
    drainage density, 132, 135–136, 138–139, 140f, 147–148, 226
    elongation ratio, 132, 135t, 136–137, 142–144, 144t, 145f, 148
    geological settings, 133
    hypsometric curve, 132, 135t, 136, 139–141, 141f
    hypsometric integral, 132, 135t, 136, 141f
    introduction, 132–133
    lithology map, 147f
    materials and methods, 135–137
    morphotectonic indices, 132, 136, 139–146, 148
    results, 137–146
    river sinuosity, 137, 145, 145t, 147
    SI values, 136, 142, 142t, 143f, 146–147
    stream length, 132, 136–138, 139t, 140f, 147–148
    stream-length gradient index (Hack's Index), 132, 135t, 136–137, 142
    stream number, 132, 135–137, 135t, 138t, 139f
    stream order, 132, 135, 135t, 137–138, 138f, 139f, 140f, 145f, 148
    study area, 133–134, 134f
    tectonic setting, 133
    watersheds and classes, 141t
Morphotectonic indices, 132, 136, 139–146, 148
Mosbahi, M., 225
Mueller, J. E., 135t, 137
Müller, F., 38–39, 126
Multi-layer perceptron (MLP), 83
Munsiari Thrust, 252, 254f, 257f

## N

Nag, S., 135t
Nash–Sutcliffe efficiency, 225, 228, 229t
NAT, *see* North Almora Thrust (NAT)
n-dimensional visualizer, 256–258, 257f, 258t
NDVI, *see* Normalized difference vegetation index (NDVI)
NDWI, *see* Normalized difference water index (NDWI)
Near infrared (NIR), 3f, 4–5, 36, 36t, 54f, 57, 90, 94f, 96, 295
    visible (VNIR), 14, 43, 253, 255, 259
Normalized difference vegetation index (NDVI), 1–2, 3f, 5, 7–8, 8f, 54, 56–57, 175, 177–178, 183–186, 183t, 184f, 225, 295, 296t, 300
    Kuttanad, 8f
Normalized difference water index (NDWI), 1–2, 3f, 4–8, 54
    Kuttanad, 7f
Normalized root mean square error (NRMSE), 174, 185t, 186
North Almora Thrust (NAT), 252–253, 254f, 257f, 264f, 267f, 269f, 270, 271f, 273
North India, fog over, *see* Fog over North India
NSE, 174, 185t
Nutrient concentration, 243–249, 246f, 247f, 247t, 248t
    abstract, 243
    areas, 246t
    data sources, 245
    distribution, 246
    introduction, 244
    methodology, 245
    study area, 244, 245f
    trend, rate, and magnitude, 247t
Nye, J. F., 41–42

## O

OAT approach, *see* One-factor-at-a-time (OAT) approach
OLI, *see* Operational land imager (OLI)
Olorunfemi, J. F., 13
One-factor-at-a-time (OAT) approach, 225, 228, 230
*Open area*, 55–56, 58–59, 58f, 59f, 60f, 61, 62f, 63, 63f, 64f, 65–67, 65f, 66f, 67f, 68f

# Index

OpenStreetMap, 83, 288
Operational land imager (OLI), 3t, 54t, 55t, 57, 125t, 276, 283
Ostrem, G., 33
Outer Himalaya and Ramgarh Group of Gneissic, 252

## P

Pai, D. S., 212–213
Parbati sub-basin, 34f
Pareta, K., 135t
Pareta, U., 135t
Paul, F., 36t, 37t
Penman–Monteith method, 209, 227–228, 234
Philip, G., 36t
Pinter, P. J., 135t, 137
Pixel purity index (PPI), 256–258, 257f, 258f, 262
Planck's equation, 56
Plant evaporation compensation factor (EPCO), 225, 229t, 230, 233, 234f
Potential evapotranspiration (PET), 227
PPI, *see* Pixel purity index (PPI)
Predictive modeling, 73–84
    abstract, 73–74
    agent-based model, 79
    analysis and results, 83
    artificial neural networks (ANN), 36f, 75, 88–89, 91f, 93, 94f, 97, 101–102, 101f, 102t, 104t, 105
    cellular automata model, 75, 79
    conclusion, 84
    data and software used, 81
    economic model, 79
    introduction, 74
    land change modeling, 73–80, 74t, 76t–78t, 84
    land use and land cover, 82f, 83t
    Lucknow case study, 73–74, 80–84, 81f, 82f, 83t, 109, 115–119, 115f, 116f, 117t
    Markov chain model, 75
    methodology, 81
    statistical regression model, 75
    study area, 80–81, 81f, 82f
    urban land change modeling, 74, 80

## Q

Quick unsupervised anisotropic clustering (QUAC), 253, 255

## R

Racoviteanu, A. E., 36t
Rai, P. K., 135t
Ramgarh Thrust, 252–254
Randolph Glacier Inventory (RGI), 38
Random forest-based classification (RF), 88–89, 91f, 94, 97, 101–102, 102f, 104t, 105, 112
Ranzi, R., 36t
Rao, A. R., 155, 216t, 252
Ravi Basin, 28, 35, 35f
RCHR_DP, 229t, 230–231, 231f
Reddy, C. S., 80
Regions of interest (ROI), 90
Remote sensing, 41–42, 112–114, 176–177, 176t, 219f, 283t, 287–288
REVAPMN, 229t, 230, 231f, 238
RF, *see* Random forest-based classification (RF)

RGB composite, 39f
Rice model, *see* CERES-rice model in Madhya Pradesh
River, 38t, 81, 115, 142, 237, 276, 280, 286–287, 289
    basin, 152–153, 224, 226, 226f
    channel, 145, 280–281
    sinuosity, 137, 145, 145t, 147
*River*, 55–56, 61, 64, 64t
*River App*, 287, 287f
River instability, 145
RMSE, *see* Root mean square error (RMSE)
ROI, *see* Regions of interest (ROI)
Root mean square error (RMSE), 89, 93, 167, 168t, 174, 185, 214–216, 215t, 216t, 219, 219t, 229t
    normalized (NRMSE), 174, 185t, 186
    R-, 174, 186
    relative, 186
Rosgen, D. L., 135t
Rott, H., 36t
Roy, P. S., 283t

## S

Sahandi, M., 133
Sah, M. P., 36t
Sailana Florican Sanctuary, 277
SAM, *see* Spectral angle mapper (SAM)
SAT, *see* South Almora Thrust (SAT)
Scaling, 40, 40t, 43–44, 124, 126, 126t, 253
Scan line corrector (SCL), 176
Schumm, A., 135t, 137, 142, 144t
SCL, *see* Scan line corrector (SCL)
*Scrub*, 55–56, 60–61, 60f, 62f, 63
Scrubland, 11–12, 15t, 17–18, 17f, 17t, 18f, 18t, 19f, 19t, 20, 21t, 22t, 23–25, 193f, 199f
SEBAL, *see* Surface energy balance algorithm for land (SEBAL)
Selected measures, 91f, 94–95, 99t, 100t, 101, 102t, 103t
Sen's, 155, 161t, 163t
Sensitivity analysis, 207–208, 210, 215–217, 224–226, 228, 230
Separability analysis, 88, 91–92, 91f, 97, 97t, 98t
SFF, *see* Spectral feature fitting (SFF)
Shankar, S., 135t
Sharma, C. K., 126
Shortwave infrared (SWIR), 36, 36t, 43, 54t, 253, 255, 259
Shukla, A., 37t
Shuttle radar topography mission (SRTM), 38t, 41, 46
Sidjak, R. W., 36t
Sinha-Roy, S., 135t
SLEUTH, *see* Slope, Land use, Exclusion, Urban extent, Transportation, Hillshade (SLEUTH) model
Slope, 13, 18, 25, 28, 37t, 39, 39f, 41–43, 113, 147, 227, 260
    basin, 132, 135t, 136–137, 145–146, 146f
    channel, 136, 142
    dependent estimations, 123
    gentle, 142, 144–145
    higher basin, 146
    layers, 228
    lower basin, 146
    map, 146f
    median, 155
    Sen's, 161t, 163t
    steep, 144
    stream section, 136

Slope, Land use, Exclusion, Urban extent, Transportation, Hillshade (SLEUTH) model, 76t, 79
Sl values, 136, 142, 142t, 143f, 146–147
Smith, T., 38t
Snout, 27, 29–30, 32, 32f, 34, 125
Soil-adjusted vegetation index (SAVI), 183, 183t, 295, 297f
Soil and water assessment tool (SWAT), 224–228, 229t, 231–232, 234–236, 235f, 236f, 237f, 238–239
Soil Science Society of Sri Lanka and Canadian Society of Soil Science (SRICANSOL), 195, 197, 201t
Soil vegetation atmosphere transfer (SVAT), 174–175, 177
SOL_AWC, 232–233, 232f, 238
SOL_K, 232–233, 232f, 238
SOL_Z, 232, 232f, 238
South Almora Thrust (SAT), 252–253, 254f, 257f, 264f, 267f, 269f, 270, 271f, 272t, 273
Space technology, 282–287
    flood mapping through active microwave sensor, 284–285
    land use and land cover mapping, 283–284
    river water level monitoring, 286–287
Sparse vegetation, 15t, 17f, 17t, 18–20, 18f, 19f, 19t, 21t, 22t, 81, 82f, 82t, 83, 83t, 91t, 94f, 97f, 98t, 98t, 99t, 100–101, 100t, 101f, 102f, 102t, 103, 103t, 115, 116f, 116t, 117t, 118t, 119
Spatial analysis, 125, 137, 169
Spatial variability, 164–167, 166f–167f, 169, 227
Spatiotemporal analysis of urban expansion, 1–8
    abstract, 1
    flowchart of methodology in study, 3f
    image classification, 5
    introduction, 2
    land use changes, 6t
    map of LST of Kuttanad, 7f
    map of LULC of Kuttanad, 6f
    map of NDVI of Kuttanad, 8f
    map of NDWI of Kuttanad, 7f
    map of study area, 4f
    materials and methods, 2–6
    mono-window algorithm for estimation of land surface temperature, 5–6
    normalized difference vegetation index, 5
    normalized difference vegetation index and land surface temperature, 7–8
    normalized difference water index (NDWI), 1–2, 3f, 4–8, 7f, 54
    normalized difference water index and land use/land cover, 6–7
    preprocessing of image, 5
    result and discussions, 6–8
    study area, 3–4
    thematic mapper, 3t
Spectral angle mapper (SAM), 252, 259–263, 261f, 262t, 264f, 265t, 266t, 268, 269f, 270, 270t, 272–273
Spectral feature fitting (SFF), 252, 259, 261–263, 262t, 265t, 266t, 268, 272–273
Spectral mapping methods of mineral aggregates using Hyperion data in Uttarakhand, 251–273, 267f
    abstract, 252–254
    bad band designation, 255
    binary encoding, 252, 259–260, 269f
    class distribution result, 272t
    data dimension reduction, 256–258
    endmember or pure pixel extraction, 256, 258t, 263f, 264f, 265t, 268t, 270t
    endmember spectral library, 259
    field spectra data, 270, 271f
    geological interpretation, 273
    geological setting, 253
    introduction, 252–253
    lithological map and thrust zones, 254f
    minimum noise fraction transformation, 256, 257f
    n-dimensional visualizer, 257–258, 257f, 258t
    preprocessing data, 254–255, 256t
    radiometric calibration, 255–256
    spectral analysis, 259–261
    spectral analyst, 259–261
    spectral angle mapping, 260–261
    spectral library, 259
    spectral mapping, 266t, 272t
    weights and scores, 261–262
    weight scores of different mapping methods and results, 262–272, 262t, 263f, 264f, 269f
SPOT5 DEM, 43, 46
SRICANSOL, see Soil Science Society of Sri Lanka and Canadian Society of Soil Science (SRICANSOL)
Srivastava, P., 256t
SRTM, see Shuttle radar topography mission (SRTM)
Statistical regression model, 75
Strahler, A. N., 132, 135–137, 135t, 139
Stream length, 132, 136–138, 139t, 140f, 147–148
Stream-length gradient index (Hack's Index), 132, 135t, 136–137, 142
Stream number, 132, 135–137, 135t, 138t, 139f
Stream order, 132, 135, 135t, 137–138, 138f, 139f, 140f, 145f, 148
Support vector machines (SVMs), 88–89, 91f, 92–93, 97, 98f, 99t, 100–105, 100t, 104t, 112
Surface energy balance algorithm for land (SEBAL), 46, 173–175, 177, 181, 185–187, 185t, 187f, 225
SVAT, see Soil vegetation atmosphere transfer (SVAT)
SVMs, see Support vector machines (SVMs)
SWAT, see Soil and water assessment tool (SWAT)

T

Tectonic setting, 133
Template method, 44–45
Theil, H., 155
TM4/TM5 ratio image, 34, 37t, 39f
TNR, see True negative rate (TNR)
TPR, see True positive rate (TPR)
Transformed divergence method, 88–89, 92
Troiani, F., 135t
True negative rate (TNR), 94–95, 99t, 100t, 102t, 103t
True positive rate (TPR), 94–95, 99t, 100–104, 100t, 102t, 103t

U

UHI, see Urban heat island (UHI)
UNESCO, 38
    World Heritage Site, 276
Universal transverse mercator (UTM), 15, 55, 125, 245, 283

Urban heat island (UHI), 52, 54, 57, 66
Urbanization, 25f
   developing and developed nations, 110–111
   remote sensing, geographic information system, and spatial metrics, 112–114
   spatial metrics, 113t–114t; *see also* Spatiotemporal analysis of urban expansion
Urbanization in India, 51–69
   abstract, 51
   basic information on demography and climate, 53t
   Chennai, 58–60, 58f, 59f, 60f
   data, 54–58, 54t
   Delhi, 60–63, 60f, 62f
   introduction, 51–52
   Kolkata, 63–64, 63f, 64f, 65f
   land surface temperature estimation, 56–58
   land use and land cover mapping, 55–56
   location of four cities, 53f
   LST maps, 67f, 68f
   LULC classification, 55t
   Mumbai, 64–68, 66f, 68f
   results, 58–68
   study area, 52–54
Urbanization in Lucknow City, 109, 115–119
   abstract, 109
   analysis and results, 118–119
   data and software used, 115
   growth from 2000 to 2017, 117f, 117t
   introduction, 109–110
   land use/land cover, 116f, 116t, 118t
   linkage between growth, ecology, and growth management, 111–112
   methodology, 115–116
   phenomenon, 110
   spatial metrics, 118f
   study area, 115, 115f
Urban land definition, 116
Urban land change modeling, 74, 80
U.S. Geological Survey (USGS), 4, 55, 253
UTM, *see* Universal transverse mercator (UTM)

## V

Vaikrita Thrust, 252, 254f, 257f
Vegetation, 1–2, 6f, 8, 15t, 37t, 39f, 139, 174, 183, 237, 244, 248, 276, 280, 282
   index, 4–5, 7–8, 295, 296f, 297f; *see also* Dense vegetation; Normalized difference vegetation index (NDVI); Soil-adjusted vegetation index (SAVI); Soil vegetation atmosphere transfer (SVAT); Sparse vegetation
*Vegetation*, 55–56, 58–61, 58f, 59f, 60f, 63–67, 63f, 64f, 65f, 66f, 67f, 68f
Velavador National Park, 277
Vincent, C., 46
Visible near infrared (VNIR), 14, 43, 253, 255, 259
Volume of glaciers and glacierets, 123–129, 128t
   abstract, 123
   area change and surface area, 127t
   catchment and glacier areas, 125, 127f
   data and methods, 125–126
   ice reserved in Stok and Matoo village catchments, 127, 128f
   introduction, 123–124
   results and discussion, 127–128
   satellite data, 125t
   scaling parameters, 126t
   study region, 124–125, 124f, 124t
   thickness and volume estimation, 125–126
   total glaciated area in Stok and Matoo village catchments, 127
Volumetric estimation, 42
   glacier flow mechanics method, 42
   remote sensing–based methods, 42
von Thünen, J. H., 80

## W

*Water*, 55–56, 58f, 59, 59f, 60f, 61, 62f, 64–66, 66f, 67f, 68f
Water body, 15t, 18t, 19t, 198t, 284, 284f, 284t
Water quality, 52, 224–225, 244–245, 248–249, 281–282
Water resources board (WRB), 191–192, 194–195, 200t, 204
Weber, A., 80
Wetland, 7, 55–56, 63, 68, 77t, 111; *see also* Land use and land cover in Kabar Tal wetland, Begusarai (Bihar)
*Wetland vegetation*, 55, 63, 63f, 64f, 65f
Wetland water, 244, 248–249
*Wetland water*, 55–56, 63–64, 63f, 64f, 65f
Wheate, R. D., 36t
Wheeler, M. C., 210, 212
Williams, M. W., 33
WRB, *see* Water resources board (WRB)
Wu, K., 225

## Z

Zare, M., 80
Z-score, 155–156, 165, 166f, 167
Z-test, 89, 95–96, 104–105, 104t